BIOMEDICAL APPLICATIONS OF VIBRATION AND ACOUSTICS FOR IMAGING AND CHARACTERISATIONS

Edited by

Ahmed Al-Jumaily
Auckland University of Technology

Mostafa Fatemi
Mayo Clinic College of Medicine

ASME PRESS

Library of Congress Cataloging-in-Publication Data

Biomedical applications of vibration and acoustics for imaging and characterisations / edited by Mostafa Fatemi, Ahmed Al-Jumaily.
p. ; cm.
Includes bibliographical references.
ISBN 978-0-7918-0273-1
1. Ultrasonic imaging. 2. Acoustic imaging. I. Fatemi, Mostafa. II. Al-Jumaily, Ahmed. [DNLM: 1. Elasticity Imaging Techniques. 2. Acoustics. 3. Vibration. WN 208 B615 2008]

RC78.7.U4.B565 2008
616.07'543--dc22

2007052818

TABLE OF CONTENTS

PREFACE

This book introduces the reader to the applications of vibration and acoustics in developing various biomedical imaging and characterization (BIC) techniques. The primary objective of the book is to compile the latest research topics in BIC which are currently under investigation.

This book is the fruit of hard work by a number of world leaders in the field who have summarized their work in thirteen chapters, which are grouped in two parts.

The first part covers imaging applications. This part includes seven chapters on the use of ultrasound radiation forces for imaging and one chapter on magnetic resonance elastography. This part includes fundamental concepts, theory, modelling, experimental methodologies, and clinical applications.

The second part explores a number of tissue characterization methods based on vibration and acoustics. These include methods for characterization of vessels, soft tissues, and bone.

This book (the first of two) is a collection of invited chapters as well as chapters built around selected papers presented at previous Symposia on Vibrations and Acoustics in Biomedical Applications at IMECE05,[1] IMECE06,[1] and IDET05,[2] all parts of ASME annual conferences.

The content and format of this book have been designed to benefit a variety of audience, including:

1. Academic educators teaching upper-level graduate courses in BIC.
2. Graduate and undergraduate students in biomedical, electrical and mechanical engineering wanting to learn about state-of-the-art BIC.
3. Researchers interested in learning relevant significant principles, theory, modelling, experimental and clinical applications of BIC.
4. Medical practitioners who are involved in using BIC techniques.
5. Industrial professionals wishing to understand the foundation, fundamentals, and applications of BIC. This includes industries in the field of ultrasound, magnetic resonance imaging, and tissue evaluation.

The editors' wish is that this volume will interest and intrigue its readers with the desire to promote further discoveries inspired by the ideas and procedures outlined here. This, in turn, will lead to the advancement of the filed of biomedical imaging and tissue characterization based on vibration and acoustics.[1]

[1] International Mechanical Engineering Congress and Exposition

[2] International Design Engineering Technical Conference

EDITORS' INTRODUCTION

Viscoelastic parameters of tissue are closely related to tissue pathology. For this reason, techniques that are aimed at estimation of tissue viscoelasticity are of great interest in medical diagnosis.

Research on imaging and characterization of tissue viscoelasticity has grown rapidly in the past decade. In particular, those methods that are based on vibration and acoustics have attracted a great number of investigators, clinicians, and industries.

The primary objective of this book is to compile the latest research topics on biomedical imaging and tissue characterization techniques that utilize vibration and acoustics.

This book includes two parts. The first part is dedicated to imaging, which is comprised of eight chapters. The first seven chapters in this part are focused on methods that utilize acoustic radiation force. These chapters include methods, such as Vibroacoustography, Dual Radiation Force, Acoustic Radiation Force Imaging, and Harmonic Motion Imaging. The eighth chapter in this part is on Magnetic Resonance Elastography.

The second part is dedicated to applications of vibration and acoustic in tissue characterization. This part contains five chapters. The first two chapters are on characterization of arterial vessels, using either pressure waves or radiation force of ultrasound. The next two chapters are focused on tissue motion detection and estimation of tissue viscoelasticity. The last chapter in this part is on characterization of bone using elastic waves.

Since most methods presented in this book are based on ultrasound, it deemed appropriate to include an appendix on ultrasound bioeffects as well as safety standards and guidelines for practice of diagnostic ultrasound.

The Editors hope that the readers will find this text a valuable asset in keeping them abreast of the latest techniques in applications of vibration and acoustics in imaging and tissue characterization. This book will appeal not only to fellow researchers, but also to practitioners, lecturers and students in this exciting and vital field of study.

Ahmed Al-Jumaily, PhD
Institute of Biomedical Technologies
Auckland University of Technology

Mostafa Fatemi, PhD
Biophysics and Biomedical Engineering
Mayo College of Medicine

EDITOR'S INTRODUCTION

EDITORS' INTRODUCTION

Viscoelastic parameters of tissue are closely related to tissue pathology. For this reason, techniques that are aimed at estimation of tissue viscoelasticity are of great interest in medical diagnosis.

Research on imaging and characterization of tissue viscoelasticity has grown rapidly in the past decade. In particular, those methods that are based on vibration and acoustics have attracted a great number of investigators, clinicians, and industries.

The primary objective of this book is to compile the latest research topics on biomedical imaging and tissue characterization techniques that utilize vibration and acoustics.

This book includes two parts. The first part is dedicated to imaging, which is comprised of eight chapters. The first seven chapters in this part are focused on methods that utilize acoustic radiation force. These chapters include methods, such as Vibroacoustography, Dual Radiation Force, Acoustic Radiation Force Imaging, and Harmonic Motion Imaging. The eighth chapter in this part is on Magnetic Resonance Elastography.

The second part is dedicated to applications of vibration and acoustic in tissue characterization. This part contains five chapters. The first two chapters are on characterization of arterial vessels, using either pressure waves or radiation force of ultrasound. The next two chapters are focused on tissue motion detection and estimation of tissue viscoelasticity. The last chapter in this part is on characterization of bone using elastic waves.

Since most methods presented in this book are based on ultrasound, it deemed appropriate to include an appendix on ultrasound bioeffects as well as safety standards and guidelines for practice of diagnostic ultrasound.

The Editors hope that the readers will find this text a valuable asset in keeping them abreast of the latest techniques in applications of vibration and acoustics in imaging and tissue characterization. This book will appeal not only to fellow researchers, but also to practitioners, lecturers and students in this exciting and vital field of study.

Ahmed Al-Jumaily, PhD
Institute of Biomedical Technologies
Auckland University of Technology

Mostafa Fatemi, PhD
Biophysics and Biomedical Engineering
Mayo College of Medicine

Part 1:

IMAGING

Section 1:

ACOUSTIC AND ULTRASOUND METHODS

Chapter 1

DYNAMIC RADIATION FORCE OF ACOUSTIC WAVES

Glauber T. Silva
Centro de Pesquisa em Matemática Computacional, Universidade Federal de Alagoas, Maceió, AL 57072-970, Brazil
Temporary address: Instituto Nacional de Matemática Pura e Aplicada — IMPA, Rio de Janeiro, RJ, 22460-320, Brazil

ABSTRACT

When an acoustic wave encounters an object, a static or dynamic radiation force is generated on the object. Static radiation force, also known as radiation pressure, has been widely studied in both acoustics and electromagnetism. It is caused by the non-linear nature of wave propagation. On the other hand, dynamic radiation force might be understood as any time-dependent force caused by wave pressure in linear or non-linear approximation. The purpose of this chapter is to present a theoretical framework of dynamic radiation force exerted by acoustic waves on suspended objects in fluids. Applications of dynamic radiation include new acoustic imaging methods such as vibro-acoustography and non-destructive material evaluation. The following analysis stems from first principles based on the differential conservation equations of fluid dynamics. Throughout this chapter, non-viscous fluids and plane progressive waves are considered. In general, a theory for radiation force comprises two elements, namely, beam-forming for an acoustic source and scattering by the object target. An exact solution of the boundary-value problem for acoustic radiation force is nearly impossible. An approximate solution using perturbation theory is provided as an alternative. On this basis, radiation force is addressed in two steps. The boundary-value problem for beam-forming is solved in second-order approximation. The obtained acoustic fields form the primary waves in linear and non-linear scattering problem for the object. The primary and secondary (scattered) fields are integrated over the surface of the object yielding the radiation force. Within the theoretical description, the spectrum of dynamic radiation force up to second-order approximation is examined. We apply the theory of dynamic radiation force wielded by bichromatic plane waves on spherical targets. An extension of the theory for polychromatic planes waves is described. The theory presented in this chapter can be readily adapted for objects with other geometrical symmetry like cylinders.

1.1 INTRODUCTION

Both electromagnetic and acoustic waves exert a force on an obstacle located within the wavepath. By 1619, Kepler [1] expressed his belief that the repulsion on comet tails is caused by the pressure of light radiation. The concept of radiation pressure was devised in Maxwell's theory of electromagnetism [2]. Accordingly, light as an electromagnetic wave is followed by a static pressure, which results from the reflection or absorption of the light wave by an object. This pressure is proportional to the total mean energy of the irradiated light beam.

Lord Rayleigh [3] formulated the first theory for radiation pressure of compressional acoustic waves. He derived the time-averaged pressure exerted by a plane progressive wave upon a piston in an ideal fluid. In his result, this mean pressure is proportional to the averaged mechanical energy density of the incident harmonic wave. An almost immediate application of this theory yielded a method to measure the power of acoustic waves by assessing the displacement caused by radiation pressure on an disk [4]. In 1928, Sivian [5] proposed an extension of this technique based on measuring the displacement of a disk not due to a static radiation pressure but that of a dynamic nature. Instead of using a harmonic acoustic wave, he applied a time-modulated wave to the disk. The displacement at the modulation-frequency is also proportional to the power irradiated by the acoustic source. This application can be considered as the "cornerstone" of dynamic radiation force. In recent years, dynamic radiation force has become of practical importance for both science and engineering. It is the underlying principle in some new acoustic imaging techniques such as shear wave elasticity imaging [6] and vibro-acoustography [7].

The use of modulated acoustic waves to dynamically excite a target yielding, for instance, measurements of acoustic power in a similar manner as radiation pressure raises some questions. Is the induced dynamic force at the modulation-frequency also a type of radiation pressure? What is the difference between this force and that generated by the pressure of a harmonic acoustic wave at the same frequency upon a target? The answer to these questions is found in the non-linear nature of acoustic wave propagation. In principle, both static and dynamic radiation pressures are caused by the same physical phenomenon due to non-linearity in fluid dynamics. Furthermore, by letting the modulation frequency approximate to zero, the magnitude of dynamic radiation pressure approaches that of static radiation pressure. Thus, we can certainly extend the classical concept of radiation pressure (force) to that of oscillating (dynamic) radiation pressure. There is no fundamental difference between a force caused by a primary harmonic pressure wave and dynamic radiation force. They are both time-varying forces. However, in some imaging methods, dynamic radiation force is easier to control and focus in a small volume in space, being thus more suitable for imaging applications.

A theory for dynamic radiation force of acoustic waves in ideal fluids is presented in Refs. [8-11]. However, the analyses in Refs. [8, 9] do not take into account the parametric regime and mechanical impedance of the target. The theory for radiation force comprises two elements, namely, beam-forming for an acoustic source and scattering by an object target. An exact analytical solution the radiation force problem on an arbitrarily shaped target is nearly impossible to achieve. An approximate solution with the aid of perturbation theory is sought. In this framework the boundary-value problem for beam-forming and scattering is solved in second-order approximation. This review reports the major results so far achieved in the theory and experiments of dynamic radiation force of acoustic waves in fluids.

1.2 HYDRODYNAMIC EQUATIONS

Consider an unbounded medium composed of a homogeneous isotropic fluid. The analysis is limited to a fluid in which thermal conductivity and viscosity are neglected — the so-called ideal fluid. The medium is described by the following fields: pressure p, density ρ, specific entropy s, and particle velocity $\mathbf{v}$. The velocity field is given as the gradient of a potential function ϕ, i.e., $\mathbf{v} = -\nabla\phi$. We assume that the velocity potential is a complex function, i.e., $\phi \in \mathbf{C}$. All fields previously introduced are given in terms of Eulerian coordinates. They are functions of the position vector $\mathbf{r}$ and time t. The ambient values of the acoustic fields are $p = p_0$, $\rho = \rho_0$, and $\mathbf{v} = (0,0,0)$, with $p_0, \rho_0 > 0$. The quantity $p - p_0$ is the excess of pressure in the medium. When effects of gravity can be neglected, the dynamics of an ideal fluid are described by the set of non-linear partial differential equations [12]

$$\frac{\partial \rho}{\partial t} + \nabla \cdot (\rho \mathbf{v}) = 0 \quad (1-1)$$

$$\frac{\partial \mathbf{v}}{\partial t} + (\mathbf{v} \cdot \nabla)\mathbf{v} = -\frac{\nabla p}{\rho} \quad (1-2)$$

$$p = p(\rho, s) \quad (1-3)$$

where the symbol means the scalar product. The conservation of fluid mass is represented by Eq. (1-1). Equation (1-2) is the momentum conservation equation, while Eq. (1-3) is the adiabatic equation of state of the fluid. Taylor's expansion of the equation of state for constant entropy $s = s_0$ is

$$p - p_0 = A\left(\frac{\rho - \rho_0}{\rho_0}\right) + \frac{B}{2}\left(\frac{\rho - \rho_0}{\rho_0}\right)^2 + \cdots \quad (1-4)$$

where $A = \rho_0(\partial p/\partial \rho)_{s0} = \rho_0 c_0^2$ and $B = \rho_0^2(\partial^2 p/\partial \rho^2)_{s0}$. The adiabatic speed of sound for a wave of infinitesimal amplitude is denoted by c_0. The ratio B/A gives a measure on how non-linear is the response of the pressure to variations in fluid density. Note that even for a linear pressure-density relation ($B = 0$) , wave propagation still a non-linear phenomenon according to Eqs. (1-1) and (1-2).

An exact solution of Eqs. (1-1) to (1-3) for scattering problems, given proper boundary conditions, is a difficult task and nearly impossible to achieve. Instead, approximate solutions up to second-order for radiation force problems are sought. The approximate solution can be obtained by using regular perturbation theory. In this case, it is necessary to find a relevant parameter, which encloses the characteristics of the host fluid and the propagating wave. The solution is given as a series expansion on this parameter.

In order to compare the advection term $(\mathbf{v} \times \nabla)\mathbf{v}$ with the linear term $\partial \mathbf{v}/\partial t$ in Eq. (1-2), it is assumed that an acoustic harmonic wave with particle velocity magnitude v_0, frequency ω_0, and wavenumber k_0 is propagating in the medium, with v_0, ω_0, $k_0 > 0$. Thus, $\partial \mathbf{v}/\partial t \approx \omega_0 v_0$ and $(\mathbf{v} \cdot \nabla)\mathbf{v} \approx k_0 v_0^2$. The ratio of these terms is $\varepsilon = v_0/c_0$ with $c_0 = \omega_0/k_0$ being the adiabatic speed of sound in the fluid. The quantity is known as the Mach number. It gives a measure on how non-linear effects might be relevant in the wave propagation process. The linear approximation is valid when $\varepsilon \to 0$; in other words, when the wave has infinitesimal amplitude. In any other case, the wave is of finite amplitude.

Deviation from a linear regime of acoustic wave propagation gives rise to some interesting physical phenomena. Among them are (static) radiation pressure, and harmonic and subharmonic generation. The effects of non-linear terms in Eqs. (1-1) to (1-3) can accumulate as the wave propagates or as the time evolves. Such accumulation depends also on energy dissipation and dispersion of the wave.

We can expand the velocity potential in terms of the Mach number as follows

$$\phi = \varepsilon\phi^{(1)} + \varepsilon^2\phi^{(2)} + O(\varepsilon^3) \quad (1-5)$$

The constant term in the expansion is arbitrarily set to zero. It follows imediately from Eq. (1-5) that in a second-order approximation we have $\mathbf{v} = \varepsilon\mathbf{v}^{(1)} + \varepsilon^2\mathbf{v}^{(2)}$, where $\mathbf{v}^{(n)} = -\nabla\phi^{(n)}$ for $n = 1, 2$. The expansion for the pressure p in terms of the velocity potentials is less obvious. We substitute $\mathbf{v} = -\nabla\phi$ in Eq. (1-2) yielding

$$\nabla\left[\frac{\partial \phi}{\partial t} + \frac{\|\nabla\phi\|^2}{2}\right] = -\frac{\nabla p}{\rho} \quad (1-6)$$

where the brackets $\|\cdot\|$ means the magnitude of a vector. Furthermore, given $\mathbf{a} \in \mathbb{C}^3$, the magnitude squared of this vector is $\|\mathbf{a}\|^2 = \mathbf{a} \cdot \mathbf{a}^*$.

It is now useful to introduce the specific enthalpy h of the fluid as through the relation

$$dh = T ds + \frac{dp}{\rho} \tag{1-7}$$

where T and s are the temperature and specific entropy of the system. Since thermal effects are neglected the wave propagation is considered to be adiabatic. Therefore, $ds = 0$ and $dh = dp/\rho$, which implies that $\nabla h = \nabla p/\rho$. We thus recognize that

$$h = -\frac{\partial \phi}{\partial t} - \frac{\|\nabla \phi\|^2}{2} \tag{1-8}$$

Note that to ensure that enthalpy vanishes at infinity, we set the integration constant, which may depend on time, to zero. The expansion of the excess of pressure in terms of the system enthalpy around the equilibrium state is

$$p - p_0 = \left.\frac{\partial p}{\partial h}\right|_{s=s_0} h + \left.\frac{\partial^2 p}{\partial h^2}\right|_{s=s_0} h^2 + O(h^3) \tag{1-9}$$

where s_0 is entropy in the fluid (constant). From Eq. (1-7), we have $\rho = (\partial p/\partial h)_{s0}$ and $(\partial^2 p/\partial h^2)_{s0} = (\partial \rho/\partial h)_{s0} (\partial p/\partial h)_{s0} = \rho/c_0^2$, for which the relation used was $c_0^2 = (\partial p/\partial \rho)_{s0}$. By substituting Eq. (1-8) into Eq. (1-9) along with Eq. (1-5), the desired expansion for the excess of pressure is obtained.

$$p - p_0 = p^{(1)} + p^{(2)} + O(\varepsilon^3) \tag{1-10}$$

where

$$p^{(1)} = \varepsilon \rho_0 \frac{\partial \phi^{(1)}}{\partial t} \tag{1-11}$$

is the linear pressure and

$$p^{(2)} = \varepsilon^2 \rho_0 \left[\frac{\partial \phi^{(2)}}{\partial t} - \frac{\left\|\nabla \phi^{(1)}\right\|^2}{2} + \frac{1}{2c_0^2} \left(\frac{\partial \phi^{(1)}}{\partial t} \right)^2 \right] \tag{1-12}$$

From the hydrodynamic equations for ideal fluids it can be shown that $\phi^{(1)}$ and $\phi^{(2)}$ satisfy the following wave equations [12]

$$\left(\nabla^2 - \frac{1}{c_0^2} \frac{\partial^2}{\partial t^2} \right) \phi^{(1)} = 0 \tag{1-13}$$

$$\left(\nabla^2 - \frac{1}{c_0^2} \frac{\partial^2}{\partial t^2} \right) \phi^{(2)} = -\frac{1}{c_0^2} \frac{\partial}{\partial t} \left[\frac{1}{2} \left(\nabla^2 - \frac{1}{c_0^2} \frac{\partial^2}{\partial t^2} \right) \phi^{(1)2} + \frac{\gamma}{c_0^2} \left(\frac{\partial \phi^{(1)2}}{\partial t} \right) \right], \tag{1-14}$$

where $\gamma = 1 + B/A$. Equation (1-13) is the well-known linear wave equation. Perturbation theory yields Eq. (1-14) as an inhomogeneous linear wave equation. However, the source term in the right-hand side of Eq. (1-14) is non-linear on $\phi^{(1)}$. It also depends on the non-linearity of the fluid, i.e., B/A parameter. In water $B/A \approx 5$ at room temperature.

In imaging applications of dynamic radiation force, beam-forming represents an important issue in designing image systems. For beamforming purposes it is assumed that the beam source is a flat piston described by a closed region $D \subset \mathbf{R}^2$. In Cartesian coordinates, D is considered to be in the xy plane. The beam is generated by linearly vibrating the piston. It subsequently propagates in the half-space $z > 0$. The analysis is restricted to a rigid baffle in which outside D the applied vibration vanishes. There is no induced velocity on the piston in second- or higher-order approximation. Therefore, the boundary conditions for Eqs. (1-13) and (1-14) are

$$\left.\frac{\partial \phi^{(1)}}{\partial z}\right|_{z=0} = \begin{cases} -\varepsilon c_0 g(t), \ t \in \mathbf{R}^3, \ (x,y) \in D \\ 0, \ (x,y) \notin D \end{cases} \tag{1-15}$$

$$\left.\frac{\partial \phi^{(2)}}{\partial z}\right|_{z=0} = 0 \tag{1-16}$$

where $g(t)$ is a dimensionless excitation function. For an induced harmonic vibration on the source, $\phi^{(1)}$ satisfies the Sommerfeld radiation condition alongside with the potential $\phi^{(2)}$. Hence, Eqs. (1-13) and (1-14) with boundary conditions (1-15) and (1-16) have their respective unique solutions.

1.3 DYNAMIC RADIATION FORCE

Consider an acoustic wave striking a object occupies a region $\Omega \subset \mathbb{R}^3$ and has boundary surface $\partial\Omega_0$ at rest. The surface of the object may be deformed and dislocated. The surface of the object at the time t is denoted by $\partial\Omega$. The instantaneous net force $\mathbf{f}$ acting on the object is obtained by integrating Eq. (1-2) on the volume of the object. Since the ambient pressure p_0 does not contribute to this force, we can replace p by $p - p_0$ and using the Gauss' integral theorem, the following is obtained.

$$\mathbf{f}(t) = -\int_{\partial\Omega} (p - p_0)\mathbf{n}\mathrm{d}S \tag{1-17}$$

where $\mathbf{n}$ is the outward normal unit vector on $\partial\Omega$ and $\mathrm{d}S$ is the area element.

The analysis of dynamic radiation force for fluids in which B/A is not very large (<100) can be carried out in second-order approximation. If this is not the case, higher-order terms may become relevant in the analysis as well. Thus, substituting Eq. (1-9) into Eq. (1-17) with the aid of Eqs. (1-11) and (1-12) gives

$$\mathbf{f}(t) = -\varepsilon\rho_0 \left\{ \int_{\partial\Omega} \frac{\partial \phi^{(1)}}{\partial t}\mathbf{n}\mathrm{d}S + \varepsilon \int_{\partial\Omega_0} \left[\frac{\partial \phi^{(2)}}{\partial t} + \frac{1}{2c_0^2}\left(\frac{\partial \phi^{(1)}}{\partial t}\right)^2 \right.\right.$$
$$\left.\left. - \frac{1}{2}\left\|\nabla\phi^{(1)}\right\|^2 \right]\mathbf{n}\mathrm{d}S \right\} \tag{1-18}$$

The transport theorem for $\phi^{(1)}$ reads

$$\frac{\mathrm{d}}{\mathrm{d}t}\int_{\partial\Omega} \phi^{(1)}\mathbf{n}\mathrm{d}S = \int_{\partial\Omega} \frac{\partial \phi^{(1)}}{\partial t}\mathbf{n}\mathrm{d}S - \int_{\partial\Omega_0} (\mathbf{n}\cdot\mathbf{v}^{(1)})\mathbf{v}^{(1)}\mathrm{d}S \tag{1-19}$$

Substituting this relation into Eq. (1-18) gives

$$\mathbf{f}(t) = -\varepsilon\rho_0 \left\{ \frac{\mathrm{d}}{\mathrm{d}t}\int_{\partial\Omega} \varphi^{(1)}\mathbf{n}\mathrm{d}S + \varepsilon\int_{\partial\Omega_0}\left[\frac{\partial\varphi^{(2)}}{\partial t} + \frac{1}{2c_0^2}\left(\frac{\partial\varphi^{(1)}}{\partial t}\right)^2 - \frac{1}{2}\left\|\nabla\varphi^{(1)}\right\|^2\right]\mathbf{n}\mathrm{d}S + \varepsilon\int_{\partial\Omega_0}(\mathbf{n}\cdot\mathbf{v}^{(1)})\mathbf{v}^{(1)}\mathrm{d}S\right\}$$

$$\mathbf{f}(t) = -\varepsilon\rho_0 \left\{ \frac{\mathrm{d}}{\mathrm{d}t}\int_{\partial\Omega} \phi^{(1)}\mathbf{n}\mathrm{d}S + \varepsilon\int_{\partial\Omega_0}\left[\frac{\partial\phi^{(2)}}{\partial t} + \frac{1}{2c_0^2}\left(\frac{\partial\phi^{(1)}}{\partial t}\right)^2 - \frac{1}{2}\left\|\nabla\phi^{(1)}\right\|^2\right]\mathbf{n}\mathrm{d}S + \varepsilon\int_{\partial\Omega_0}(\mathbf{n}\cdot\mathbf{v}^{(1)})\mathbf{v}^{(1)}\mathrm{d}S\right\} \quad (1-20)$$

This equation describes all forces acting on the object in a second-order approximation. The first term in the right-hand side of Eq. (1-20) is related to the linear pressure acting on the moving surface $\partial\Omega$. The second term in this equation depends on the second-order excess of pressure exerted on the fixed surface $\partial\Omega_0$. The last term is related to the momentum flux across the surface $\partial\Omega_0$.

Analysis of the force on the object is restricted to that caused by a bichromatic acoustic wave with primary frequencies $\omega_{21} < \omega_1 < \omega_2$, where $\omega_{21} = \omega_2 - \omega_1$. Suppose that the primary and the difference-frequency waves that hit the object are scattered. According to Eqs. (1-13) and (1-14), each wave can separately be considered linear. Therefore, the superposition principle ensures that the total velocity potential in the fluid for each wave is the sum of incident and scattered contributions. In the radiation force problem, the contribution of the scattered wave must be considered because the work done by radiation force depends on both incident and scattered energies.

The velocity potentials of the primary and the difference-frequency waves including the contribution of both incident and scattered waves are denoted, respectively, by

$$\begin{aligned} \phi_1^{(1)} &= \hat{\phi}_1 e^{-i\omega_1 t} \\ \phi_2^{(1)} &= \hat{\phi}_2 e^{-i\omega_2 t} \\ \phi_{21}^{(2)} &= \hat{\phi}_{21} e^{-i\omega_{21} t} \end{aligned} \quad (1-21)$$

where i is the imaginary unit and $\hat{\phi}_m$ (m =1, 2) and $\hat{\phi}_{21}$ complex functions.

Static radiation force is given as the time-average of Eq. (1-20) over a long period. It corresponds to the DC component of the instantaneous force acting on the object. The first two terms in the right-hand side of this equation average zero. The remaining terms in the equation are squared complex functions. Care should be practiced in taking the square of these functions because (static) radiation force is a real-value quantity. In fact, observables correspond to the real part of acoustic fields in Eqs. (1-1) to (1-3). The time average of the instantaneous force $\mathbf{f}(t)$ corresponds to the DC component in the spectrum of this force. Therefore, static radiation force on the object is formed by the sum of the contribution of each frequency ω_1 and ω_2 as $\langle\mathbf{f}\,(t)\rangle = \langle\mathbf{f}_1\rangle + \langle\mathbf{f}_2\rangle$ with

$$\langle\mathbf{f}_m\rangle = -\frac{\varepsilon^2\rho_0}{2}\int_{\partial\Omega_0}\left[\frac{1}{2}\left(k_m^2\left|\hat{\phi}_m\right|^2 - \left|\nabla\hat{\phi}_m\right|^2\right)\mathbf{n} + \left(\frac{\partial\hat{\phi}_m}{\partial n}\right)\nabla\hat{\phi}_m^*\right]\mathrm{d}S, \quad m = 1,2 \quad (1-22)$$

The brackets ⟨⟩ mean time average over a long period, $\partial/\partial n \equiv (\mathbf{n} \cdot \nabla)$, and the symbol * signifies the complex conjugate. To derive Eq. (1-22), the relation $\langle \mathrm{Re}^2\{\hat{g}e^{-i\omega t}\}\rangle = |\hat{g}|^2/2$ is employed with $\hat{g}$ being a complex function, where Re means the real part of a complex variable. It is straightforward to generalize Eq. (1-22) when there are more than two different primary frequencies. Each individual contribution is added to provide the total static radiation force.

In order to derive the dynamic radiation force formula, let F represent the Fourier transform from time to the frequency space. The inverse transform is denoted by F^{-1}. Let $\hat{g}_1$ and $\hat{g}_2$ be complex functions. The following identity holds

$$\mathrm{F}^{-1}\{\mathrm{F}[\mathrm{Re}\{\hat{g}_1 e^{-i\omega_1 t}\}\mathrm{Re}\{\hat{g}_2 e^{-i\omega_2 t}\}]_{\omega=\omega_{21}}\} = \frac{1}{2}\mathrm{Re}\{\hat{g}_1\hat{g}_2^* e^{-i\omega_{21} t}\} \quad (1-23)$$

Using this relation in Eq. (1-19), it is found that the dynamic radiation force on the object is given by

$$\mathbf{f}_{21}(t) = \mathrm{F}^{-1}\left\{\mathrm{F}\left[\mathbf{f}(t)\right]_{\omega=\omega_{21}}\right\} = \mathrm{Re}\left\{\hat{\mathbf{f}}_{21} e^{-i\omega_{21} t}\right\} \quad (1-24)$$

where

$$\hat{\mathbf{f}}_{21} = -\varepsilon^2 \rho_0 \int_{\partial\Omega_0} \left\{ \left[i\omega_{21}\hat{\phi}_{21} + \frac{1}{2}\left(k_1 k_2 \hat{\phi}_1 \hat{\phi}_2^* - \nabla\hat{\phi}_1 \cdot \nabla\hat{\phi}_2^*\right)\right]\mathbf{n} + \frac{1}{2}\sum_{\substack{\ell,m\in\{1,2\}\\ \ell\neq m}} \left(\frac{\partial\hat{\phi}_\ell}{\partial n}\right)\nabla\hat{\phi}_m^* \right\} \mathrm{d}S - i\varepsilon^2\rho_0\omega_{21}\mathrm{F}\left[\int_{\partial\Omega}\phi^{(1)}\mathbf{n}\mathrm{d}S\right]_{\omega=\omega_{21}} \quad (1-25)$$

Note that dynamic radiation force is characterized by both magnitude and phase.

The first integral in Eq. (1-24) resembles that in Eq. (1-12), except for the presence of the potential ϕ_{21}. It characterizes a type of parametric amplification underlaid in Eq. (1-20). The farther the object to the acoustic source or the higher the difference-frequency, the larger the dynamic radiation force. This happens because the energy of the primary frequency wave is pumped to the difference-frequency component. In actual fluids, this amplification is limited by energy attenuation, dispertion, and geometric spreading of the beam.

The total force due to radiation pressure exerted on the object including up to ω_{21} component is given by

$$\mathbf{f}_{\mathrm{rad}}(t) = \langle \mathbf{f}_1(t)\rangle + \langle \mathbf{f}_2(t)\rangle + \hat{\mathbf{f}}_{21} e^{-i\omega_{21} t} \quad (1-26)$$

Let the total irradiated power by the source be P_0 constant. Referring to Eq. (1-22) we find that $\|\langle \mathbf{f}_{\mathrm{rad}}(t)\rangle\| \propto P_0 S$, where S is the projection area of the incident wave onto the boundary surface of the object. Furthermore, from Eqs. (1-22) and (1-25) we see that

$$\left\|\hat{\mathbf{f}}_{21}\right\| \underset{\omega_{21}\to 0}{\approx} \|\langle \mathbf{f}_m \rangle\|, \; m = 1,2 \quad (1-27)$$

This means that the magnitude of dynamic radiation force approaches its static counterpart as the difference-frequency becomes smaller.

1.4 ACOUSTIC SCATTERING

The road to dynamic radiation force on a target is not complete only with beam-forming from an acoustic source. It is necessary to account for the scattering problem on the target.

The scattering problem for a bichromatic wave has to be solved in the second-order approximation. An analytical solution for arbitrarily shaped targets is nearly impossible even for circular or rectangular pistons. A possible method is to employ numerical techniques such as finite difference schemes or finite element method.

We restrict our analysis to a simpler case for which a plane progressive wave is generated on xy plane and the target is a solid elastic sphere. The sphere of radius a is placed in the z-axis at the distance z_0 from the acoustic source. The density of the sphere is ρ_1 and its compressional and shear speed of sound are denoted, respectively, by c_c and c_s. To further simplify the problem, we consider $a = z_0$. Referring to Eq. (1-15), the following equation gives the biharmonic excitation at the acoustic source

$$\left.\frac{\partial \phi^{(1)}}{\partial z}\right|_{z=0} = -\varepsilon c_0(\sin \omega_1 t + \sin \omega_2 t) \qquad (1-28)$$

Thus, the solution of Eq. (1-13) is

$$\phi^{(1)} = c_0 \mathrm{Re}\left\{\frac{1}{k_1}e^{-i(\omega_1 t - k_1 z)} + \frac{1}{k_2}e^{-i(\omega_2 t - k_2 z)}\right\} \qquad (1-29)$$

where $k_1 = \omega_1/c_0$, and $k_2 = \omega_2/c_0$. For the dynamic radiation force problem, the focus is on the scattering of the wave at the difference-frequency $\omega_{21} = \omega_2 \quad \omega_1$. Hence, the solution of Eq. (1-14) for this component is given by

$$\phi_{21}^{(2)} = \frac{\gamma c_0}{2k_{21}} \mathrm{Re}\left\{(k_{21} z + i)e^{-i(\omega_{21} t - k_{21} t)}\right\} \qquad (1-30)$$

where $k_{21} = \omega_{21}/c_0$.

According to Eq. (1-12) the contribution of Eq. (1-30) to the second-order pressure is given by

$$p_{21}^{(2)} = -i\gamma E_0 \mathrm{Re}\left\{(k_{21} z + i)e^{-i(\omega_{21} t - k_{21} z)}\right\} \qquad (1-31)$$

where $E_0 = \varepsilon^2 \rho_0 c_0^2/2$ is the energy density at the acoustic source. The amplitude of $p_{21}^{(2)}$ increases linearly with distance and difference-frequency. Experimental confirmation of this amplification was supplied by Ref. [13]. Such phenomenon resembles parametric amplification as known in electronics and optics. The pressure $p_{21}^{(2)}$ is within the framework of difference-frequency generation. However only this component of the second-order pressure is subject to amplification. The process of difference-frequency generation and amplification is also known as a *parametric acoustic array* [14].

The velocity potentials in terms of spherical waves in the spherical coordinates $(r, ,)$ are given by

$$\hat{\phi}_m = A_m \sum_{n=0}^{+\infty} (2n+1)i^n \left[h_n^{(2)}(x_m) + S_n(x_m)h_n^{(1)}(x_m)\right] P_n(\cos\theta), \quad m = 1,2 \qquad (1-32)$$

$$\hat{\phi}_{21} = A_{21} \sum_{n=0}^{+\infty} (2n+1)i^n \left[h_n^{(2)}(k_{21} r) + S_n(x_{21})h_n^{(1)}(k_{21} r)\right] P_n(\cos\theta) \qquad (1-33)$$

where $A_m = c_0 e^{-ik_m z_0} / (2k_m)$, $A_{21} = \gamma c_0 z_0 (k_{21} z_0 + i)e^{-ik_{21} z_0} / 4$, $x_m = k_m a$, and $x_{21} = k_{21} a$. The functions $h_n^{(1)}$ and $h_n^{(2)}$ are, respectively, the first- and second-kind spherical Hankel functions of the nth order, while P_n is the Legendre polynomial of the nth order. The quantity S_n is the modal scattering function that depends on the mechanical parameters of the material.

To find the scattering function S_n explicitly in terms of the mechanical parameters of the sphere, it is necessary to solve the boundary-value problem of the wave propagation outside and inside the sphere. In summary, the boundary conditions are: the pressure in the fluid must be equal to the radial component of the stress in the sphere; the radial component of displacement of the fluid must be equal to the normal component of displacement of the sphere; and tangential components of the stress in the sphere must vanish. These conditions yield a linear system with three equations with unique solution. Solving this system provides the scattering function as [15]

$$S_n = \det \begin{bmatrix} a_{11} & d_{12} & d_{13} \\ a_{21} & d_{22} & d_{23} \\ 0 & d_{32} & d_{33} \end{bmatrix} \det^{-1} \begin{bmatrix} d_{11} & d_{12} & d_{13} \\ d_{21} & d_{22} & d_{23} \\ 0 & d_{32} & d_{33} \end{bmatrix}. \qquad (1-34)$$

Matrix elements of this equation are given in the Appendix.

1.5 RADIATION FORCE ON SPHERES

1.5.1 Radiation Force Function

The radiation force exerted by a bichromatic plane wave on a solid elastic sphere is now considered. The analysis is restricted to waves in which $\omega_{21} < \omega_1$, otherwise there may be superposition of dynamic radiation force and force due to pressure of the primary waves.

Referring to Eq. (1-25), the second integral in the right-hand side depends on the surface boundary $\partial\Omega$, which may vary in time. The evaluation of this integral for a varying boundary surface is a self-coupled problem. It is necessary to calculate the force on the object in first order by integrating the linear pressure on the boundary of the object. However, the pressure may deform or move the boundary of the object. An approximate calculation of this term for a rigid spherical target can be determined.

Consider a sphere of radius a and density ρ_1 in a fluid. The sphere is free to move and without perturbation it has zero net force. The origin of the coordinate system coincides with the center of the sphere at rest. It is assumed that the sphere oscillates only along the z-axis without deformation. This means that only the normal vector **n** varies in time. We should expect such variation is a function of the position of the center of the sphere denote by $z_c(t)$. Thus $\mathbf{n} = \mathbf{n}(x/a, y/a, z/a - \tilde{z})$, where $\tilde{z} = z_c/a$. By expanding **n** for small variations of $\tilde{z}$, we obtain

$$\mathbf{n}(\tilde{z}) = \mathbf{n}(0) + \tilde{z}\mathbf{n}'(0) + \cdots, \quad |\tilde{z}| \ll 1 \qquad (1-35)$$

where the symbol $'$ means the derivative of the function with respect to its argument. One recognizes that $\mathbf{n}(0) = (x/a, y/a, z/a) = \mathbf{e}_r$ and $\mathbf{n}'(0) = -\mathbf{e}_z$, where $\mathbf{e}_r$ is the radial unit vector and $\mathbf{e}_z$ is the unit vector along the positive z direction. The relative displacement of the center of the sphere due to the primary waves is given by the real part of

$$\tilde{z} = \hat{z}_1 e^{-i\omega_1 t} + \hat{z}_2 e^{-i\omega_2 t} \qquad (1-36)$$

where the quantities $\hat{z}_1$ and $\hat{z}_2$ are the amplitudes of the displacement at the frequencies ω_1 and ω_2, respectively. Now, combining Eqs. (1-35) and (1-36) into Eq. (1-25) yields

$$\mathrm{F}\left[\int_{\partial\Omega} \phi^{(1)}\mathbf{n}dS\right]_{\omega=\omega_{21}} \approx \mathbf{e}_z \int_{\partial\Omega_0} \mathrm{F}\left[\phi^{(1)}\left(\hat{z}_1 e^{-i\omega_1 t} + \hat{z}_2 e^{-i\omega_2 t}\right)\right]_{\omega=\omega_{21}} dS \qquad (1-37)$$

To obtain the amplitudes $\hat{z}_1$ and $\hat{z}_2$, the definition of mechanical impedance of an oscillating object can be employed. Consider that an oscillating force at frequency ω induces a vibration on an object. The relation between the amplitude of the applied force and the induced velocity of the object is the mechanical impendace of the object $Z(\omega)$. Hence, the oscillation amplitudes can be obtained from the relation

$$\hat{z}_m = \frac{i}{a\omega_m Z_m} \int_{\partial\Omega_0} \hat{p}_m^{(1)} \mathrm{d}S, \quad m = 1,2 \tag{1-38}$$

where $\hat{p}_m^{(1)}$ is the amplitude of the linear pressure, which can be calculated from Eq. (1-11). The mechanical impedances of the sphere immersed in a lossless fluid are given by [16]

$$Z_m = i\frac{4}{3}\pi a^2 c_0 x_m \left[\rho_0 \frac{2 + x_m^2 - i x_m^3}{4 + x_m^4} + \rho_1\right], \quad m = 1,2 \tag{1-39}$$

Now, the dynamic radiation force on the spherical target can be found as follows. The potential functions given in Eqs. (1-32) and (1-33) are substituted into Eq. (1-25) along with Eqs. (1-37) and (1-38). Accordingly, the dynamic radiation force on the sphere is given by

$$\mathbf{f}_{21}(t) = \pi a^2 E_0 \hat{Y}_{21} e^{-i(\omega_{21}t - k_{21}z_0)} \mathbf{e}_z \tag{1-40}$$

where $\hat{Y}_{21}$ is called radiation force function. This equation might be understood as follows. The term $\pi a^2 E_0$ corresponds to how much energy is projected on the cross section area of the target. The function $\hat{Y}_{21}$ carries information on the incident beam, the mechanical parameters of the sphere as well as its scattering properties. The sphere vibrates in the z direction around the equilibrium similar to a fluid particle in a plane wave located at z_0.

The radiation force function might be expressed as the sum of the following components

$$\hat{Y}_\mathrm{i} = -i\pi a^2 \rho_0 c_0 \left(Z_1^{-1*} + Z_2^{-1}\right) \frac{x_{21}}{x_1 x_2} R_0(x_2) R_0^*(x_1) \tag{1-41}$$

$$\hat{Y}_\mathrm{p} = -\frac{i\gamma z_0}{a} x_{21} R_1(x_{21}) \tag{1-42}$$

$$\begin{aligned}\hat{Y}_\mathrm{s} = -\sum_{n=0}^{+\infty} \frac{n+1}{x_2 x_1} &\{[x_2 x_1 - n(n+2)]\left[R_n(x_2)R_{n+1}^*(x_1) + R_{n+1}(x_2)R_n^*(x_1)\right] \\ &- n[x_1 R_n(x_2) R_{n+1}'^*(x_1) \\ &+ x_2 R_{n+1}'(x_2) R_n^*(x_1)] + (n+2)\left[x_2 R_n'(x_2) R_{n+1}^*(x_1) + x_1 R_{n+1}(x_2) R_n'^*(x_1)\right] \\ &+ x_2 x_1 [R_n'(x_2) R_{n+1}'^*(x_1) + R_{n+1}'(x_2) R_n'^*(x_2)]\}\end{aligned} \tag{1-43}$$

where $R_n(x) = i^n[h_n^{(2)}(x) + S_n(x)h_n^{(1)}(x)]$ and $R'_n(x) = i^n[h_n^{(2)\prime}(x) + S_n(x)h_n^{(1)\prime}(x)]$. The function $\hat{Y}_{21}$ depends on the size factors x_1, x_2, and x_{21} with the constraint $x_{21} < x_1 < x_2$.

The function $\hat{Y}_i$ is proportional to the inverse of the mechanical *impedance* of the sphere at the primary frequencies. When $x_{21} \ll x_1, x_2$ the contribution of $\hat{Y}_i$ to dynamic radiation force function may be disregarded. This hypothesis can be safely assumed for transducers operating in MHz range and objects in millimeter scale. It is noteworthy when both x_1 and x_2 approach to $x_{21} \ll 1$ then

$$|\hat{Y}_\mathrm{i}| \underset{x_1, x_2 \to x_{21}}{\sim} \frac{\rho_0}{(\rho_0/2 + \rho_1)x_{21}^4} \tag{1-44}$$

This divergent behavior cannot take place in the actual radiation force phenomenon because it violates the principle of conservation of energy. In other words, dynamic radiation force would be infinitely large as x_1, x_2, and x_{21} approach to zero. However, to calculate the function $|\hat{Y}_i|$, it was hypothesized that the sphere develops small oscillation compared to its radius, i.e., $|\tilde{z}| \ll 1$. From Eqs. (1-38) and (1-39) we see that

$$\lim_{x_1, x_2 \to 0} |\tilde{z}| = +\infty \qquad (1-45)$$

and thus the hypothesis is not valid. Therefore, the conclusion must be that the model is still valid for small values of x_1 and x_2 only if $|\tilde{z}|$ remains much less than the unit. From Eq. (1-39), it can be seen that this condition is satisfied if $x_m Z_m \gg \varepsilon\pi a^2 \rho_0 c_0$, $m = 1,2$. For a sphere in the millimeter range in water and x_m 0.1, it is found that $\varepsilon = 0.4$. The divergent behavior of $|\hat{Y}_i|$ happens because damping in the oscillation of the sphere has not been considered.

The contribution of function $\hat{Y}_P$ describes the *parametric* regime of dynamic radiation force. It has a dominant role whenever the size factor x_{21} is not a small quantity. It also depends linearly on the parameter and the ratio z_0/a. This function has the following asymptotic behavior

$$|\hat{Y}_P| \underset{x_{21}\to+\infty}{\sim} \frac{2z_0\gamma}{a}\left(1 + \frac{1}{x_{21}}\right) \qquad (1-46)$$

The amplification of the radiation force with distance does not increase indefinitely. Two aspects of the amplification should be clarified. First, the distance z_0 should be within the preshock wave range, otherwise the conservation Eqs. (1-1) to (1-3) are no longer valid. Second, diffraction and thermoviscous effects may change the dependence of radiation force with distance to the acoustic source (see Ref. [17]).

The function $\hat{Y}_s$ has a prominent contribution to the dynamic radiation force when $x_{21} \ll 1$ and $x_{21} \ll x_1, x_2$. This function behaves similarly to the *static* radiation force function obtained in Ref. [18]. In other words, the asymptotic approximation of dynamic to static radiation force given in Eq. (1-27) may be rewritten as follows

$$|\hat{Y}_{21}| \underset{x21\to 0}{\sim} Y_{st} \qquad (1-47)$$

where Y_{st} is the static radiation force function for a sphere as given in Ref. [18].

It is now possible to analyze the spectrum of the radiation force function $\hat{Y}_{21}$ for a stainless steel sphere immersed in water. The parameters that characterize the sphere are $a = 10^{-3}$ m, $\rho_1 = 7670$ kg/m^3, $c_c = 5240$ m/s, and $c_s = 2978$ m/s. It is assumed that the incident beam is a time-modulated wave with carrier suppressed. The angular carrier frequency is denoted by ω_0, while ω_{21} is the angular modulation frequency. The primary frequencies of the beam are related as follows $\omega_1 = \omega_0 - \omega_{21}/2$ and $\omega_2 = \omega_0 + \omega_{21}/2$. This corresponds to the following relation between the size factors: $x_1 = x_0 - x_{21}/2$ and $x_2 = x_0 + x_{21}$, where $x_0 = \omega_0 a/c_0$.

Figure 1-1, plots $\hat{Y}_{21}$ as a function of x_0 (0.1,20]. This interval is partitioned into 1024 intervals. Both magnitude and phase are illustrated. For $x_{21} = 10^{-6}$, the magnitude of dynamic radiation force is close to that of static radiation force. Furthermore, considering $|\hat{Y}_{21}|$ ~ 1.5, resonances are observed in agreement with calculations for static force in Ref. [18]. In this case, the parametric contribution is nearly non-existent. For $x_{21} = 0.05$ and 0.1, an increase in the magnitude is observed specifically when $x_0 < 1$ as described by Eq. (1-47). In the region $x_0 > 2$, then $|\hat{Y}_{21}| \sim 3.2, 9.5$, for $x_{21} = 0.05, 0.1$, respectively. The difference on the magnitude of the shown functions can be accounted for by parametric interaction of the primary waves.

Figure 1-2 shows the spectrum of the radiation force function varying with $x_{21} \in [0, 20]$. Again the interval is divided into 1024 partitions. Considering $x_1, x_2 > 1$, contributions from $\hat{Y}_i$ and $\hat{Y}_s$ to the radiation force function are negligible. In the region $x_{21} > 3$, $|\hat{Y}_p|$ is as

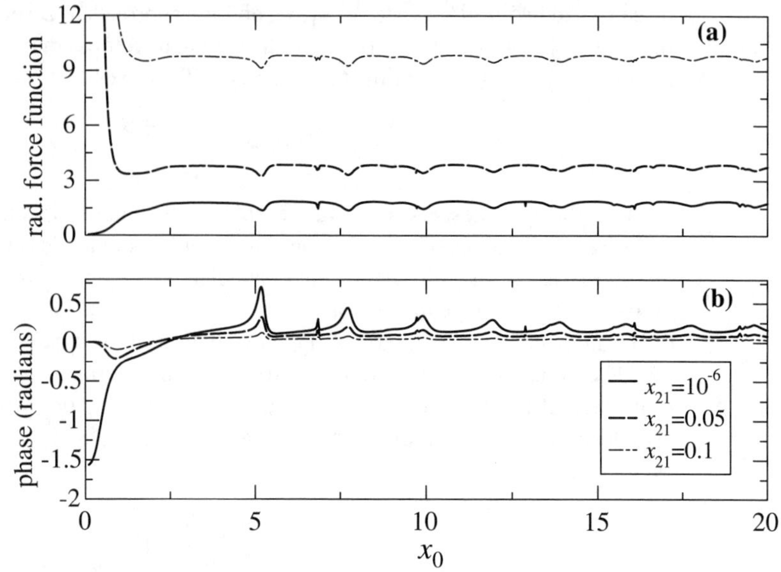

Figure 1-1. Radiation force function for the stainless steel sphere varying with the size factor x_0. (a) Magnitude. (b) Phase

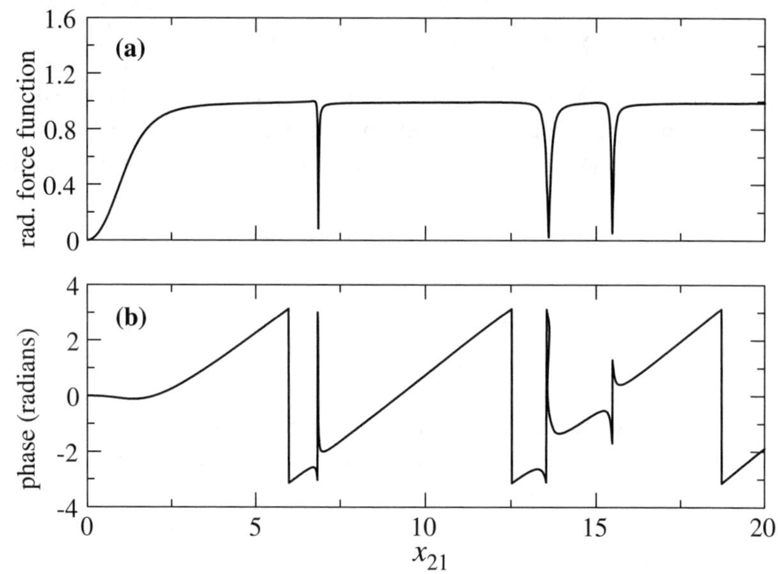

Figure 1-2. Normalized radiation force function for the stainless steel sphere varying with the size factor x_{21}. (a) Magnitude. (b) Phase

much as 60 dB higher than the magnitude of $|\hat{Y}_s|$, which is of the order of static radiation force function. Resonances are clearly found in the spectrum of $|\hat{Y}_{21}|$ followed by jumps in the phase. These resonances are related to that observed for dipole mode in the Resonant Scattering Theory (RST) [15].

1.5.2 Experimental Achievements

Validation of the theory presented here was performed in a set of experiments in fluids. In one experiment [19], both static and dynamic radiation force were measured on a stainless steel sphere of radius $a = 0.64$ mm for difference-frequencies, namely, 100 Hz, 200 Hz, and 400 Hz. The sphere was suspended through a bifilar pendulum in a waterbath. The primary frequencies were about 930 kHz. The sphere size factors for the difference-frequencies in increasing order are $x_{21} = 2.6 \cdot 10^{-4}, 5.3 \cdot 10^{-4}$, and 10^{-3}. For the primary frequencies x_1 $x_2 = 2.45$. The measured magnitude of both static and dynamic radiation forces in all situations were around 3.50 ∝N with relative difference less than 5.6%. This result supports the statement that dynamic and static radiation force magnitudes are alike when x_{21} 0 as described in Eq. (1-47).

In another experiment [11], parametric regime in dynamic radiation force was investigated. An acrylic sphere of radius 1.60 mm and density 1190 kg/m^3 was used as a target. A 22-mm flat circular transducer was driven with a suppressed-carrier amplitude-modulated signal. The carrier frequency was set to 2.25 MHz, while the modulation-frequency ranged from 10 kHz to 240 kHz. Experiments were performed in water and ethyl alcohol whose speed of sound are 1500 m/s and 1100 m/s, respectively. Measurement were done employing a laser vibrometer. The density of water and alcohol are 1000 kg/m^3 and 700 kg/m^3. The sphere was placed at 200 mm away from the transducer. The non-linear parameter for both water and alcohol are = 6,11, respectively. In this experiment, the size factors are x_1 $x_2 = 15.1, 20.5$ for water and alcohol, respectively. The difference-frequency are $x_{21} \in [6.7 \times 10^{-6}, 1.61]$ for water and $x_{21} \in [9.1 \times 10^{-5}, 2.19]$ for alcohol. Figure 1-3 exhibits a comparison between the measured dynamic radiation force and theoretical predictions. An amplification of dynamic radiation force for frequencies above 3 kHz can be seen. The theory predicts a gain of 52 (44) dB for alcohol (water) as the difference frequency goes from 10 Hz to 240 Hz. The observed amplification gains are 80 (66) dB for alcohol (water).

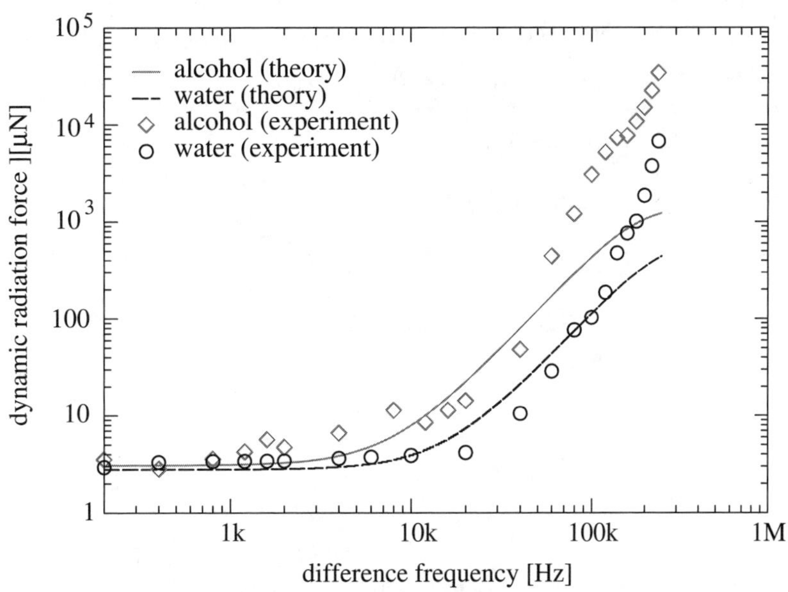

Figure 1-3. Theoretical and experimental results for the magnitude of the dynamic radiation force exerted by a bichromatic acoustic beam on an acrylic sphere varying with the difference-frequency. Measurements are taken in alcohol and water

Differences between theoretical and experimental results might be related to diffraction and thermoviscous effects of the fluid. Thermoviscous effects may produce streaming in boundary layers surrounding the sphere and nearby the walls of the tank yielding more force on the sphere. A theory with thermoviscous effects requires a full solution of the scattering problem using the Navier-Stokes equations.

1.5.3 Multifrequency Radiation Force

Suppose that instead of a bichromatic primary wave hitting a target there is a polychromatic wave with N-frequency components. It is expected that a radiation force arises on the target at the modulation frequencies generated by the incident wave. Stemming from Eq. (1-17), it has been demonstrated that the multifrequency radiation force due to a N-frequency beam is given by [20]

$$\mathbf{f}_{\text{rad}}(t) = \sum_{\substack{m,n=1 \\ m>n}}^{N} \left(\langle \mathbf{f}_m(t) \rangle + \hat{\mathbf{f}}_{mn} e^{-i\omega_{mn} t} \right), \qquad (1-48)$$

where $\omega_{mn} = \omega_m - \omega_n > 0$, $\langle \mathbf{f}_m(t) \rangle$ is the static radiation force due to each frequency ω_m, and $\hat{\mathbf{f}}_{mn}$ is the complex amplitude of the dynamic radiation force given in Eq. (1-40). The restriction $m > n$ yields $N(N-1)/2$ frequency components for dynamic radiation force. Multifrequency radiation force has been applied in an imaging method called multifrequency vibro-acoustography [21]. In this technique, it is possible to acquire an image for each frequency component of dynamic radiation force. Thus, one can obtain up to $N(N-1)/2$ images simultaneously. However, each frequency component has $2/(N^2 - N)$ fraction of the total power yielded to the transducer. Therefore, the signal-to-noise ratio in the acquired images may decrease with the inverse of N^2.

Appendix

The elements of the matrices in Eq. (1-34) are:

$$a_{11} = -(\rho_0/\rho_1) x_s^2 h_n^{(2)}(x)$$

$$a_{21} = x h_n^{(2)\prime}(x)$$

$$d_{11} = (\rho_0/\rho_1) x_s^2 h_n^{(1)}(x)$$

$$d_{12} = [2n(n+1) - x_s^2] j_n(x_c) - 4x_c j_n'(x_c)$$

$$d_{13} = 2n(n+1)[x_s j_n'(x_s) - j_n(x_s)]$$

$$d_{21} = -x h_n^{(1)\prime}(x)$$

$$d_{22} = x_c j_n'(x_c)$$

$$d_{23} = n(n+1) j_n(x_s)$$

$$d_{32} = 2[j_n(x_c) - x_c j_n'(x_c)]$$

$$d_{33} = 2x_s j_n'(x_s) + [x_s^2 - 2n(n+1) + 2] j_n(x_s)$$

where $x_c = c_0 x/c_c$, $x_s = c_0 x/c_s$, and $j_n(\cdot)$ is the spherical Bessel function of the nth order.

REFERENCES

[1] De Mairan, 1754, *Trait Physique et Historique de l'Aurore Borale*. Paris, France. As cited by Nichols, E. E., and Hull, G. F., Physical Review (Series I) 17, 26 (1903).

[2] Maxwell, J. C., 1904, *A Treatise in Electricity and Magnetism*, Oxford University, Oxford, England, Vol. 2, 3rd ed.

[3] Rayleigh, L., 1902, "On the Pressure of Vibrations," *Philosophical Magazine*, **3,** pp. 338–346.

[4] Stewart, G. W., and Stiles, H., 1913, "A Method of Producing Relative Sound Intensity and a Test of the Rayleigh Disk," *Physical Review*, **1**, pp. 309–315.

[5] Sivian, L. J., 1928, "A Modification of the Rayleigh Disk Method for Measuring Sound Intensities," *Philosophical Magazine*, **5**(7), pp. 615–620.

[6] Sarvazyan, A. P., Rudenko, O. V., Swanson, S. D., Fowlkes, J. B., and Emelianov, S. Y., 1998, "Shear Wave Elasticity Imaging: A New Ultrasonic Technology of Medical Diagnostics," *Ultrasound in Medicine and Biology*, **24**(9), pp. 1419–1435.

[7] Fatemi, M., and Greenleaf, J. F., 1998, "Ultrasound-Stimulated Vibro-Acoustic Spectrography," *Science*, **280**, pp. 82–85.

[8] Mitri, F. G., and Chen, S., 2005, "Theory of Dynamic Acoustic Radiation Force Experienced by Solid Cylinders," *Physical Review E*, **71**(1), p. 016306.

[9] Silva, G. T., Chen, S., Greenleaf, J. F., and Fatemi, M., 2005 (May), "Dynamic Ultrasound Radiation Force in Fluids," *Physical Review E*, **71**, p. 056617.

[10] Silva, G. T., 2006, "Dynamic Radiation Force of Acoustic Waves on Solid Elastic Spheres," *Physical Review E*, **74**(2), p. 026609.

[11] Silva, G. T., Chen, S., and Viana, L. P., 2006, "Parametric Amplification of the Dynamic Radiation Force of Acoustic Waves in Fluids," *Physical Review Letters*, **96**(23), p. 234301.

[12] Heaps, H. S., 1962, "Waveform of Finite Amplitude Derived from Equations of Hydrodynamics," *Journal of the Acoustical Society of America*, **34**(3), pp. 355–355.

[13] Thuras, A. L., Jenkins, R. T., and O'Neil, H. T., 1935 (January), "Extraneous Frequencies Generated in Air Carrying Intense Sound Waves," *Journal of the Acoustical Society of America*, **6**, pp. 173–180.

[14] Westervelt, P. J., 1963 (April), "Parametric Acoustic Array," *Journal of the Acoustical Society of America*, **35**(4), pp. 535–537.

[15] Gaunaurd, G. C., and Uberall, H., 1983 (January), "RST Analysis of Monostatic and Bistatic Acoustic Echos from Elastic Sphere," *Journal of the Acoustical Society of America*, **73**(1), pp. 1–12.

[17] Tjøtta, J. N., and Tjøtta, S., 1981, "Nonlinear Equations of Acoustics, with Application to Parametric Acoustic Arrays," *Journal of the Acoustical Society of America*, **69**(6), pp. 1644–1652.

[18] Hasegawa, T., and Yosioka, K., 1969, "Acoustic Radiation Force on a Solid Elastic Sphere," *Journal of the Acoustical Society of America*, **46**, pp. 1139–1143.

[19] Chen, S., Silva, G. T., Kinnick, R. R., Greenleaf, J. F., and Fatemi, M., 2005 (May), "Measurement of Dynamic and Static Radiation Force on a Sphere," *Physical Review E*, **71**, p. 056618.

[20] Silva, G. T., Urban, M. W., and Fatemi, M., 2007, "Multifrequency Radiation Force of Acoustic Waves," *Physica D*, **232**, pp. 48–53.

[21] Urban, M. W., Silva, G. T., Fatemi, M., and Greenleaf, J. F., 2006, "Multifrequency Vibro-Acoustography," *IEEE Transactions on Medical Imaging*, **25**, pp. 1284–1295.

Chapter 2

NUMERICAL MODELING FOR ASSESSMENT AND DESIGN OF ULTRASOUND VIBROACOUSTOGRAPHY SYSTEMS

Alison E. Malcolm[1], Fernando Reitich[2], Jiaqi Yang[2], Mostafa Fatemi[3], and James F. Greenleaf[3]
[1]Department of Earth Sciences, Utrecht University, Utrecht, The Netherlands
[2]Department of Mathematics, University of Minnesota, Minneapolis, MN 55455, USA
[3]Mayo Clinic College of Medicine, Rochester, MN 55905, USA

ABSTRACT

Ultrasound vibroacoustography is a new medical imaging technique that relies on the non-linear interaction of high-frequency fields to produce a low-frequency acoustic emission. Combining the resolution obtained by using a high-frequency focused source with the clean (free of artifacts from multiple scattering) images obtained with low-frequency fields yields a clearer image of abnormal tissue than does traditional ultrasound. A modeling strategy for this imaging modality is presented that overcomes the traditional difficulties in modeling over the range of frequencies present in these experiments. Through a combination of appropriate approximations and exact methods, it is possible to obtain images that capture the basic structure of experimental data despite the use of a simplified physical model.

2.1 INTRODUCTION

The objective of this work is to construct a complete simulation infrastructure to assist in the understanding and design of ultrasound vibroacoustography (UVA) systems such as those pioneered at the Mayo Clinic [1, 2]. A diagram of the experimental setup is shown in Fig. 2-1; this diagram also illustrates the length scales of interest in this problem. The experiment uses a spherical confocal transducer to focus two beams at slightly different frequencies on a small region within an object submerged in a water tank. The confocal transducer has two regions, which are driven at two different frequencies: the inner shell is driven at ω_1 and the outer one at ω_2. The beams are focused (assuming the sound velocity to be constant within the beam region) so that they interact constructively only within a small neighborhood of the focal point. This (non-linear) interaction results in a low-frequency field that oscillates at the difference frequency, $\delta\omega = \omega_2 - \omega_1$. This field is recorded by using a hydrophone, and an image is produced by plotting its amplitude as a function of the focus position of the source. Although there are a few studies that investigate the modeling of similar experiments (see, e.g., [3, 4]), there does not, to the authors' knowledge, exist a comprehensive strategy for modeling UVA.

The lack of such comprehensive simulation models is largely due to the significant challenges that the experiment presents from a computational perspective. First, as illustrated in Fig. 2-1, the field excited by the transducer is a high-frequency field in that it propagates on the order of a hundred wavelengths before focusing. However, asymptotic high-frequency models (e.g., ray acoustics) are not suitable for their representation, as the wavelength is not significantly smaller than the size of the object to be imaged. Modeling a field at this frequency scale with classical full-wave methods (finite differences, finite element, integral equation solvers), on the other hand, requires the resolution of the wavelength and, thus, would lead to inordinate computational costs. To avoid these limitations, a parabolic approximation [4-7] is used, which significantly reduces the cost while preserving accuracy in the directions of interest. In this approximation only waves traveling in a particular (though arbitrarily defined) direction are propagated and the resulting one-way wave pseudodifferential equation is amenable to fast solution via fast Fourier transforms (FFTs). In the present case, the use of this approximation is justified by the focused nature of the source fields and the irrelevance of the reflected fields to the recorded data. To verify this justification, this field is also computed using an integral approach as this approach gives the exact wavefield at a higher computational cost.

In addition to the difficulties associated with the acoustic source as described above, the modeling of vibroacoustics is further complicated by the need to incorporate non-linear effects. Indeed, as discussed above, these effects constitute the core of the imaging technique, as it is their propagation that is recorded in the formation of the image. Here a simple non-linear model of this interaction is reviewed [8, 9] that explains the emergence of a field at the difference frequency and can be easily incorporated into the simulation framework.

Finally, the propagation of the low-frequency field excited by this non-linear interaction must also be computed accurately. Although it would be possible to do this with the parabolic approximation, this is not ideal as the field is truly omni-directional and the hydrophone is typically located far from the focal point. An additional complication to the propagation of the low-frequency field is that the size of the focal area is governed by the high-frequency field and so must be discretized on a grid appropriate for this frequency; this is a very fine

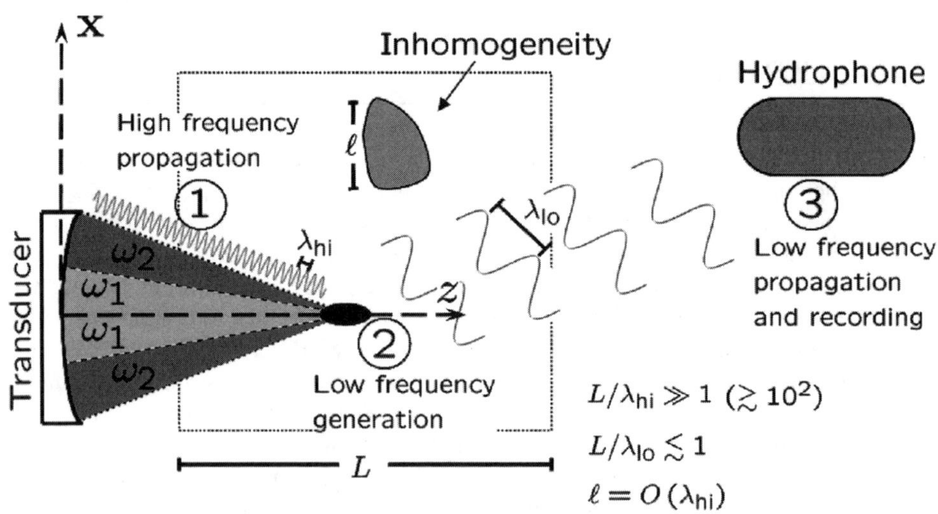

Figure 2-1. Schematic of the experimental setup, as described in [14, 15]

mesh at the difference frequency. Whichever method is chosen, it must be able to accommodate this additional length scale to obtain an accurate solution. The low-frequency nature of this field does, however, permit the use of a variety of methodologies designed to solve such problems. As detailed below, a numerical scheme based on integral equations bypasses the need for artificial absorbing boundaries and is also amenable to accelerated evaluations (e.g., via FFTs [10, 11], fast multipole methods [12], etc.). Two methods are discussed for computing this low-frequency field; the first is a surface integral formulation, suitable for the treatment of (an arbitrary number of) piecewise homogeneous inclusions. The precise formulation is detailed below, as are the quadrature rules [13] and evaluation procedures used in this initial implementation. The second method is a Lippmann-Schwinger volume integral approach allowing for the computation of the scattered field from an arbitrary number

TABLE 2-1. Table of symbols

ϕ	First-order velocity potential, high frequency
ϕ_j	Subscript j indicates that the field is at frequency ω_j
$c(x)$	Acoustic velocity
ω	Angular frequency
λ_{hi}	Wavelength at high frequency
$\delta\omega$	Difference frequency $\omega_2 - \omega_1$
λ_{lo}	Wavelength at difference frequency
ρ	Density
p	Pressure
γ	Specific heat ratio of the medium
ψ	Second-order velocity potential
$\vec{v}$	Wavefield velocity
$\vec{T}$	Stress tensor
φ	Scalar potential for elasticity
$\vec{\Phi}$	Vector potential for elasticity
A	Square-root of wave operator
P	Matrix to diagonalize wave operator
$\hat{\cdot}$	Temporal Fourier transform
ϕ_+, ϕ_-	Velocity potential for waves in positive and negative z directions
$\lvert\cdot\rvert$	Modulus of a complex number
$\lVert\cdot\rVert$	L_2-norm of a vector
k_x	Wavenumber in x-direction
k_z	Wavenumber in z-direction
z	Preferential direction for parabolic approximation
$x = (x_1, x_2)$	Orthogonal directions for parabolic approximation
F	Fourier transform in x variables
D	Inclusions for integral methods
∂D	Boundary of inclusions for surface integral method
D^c	Area outside inclusions for integral methods
G	Green's function
ψ_e, ψ_i	Potential inside and outside inclusion
ψ^{inc}, ψ^s	Incident and scattered fields
f	Source function of low-frequency waves
$\frac{\partial}{\partial n_y}$	Normal derivative with respect to y
c_0	Constant background velocity
B	Region in which $c_0 - c(x)$ is non-zero
G	Spherically averaged Green's function
λ, μ, ρ	Inclusion parameters for Lippmann-Schwinger elastic integral equation
λ_0, μ_0, ρ_0	Background parameters for Lippmann-Schwinger elastic integral equation
$\vec{U}$	Vector displacement for elastic Lippmann-Schwinger integral method
F	Source term for elastic Lippmann-Schwinger integral method

of localized inclusions not assumed to be homogeneous. The implementation of this second approach in three dimensions is discussed and both the high-frequency and low-frequency fields are computed with this technique. Using this result allows investigation of the errors made when assuming two, rather than three-dimensional wave propagation.

This chapter is organized as follows. Section 2.2 describes the mathematical model of the experiment and gives additional information about the experiment itself. The following section (Section 2.3) details the numerical procedures used to approximate the solutions to the equations derived in Section 2.2. Section 2.4 then presents some preliminary numerical results that improve the understanding of the underlying physics while also providing guidance on possible means to improve the quality of the overall technique. Finally, Section 2.5 provides some conclusions and a discussion of possible future directions of research along the lines presented here.

2.2 MATHEMATICAL MODEL

Ultrasound vibroacoustography relies on the radiation force, excited by the two coupled high-frequency ultrasound beams, to generate the low-frequency signal that is recorded by the hydrophone to form the image. A possible strategy for modeling the experiment is thus to first estimate the radiation force, as discussed in [14-16], generated by the transducer and then calculate the field resulting from this force. The experiment is easily broken down into two coupled wave propagation problems, however, which allow the extensive machinery developed for the numerical solution of wave propagation problems to be applied. This section briefly outlines the appropriate theory for this approach. Further details can be found in, e.g., [8, 9] for the acoustic case, [17] for the elastic case, and [18] for the viscoelastic case.

The two wavefields excited by the transducer are standard ultrasound beams, which are assumed to propagate through a fluid. Thus, they obey the scalar wave equation

$$c(x)^2\Delta\phi_j - \partial_t^2\phi_j = 0, \qquad j = 1, 2 \tag{2-1}$$

where ϕ is the velocity potential ($\vec{v} = \nabla\phi$, where $\vec{v}$ is the velocity), c is the acoustic velocity of the background model (c = 1.5 mm/μs in water), j = 1, 2 is used to distinguish the beam at ω_1 from that at ω_2 and Δ denotes the Laplacian (see Table 2-1 for a complete list of symbols). Although a single beam can generate a radiation force, for this force to have a frequency dependence, the beam must be somehow modulated. As discussed above, in ultrasound vibroacoustography the modulation comes from the non-linear interaction of the two high-frequency beams. Specifically (see, e.g., [8, 9]), these interactions arise from higher-order terms in the Euler model, whose lowest-order term results in Eq. (2-1). More precisely, from the Euler equations

$$\partial_t v + (v \cdot \nabla)\, v = -\rho^{-1}\nabla p \tag{2-2}$$

the equation of continuity

$$\partial_t\rho + \nabla \cdot (\rho\vec{v}) = 0 \tag{2-3}$$

and the isentropic assumption

$$p = p_0\left(\frac{\rho}{\rho_0}\right)^{\gamma} \tag{2-4}$$

a perturbation expansion is made about a constant, quiescent state with pressure and density $p \equiv p_0$ and $\rho \equiv \rho_0$. As is well-known, the first-order perturbations satisfy the wave equation

(2-1). The second-order effects lead (upon substituting Eq. (2-4) for p into Eq. (2-2), solving for ρ / ρ_0 in the latter, and again substituting this into (2-3)) leads to the equation

$$\partial_t^2 \psi - c^2 \Delta \psi = \partial_t \left[|\nabla \phi|^2 + \frac{\gamma - 1}{2c^2} |\partial_t \phi|^2 \right] \tag{2-5}$$

for a (second-order) velocity potential ψ, so that the overall velocity model takes the form

$$\vec{v} = \nabla \phi + \nabla \psi \tag{2-6}$$

Further details of this derivation are given in [8, 9] (see Table 2-1).

Equation (2-5) provides the means for the creation of the secondary field at the difference frequency. In the experiment modeled here, ϕ in (2-5) is the sum of two fields at slightly different frequencies, cf. (2-1),

$$\phi = \phi_1 + \phi_2 \tag{2-7}$$

and, mathematically, the squared terms on the right-hand side of Eq. (2-5) generate low-frequency components from the cross terms $\phi_1 \phi_2^*$ and $\phi_1^* \phi_2$, where * denotes complex conjugation. In this way, the right-hand side of Eq. (2-5) models the effective radiation force the system is subjected to. Together, Eqs. (2-1) and (2-5) provide a complete acoustic model of the experiment.

The complete acoustic model is only a starting point, however; a more accurate representation would incorporate elastic (or even viscoelastic) effects, as discussed in [18]. This is particularly important because of the interest in recovering shear moduli; because these moduli are typically several orders of magnitude smaller than the bulk modulus, their recovery (both physical and virtual) must be from shear waves. From a modeling perspective, the derivation of the elastic non-linear, second-order, low-frequency waves generated by the interaction of high-frequency forcings follows largely that described above for compressional waves. Indeed, while the continuity equation (2-3) is unchanged, the equation of motion (2-2) generalizes to

$$\partial_t \vec{v} + (\vec{v} \cdot \nabla)\vec{v} = -\rho^{-1} \nabla \cdot \vec{T} \tag{2-8}$$

where $\vec{T}$ is the Cauchy stress tensor. The field now is no longer irrotational, but it can be decomposed into

$$\vec{v} = \nabla \varphi + \nabla \times \vec{\Phi} \tag{2-9}$$

with the aid of an additional vector potential $\vec{\Phi}$. Finally, the equation of state (2-4) can either be retained as in [18] or be replaced by a linear stress-strain relation holding in the (narrow) frequency band of operation. The second option leads to a forcing term that is analogous to that in (2-5) with $\gamma = 1$.

2.3 NUMERICAL MODEL

This section describes the three stages of the numerical procedure. The first subsection describes the parabolic approximation in the context of this problem, showing examples of the fields modeled with this method. The parabolic approximation is used to propagate the high-frequency field (labeled 1 in Fig. 2-1). Following this, the second subsection develops the necessary structure to model the low-frequency wavefield (3 in Fig. 2-1) with an appropriate integral equation method. Finally, the third section briefly describes and illustrates the computation of the interaction/coupling term on the right-hand side of Eq. (2-5); this is labeled 2 in Fig. 2-1.

2.3.1 High-Frequency Propagation — Parabolic Approximation

Several different types of parabolic approximations are used extensively in exploration seismology (e.g., [6, 19]), ocean acoustics (e.g., [7, 20]), ultrasound [4] as well many other applications in which a wavefield is propagated along a preferred direction. As noted above, this class of approximations defines a preferred axis of wave propagation, here denoted by z, and propagates waves in only one direction along this axis. A parabolic approximation is ideally suited for modeling the high-frequency field propagating from the transducer to the focus point because, quite generally, the aperture is not large, the field propagates only away from the transducer, and any reflected signal is not relevant to the final result (assuming relatively small velocity contrasts). Jensen et al [20] (Chapter 6) give an overview of several methods of deriving parabolic approximations; here, the chosen method is a diagonalization procedure as this method is comparatively intuitive and is amenable to generalization.

Letting ω denote the time frequency, and denoting by $x = (x_1, x_2)$ the directions orthogonal to z, Eq. (2-1) can be written as

$$\partial_z \begin{pmatrix} \phi \\ \partial_z \phi \end{pmatrix} + \begin{pmatrix} 0 & -1 \\ \Delta_x + \dfrac{\omega^2}{c(x,z)^2} & 0 \end{pmatrix} \begin{pmatrix} \phi \\ \partial_z \phi \end{pmatrix} = \begin{pmatrix} 0 \\ 0 \end{pmatrix} \qquad (2-10)$$

This system of equations is now split into two parts, through a diagonalization procedure; the two parts then govern waves propagating in the positive and negative z directions, respectively. This diagonalization procedure is discussed in general in [21] and for the wave equation specifically in [19, 22]; only a summary of the basic ideas is given here without the mathematical details. The diagonalization procedure gives

$$\begin{pmatrix} 0 & -1 \\ \Delta_x + \dfrac{\omega^2}{c(x,z)^2} & 0 \end{pmatrix} = P \begin{pmatrix} iA & 0 \\ 0 & -iA \end{pmatrix} P^{-1} \qquad (2-11)$$

where A is a pseudodifferential operator with $A^2 = -\Delta_x - \dfrac{\omega^2}{c(x,z)^2}$ (see [23] for an introduction to the theory of pseudodifferential operators) and P is a matrix of pseudodifferential operators. Introducing the up- and down-going fields, ϕ_- and ϕ_+

$$\begin{pmatrix} \phi \\ \partial_z \phi \end{pmatrix} = P \begin{pmatrix} \phi_+ \\ \phi_- \end{pmatrix} \qquad (2-12)$$

the original system (2-10) reduces to

$$P^{-1}\partial_z P \begin{pmatrix} \phi_+ \\ \phi_- \end{pmatrix} + \begin{pmatrix} iA & 0 \\ 0 & -iA \end{pmatrix} \begin{pmatrix} \phi_+ \\ \phi_- \end{pmatrix} = \begin{pmatrix} 0 \\ 0 \end{pmatrix} \qquad (2-13)$$

The parabolic approximation is obtained by assuming $\phi_- = 0$ and that $P^{-1}\partial_z P \approx \partial_z$ yielding

$$\partial_z \phi_+ + iA\phi_+ = 0 \qquad (2-14)$$

As in the diagonalization of a matrix in linear algebra the normalization of P can be freely chosen. The accuracy of the approximation $P^{-1}\partial_z P \approx \partial_z$ depends on the choice of this normalization, however, and thus P is chosen to be

$$P = \begin{pmatrix} A^{-\frac{1}{2}} & A^{-\frac{1}{2}} \\ iA^{\frac{1}{2}} & -iA^{\frac{1}{2}} \end{pmatrix} \qquad (2-15)$$

because in this normalization the largest contribution to the error is off-diagonal. This means that the approximation affects only the coupling of up- and down-going waves and not forward propagation (for details, see [19]). The error caused by the assumption that $\phi_- = 0$ depends on the velocity contrast as this determines the strength of the ignored back-scattered field.

Recalling that A is a pseudodifferential operator, the solution of Eq. (2-14) (see [23], Chapter 1) is given by

$$\phi_+(z,x) = \left(\frac{1}{2\pi}\right)^2 \iint e^{i(z-z')\omega\sqrt{c(z,x)^{-2} - \|k_x\|^2}}\, \phi_+(z',x')e^{i\omega k_x\cdot(x-x')}\mathrm{d}x'\mathrm{d}k_x \qquad (2-16)$$

Equation (2-16) can be considered as an operator acting on ϕ_+ at a particular level z′ that propagates this wavefield to a new level, z.

For a general velocity model, the application of (2-16) is not straightforward because the square-root in the first exponential depends on both the space, x, and Fourier, k_x, variables. In the case where the velocity depends only on z, it simplifies to

$$\phi_+(z,x) = \mathcal{F}^{-1}_{k_x \to x}\left(e^{i(z-z')\omega\sqrt{c(z)^{-2} - \|k_x\|^2}}\, \mathcal{F}_{x\to k_x}(\phi_+(z',x'))\right) \qquad (2-17)$$

where $\mathcal{F}_{x\to k_x}$ denotes the spatial Fourier transform from x to k_x. In this model, the velocity is not independent of x, but rather takes on only a few distinct values: $c(z, x) = c_e$ outside the inclusions and $c(z, x) = c_i$ inside the inclusions. A number of approaches have been developed to deal with lateral heterogeneity within the parabolic approximation (see, e.g., [5, 20, 24]). For an initial implementation, the "phase-shift plus interpolation method" of Gazdag and Sguazzero [24] is used, which can be further simplified in this context due to the piecewise constant character of the velocity profile. Note that from (2-16) $\phi_+(z, x)$ depends on $c(z, x)$ and not on $c(z', x')$. In the case considered here, $c(z, x)$ takes on only a few values (e.g. c_e outside of an inclusion and c_i inside). Combining these two properties, the field $\phi_+(z, x)$ can be estimated by propagating the entire field (via 2-17) with both c_e and c_i and then recombining these fields in the space domain by choosing at every point (z, x) the field propagated with the corresponding velocity $c(z, x)$. The total field is then constructed in the space domain by taking, at each point, the field propagated with the appropriate velocity. The simplification of this method as compared with the phase-shift plus interpolation method is that interpolation is not needed to find the field at points with a velocity between the two propagation velocities as the velocity is piecewise constant.

The amplitude correction term, P, is derived in a manner similar to that used above to arrive at Eq. (2-16). These corrections are also applied to each of the two intermediate fields.

To simulate the field excited by the transducer using the parabolic approximation, the definition of an initial source distribution is needed. This is done by using a number of point sources arranged on a portion, defined by the aperture of the transducer, of a circle centered at the desired focus point. As the field is stepped forward in z with Eq. (2-17) the point sources at each z are added into the field propagated from previous z. This method of modeling the transducer ignores the interactions of different elements of the transducer as well as the fact that the transducer itself acts as a scatterer; a similar method of transducer modeling is discussed in [25], while [26] gives an overview of various transducer types.

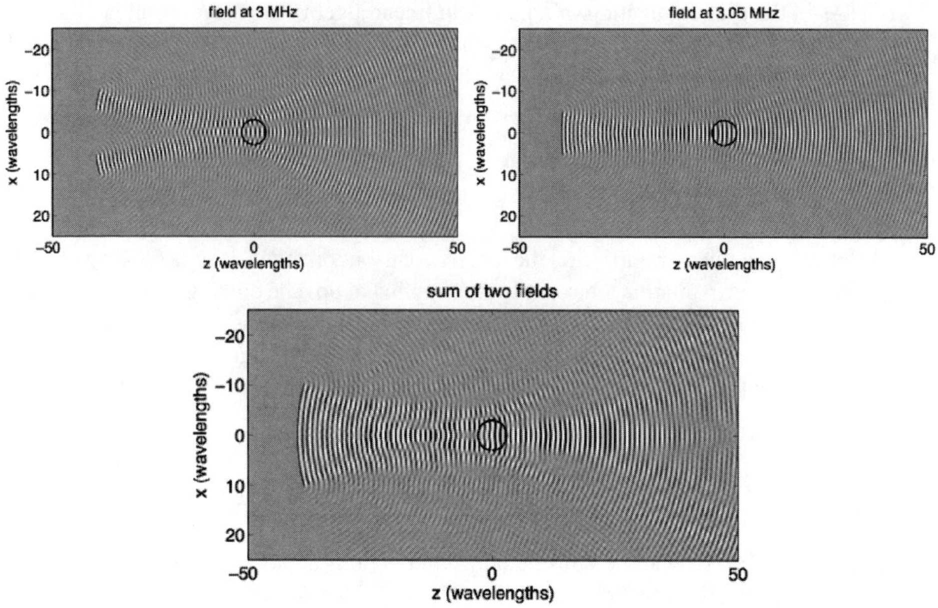

Figure 2-2. The top two figures show the real parts of the two high-frequency fields, propagated to the focus point. The bottom figure is the sum of these two fields, which is the linear interference pattern

Figure 2-2 shows the fields simulated in this way. The focus point is within the obstacle, which has a moderate 10% velocity contrast with the surroundings. With the parabolic approximation, the back-scattered field is ignored, which is a source of error; for this velocity contrast, this error is less than 1%. As is apparent from the figure depicting the sum of the fields at ω_1 and ω_2, the field is complicated by the inclusion resulting in several regions that contribute to the final image rather than a single point as would be the case for a full (360°) aperture transducer in a constant medium.

With the method described above, the dominant source of error in the parabolic approximation itself is that the back-scattered field is ignored. Figure 2-3 illustrates the effects of this approximation by contrasting it with accurate full-wave results derived from the application of the integral equation approach described in the next section. To this end, evaluations of the product $\phi_1\phi_2^*$ are compared, which mimics the behavior of the right hand side of Eq. (5). (This product is chosen here in lieu of the full right-hand side to isolate inaccuracies that arise solely from the use of an approximate one-way wave model and to avoid additional errors that may result from numerical differentiation in evaluating the exact right-hand side.) For this velocity contrast, the parabolic approximation is indeed quite accurate.

2.3.2 Low-Frequency Wavefield — Surface Integral Method

As noted in the discussion in Section 2.2, the two high-frequency beams interact to generate a low-frequency field; it is this low-frequency field that is recorded at the hydrophone. The goal of UVA is to focus this source on as small a region of tissue as possible so that the resulting source of low-frequency energy is close to a point source. Because of the focused nature of this source, the resulting low-frequency field is essentially omni-directional, ruling out the application of the parabolic approximation. As discussed in the Introduction, the use of integral equation methods for the propagation of this field is proposed because

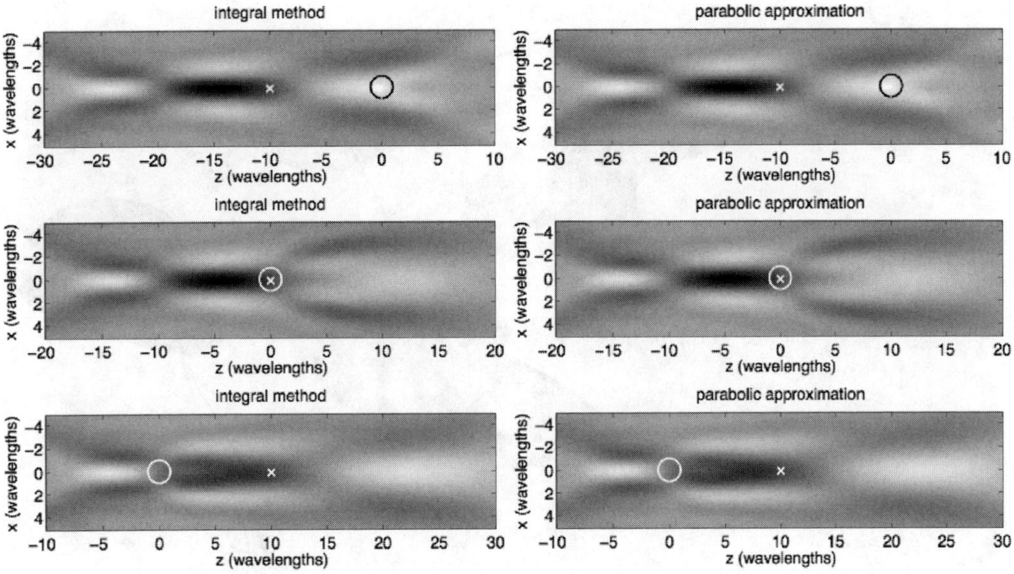

Figure 2-3. Comparison of an approximation of the right-hand side of Eq. (2-5); on the left the field is computed with the integral method of Section 2.3.2 on the right with the parabolic approximation. The circle marks the outline inclusion and the "x" marks the focus point

they compute the field accurately in all directions without the need for absorbing boundary conditions. Specifically, the equations for a surface integral method are developed as such methods require discretization only on the boundary of the inclusions greatly reducing the number of points in the integrations.

The idea behind integral equation methods is to take the differential equations describing the problem and reformulate them as (equivalent) integral equations through the use of Green's Theorem. In order to apply Green's Theorem, the background velocity model must be such that it can be divided into a background region with one or several inclusions, where the Green's function is known in each region. In the case discussed here, this is easily accomplished as the model consists of one (or more) inclusions or scatterers embedded in a background medium where both the background and inclusions have constant (though different) material parameters.

The derivation of these integral equations starts from the basic equations for the field in both the internal (within the inclusions, denoted D) and external (outside the inclusions, denoted D^c) domains (see Fig. 2-4). The field in each of these domains is split into two parts: the known incident field, which is the field in the absence of a scatterer, and the unknown scattered field, which is the difference between the incident and total fields. In other words, a solution is sought for

$$\begin{aligned} \Delta\psi_e^s + k_e^2\psi_e^s &= 0 && \text{in } D^c \\ \Delta\psi_i^s + k_i^2\psi_i^s &= 0 && \text{in } D \\ \Delta\psi_e^{inc} + k_e^2\psi_e^{inc} &= f && \text{in } D^c \\ \Delta\psi_i^{inc} + k_i^2\psi_i^{inc} &= f && \text{in } D \end{aligned} \qquad (2-18)$$

with boundary conditions that require the continuity of the field and its normal derivative across the interface along with the Sommerfeld radiation conditions at infinity. In the above, subscript "e" indicates that a quantity is external to the inclusions and subscript "i" that it is inside the inclusions, the superscript "s" denotes the scattered field and superscript "inc" denotes the incident field. The total field is given by $\psi = \psi^{inc} + \psi^s$ for both the internal

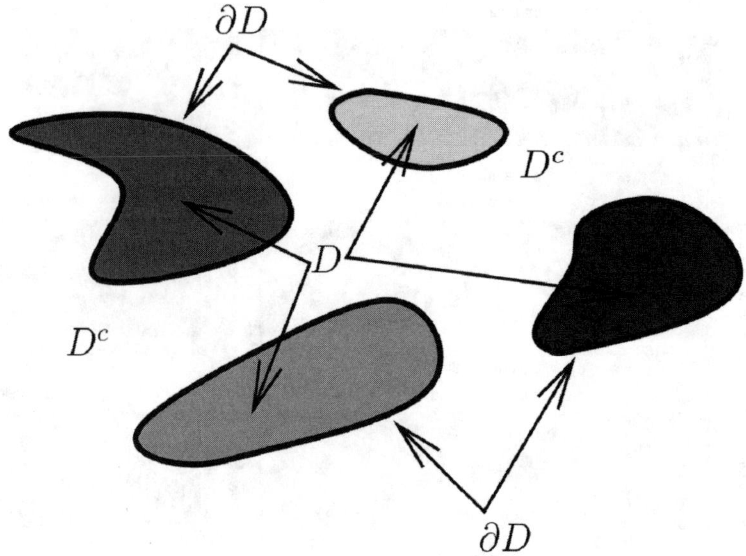

Figure 2-4. For the low-frequency field, the velocity is illustrated with the different colors; the interior, D, is the union of all the colored regions, the exterior, D^c, is the white region and the boundary, ∂D, consists of union of the black lines

and external fields and the source function is denoted by f. In what follows, knowledge of ψ^{inc} is assumed; this field is calculated using the relations

$$\psi_i^{inc}(x) = \int_D G_i(x,y) f(y)\mathrm{d}y \qquad x \in \mathbf{R}^2,$$
$$\psi_e^{inc}(x) = \int_{D^c} G_e(x,y) f(y)\mathrm{d}y \qquad x \in \mathbf{R}^2 \qquad (2-19)$$

The numerical evaluation of these integrals, along with a discussion of how f is estimated from $\phi_{1,2}$ is the subject of the next subsection.

To derive an integral equation formulation of (2-18), Green's Theorem is used to obtain

$$\psi_i^s(x) = \int_{\partial D} G_i(x,y)\frac{\partial \psi_i^s}{\partial n_y}(y) - \frac{\partial G_i}{\partial n_y}(x,y)\psi_i^s(y)\mathrm{d}y \qquad x \in D \qquad (2-20)$$

$$0 = \int_{\partial D} G_i(x,y)\frac{\partial \psi_i^{inc}}{\partial n_y}(y) - \frac{\partial G_i}{\partial n_y}(x,y)\psi_i^{inc}(y)\mathrm{d}y \qquad x \in D \qquad (2-21)$$

for the interior field, where G_i is the Green's function for the interior domain, D, defined by

$$\Delta G_i^{inc} + k_i^2 G_i^{inc} = -\delta \qquad \text{in } D \qquad (2-22)$$

with G_e defined similarly in the exterior domain, D^c. The symbol $\frac{\partial}{\partial n_y}$ and ∂D denote the normal derivative with respect to y and the boundary of D respectively. The analogous expressions for the exterior field are

$$\psi_e^s(x) = \int_{\partial D} \frac{\partial G_e}{\partial n_y}(x,y)\psi_e^s(y) - G_e(x,y)\frac{\partial \psi_e^s}{\partial n_y}(y)\mathrm{d}y \qquad x \in D^c \qquad (2-23)$$

$$0=\int_{\partial D}\frac{\partial G_e}{\partial n_y}(x,y)\psi_e^{inc}(y)-G_e(x,y)\frac{\partial \psi_e^{inc}}{\partial n_y}(y)dy \qquad x\in D^c \qquad (2-24)$$

The above expressions allow for the calculation of the field either inside or outside the inclusions from knowledge of the field on the boundary. The scattered field is not known a priori, however, and in fact the goal is to recover this scattered field, requiring an expression linking the, known, incident field on the boundary to the, unknown, scattered field also on the boundary. This relationship is obtained from the limit as $x \to \partial D$ of the above equations, resulting in

$$\frac{1}{2}\psi_i^s(x)=\int_{\partial D}G_i(x,y)\frac{\partial \psi_i^s}{\partial n_y}(y)-\frac{\partial G_i}{\partial n_y}(x,y)\psi_i^s(y)dy \qquad x\in \partial D \qquad (2-25)$$

$$\frac{-1}{2}\psi_i^{inc}(x)=\int_{\partial D}G_i(x,y)\frac{\partial \psi_i^{inc}}{\partial n_y}(y)-\frac{\partial G_i}{\partial n_y}(x,y)\psi_i^{inc}(y)dy \qquad x\in \partial D \qquad (2-26)$$

for the interior field and

$$\frac{1}{2}\psi_e^s(x)=\int_{\partial D}\frac{\partial G_e}{\partial n_y}(x,y)\psi_e^s(y)-G_e(x,y)\frac{\partial \psi_e^s}{\partial n_y}(y)dy \qquad x\in \partial D \qquad (2-27)$$

$$\frac{-1}{2}\psi_e^{inc}(x)=\int_{\partial D}\frac{\partial G_e}{\partial n_y}(x,y)\psi_e^{inc}(y)-G_e(x,y)\frac{\partial \psi_e^{inc}}{\partial n_y}(y)dy \qquad x\in \partial D \qquad (2-28)$$

for the external field. In taking the above limits, use is made of the fact that the integral containing G is a single-layer potential and the integral containing $\frac{\partial G}{\partial n_y}$ is a double-layer potential allowing us to apply the result of [13] (Theorem 2.1); further details on single- and double-layer potentials and their limiting values can be found in [27] (Chapter 2).

Summing (2-25) to (2-28) and applying the boundary conditions to the result gives

$$\psi_e(x)=\psi^{inc}(x)+\int_{\partial D}\frac{\partial(G_i-G_e)}{\partial n_y}(x,y)\psi_e(y)-(G_i-G_e)(x,y)\frac{\partial \psi_e}{\partial n_y}(y)dy \qquad (2-29)$$

recalling that $\psi_e=\psi_e^{inc}+\psi_e^s$ and defining $\psi^{inc}=\psi_i^{inc}+\psi_e^{inc}$. A similar procedure applied to the normal derivatives of the initial equations leads to

$$\frac{\partial \psi_e(x)}{\partial n_x}=\frac{\partial \psi^{inc}}{\partial n_x}(x)+\int_{\partial D}\frac{\partial^2(G_i-G_e)}{\partial n_x\partial n_y}(x,y)\psi_e(y)-\frac{\partial(G_i-G_e)}{\partial n_x}(x,y)\frac{\partial \psi_e}{\partial n_y}(y)dy \qquad (2-30)$$

The coupled system of (surface) integral equations (2-29) to (2-30) provides a complete mathematical formulation of the propagation of the low-frequency emission. Indeed, once these are solved, the field at any point in space (and, in particular, at the hydrophone) can be recovered from Eqs. (2-20) to (2-26).

The numerical solution of both the boundary system for the total field (2-29) to (2-30) and the integrals to determine the field (2-20), (2-23), on the other hand, are complicated by the singular nature of the Green's functions (which demand a careful design of quadrature formulas for accurate evaluation) and their long-range effects (which translate into full matrices upon discretization, necessitating the use of acceleration techniques for efficiency). To overcome the former difficulty in solving Eqs. (2-29) and (2-30), it is necessary to resort to spectral quadratures as defined in [13] (pp. 67–78), in the two-dimensional case shown here; an extension to three dimensions is given in [28]. Acceleration techniques that allow for fast matrix-vector multiplications are also available, for instance, with schemes based on fast multipole expansions [12, 29] or FFTs [10, 28]. For Eqs. (2-20) and (2-23), determining the value of $\psi^{s}_{i,e}(x)$ is straightforward when x is far from the boundary ∂D because at these points, G, $\frac{\partial G}{\partial n_y}$ and ψ are all smooth. When x is near the boundary, however, the singularities in G and $\frac{\partial G}{\partial n_y}$ makes the accurate evaluation of these integrals difficult. To avoid this problem, a polynomial resampling is used as suggested in [13] (Eq. 3.81). The idea behind such a method is to evaluate

$$h(x) = \int_a^b G(x,y) f(y) \mathrm{d}y \tag{2-31}$$

where G is singular at a and f is smooth through the change of variables $y = p(s)$ giving

$$h(x) = \int_a^b G(x,p(s)) f(p(s)) p'(s) \mathrm{d}s \tag{2-32}$$

enforcing that $p(a) = a$ and $p(b) = b$ and that $p'(a)$, $p''(a)$, $p^{(I)}(a) = 0$ for some order I. The composite function $G(x,p(s))p'(s)$ is smoother in s than $G(x,y)$ is in y resulting in a more rapidly converging function. From a geometrical point of view, the effect of this change of variables is to increase the density of points in the vicinity of $x = a$; this effect is stronger for larger I. The function G is assumed to be known so its evaluation on the irregular grid is straightforward. By contrast, f is not known, but is smooth, allowing for its accurate interpolation with a combination of trigonometric interpolation (to a denser, still regular, grid) and spline interpolation (to the irregular grid defined by $p(s)$). The integrals in Eqs. (2-20) and (2-23) are around closed domains, allowing a and b to be chosen in (2-32) such that a is the closest point on ∂D to x. The integral is then properly sampled resulting in an accurate value of $\psi^{s}_{e,i}(x)$.

2.3.3 Three-Dimensional Volume Integral Approach

To extend these ideas to the case where the obstacle is no longer piecewise homogeneous, the use of a method based on the solution of a suitable Lippman-Schwinger type integral equation is proposed. To investigate the errors made in assuming two-dimensional wave propagation, as was assumed previously, these integral equations are set up and solved in three dimensions. Lippmann-Schwinger type integral equations are derived by writing the basic (acoustic or elastic) differential model so that inhomogeneities are viewed as providing a source for a wave equation in a homogeneous background. Explicitly, Eq. (2-1) is written in the temporal frequency domain as

$$\Delta\hat{\phi}_j + \frac{\omega_j^2}{c(x)^2}\hat{\phi}_j = 0 \tag{2-33}$$

for the total field, $\hat{\phi}_j$, in the frequency domain; the incident field, $\hat{\phi}_j^{\text{inc}}$ satisfies

$$\Delta\hat{\phi}_j^{\text{inc}} + \frac{\omega_j^2}{c_0^2}\hat{\phi}_j^{\text{inc}} = 0 \tag{2-34}$$

Differencing these two equations gives

$$\Delta\hat{\phi}_j^s + \frac{\omega_j^2}{c_0^2}\hat{\phi}_j^s = k_{j0}^2 \frac{c(x)^2 - c_0^2}{c(x)^2}\hat{\phi}_j \qquad (2-35)$$

where $k_{j0} = \frac{\omega_j}{c_0}$; from this, by analogy to Eq. (2-19), the solution of Eq. (2-33) can be written as

$$\hat{\phi}_j(x) = \hat{\phi}_j^{\text{inc}}(x) + k_{j0}^2 \int_D G_j(x,y)\frac{c(y)^2 - c_0^2}{c(y)^2}\hat{\phi}_j(y)dy \qquad (2-36)$$

defining $G(x,y)$ as the Green's function for Eq. (2-34). To compute the incident field, $\hat{\phi}_j^{\text{inc}}$, Eq. (2-19) is solved where $f(y)$ is replaced by the set of point sources used to model the transducer, as discussed in Section 2.3.1. Assuming that the total velocity $c(x)$ consists of a background c_0 with a localized perturbation near the obstacle, it is noted that $c_0 - c(x)$ is zero outside some region, B, surrounding the obstacle. Thus, it is necessary to solve Eq. (2-36) for $\hat{\phi}_j$ only within B; from this solution, it is possible to calculate $\hat{\phi}_j$ at any position by inserting the values of $\hat{\phi}_j$ within the region B into the integral.

In the above derivation, it has been assumed that the background velocity, c_0 is constant, and also that $c(x)$ is piecewise smooth and supported (non-zero) in a limited area. This opens up the possibility, excluded by the surface integral method of the previous subsection, of modeling inclusions with variable velocity. As with the surface integral method discussed in the previous case, this method generalizes easily to the elastic case. In particular, the elastic analog of Eq. (2-36) is

$$-\rho_0(\delta\omega)^2\vec{U} = (\lambda_0 + \mu_0)\nabla(\nabla \cdot \vec{U}) + \mu_0\nabla^2\vec{U} + F(\vec{U}, \lambda, \mu, \rho) \qquad (2-37)$$

where, as above, a subscript 0 indicates a constant (background) quantity and quantities without subscript vary in space in a limited area. The notations λ, μ, are used for the Lamé constants, and $\vec{U}$ for the vector displacement. In this form, Eq. (2-37) can be solved using the Green's function for the background homogeneous medium resulting again in a (Lippmann-Schwinger) integral equation

$$\vec{U} = \vec{U}^{\text{inc}} + G * F(\vec{U}, \lambda, \mu, \rho) \qquad (2-38)$$

where the incident field, $\vec{U}^{\text{inc}}$, is computed as for the acoustic case.

As in the surface integral method of the previous subsection, the computation of the integral in Eq. (2-36) is complicated by the singularity in the Green's function when $x = y$ and the full matrices resulting from the discretization of Eq. (2-36). To overcome the singularity issue, a so-called weakening procedure is used as proposed in [30-32]. The idea behind such a procedure is to replace the Green's function by its spherical average. In other words, we replace G with

$$G(x,y) = \frac{\int_{R<\frac{\Delta y}{2}} \mathrm{G}(x, y_1 + y_1', y_2 + y_2', y_3 + y_3')\mathrm{d}y_1\mathrm{d}y_2\mathrm{d}y_3}{\int_{R<\frac{\Delta y}{2}} \mathrm{d}y_1\mathrm{d}y_2\mathrm{d}y_3} \qquad (2-39)$$

where y has been split into its Cartesian components. The advantage of this approach is that Eq. (2-39) has a closed form solution; this method has a second-order convergence rate, meaning that for large numbers of grid points, if the grid size Δy is reduced to $\Delta y/2$ the error, E, is reduced to $\frac{E}{4}$.

To solve the linear system efficiently and accurately in three dimensions, use is made of the so-called conjugate gradient FFT (CGFFT). This technique is particularly appropriate for this problem as the Green's function G depends only on the difference $x - y$ making the integral in Eq. (2-36) a convolution. In the CGFFT method, the FFT is used to evaluate this convolution at each iteration of the conjugate gradient method (see, e.g., [33] for details on conjugate gradient). Further improvements to this method can be made with appropriate pre-conditioners as discussed in [33] (Chapter 8) or [34] (Chapter 2); here, the adjoint of the Helmholtz operator is used as a pre-conditioner to make the system symmetric, which is known as the CGNR method.

By applying the above procedure at each of the two high frequencies, an estimate is obtained of both of the high-frequency fields, analogous to the results of propagating these fields with the parabolic approximation. The difference is that these solutions do not require the one-way assumption of the parabolic approximation and that the computations have been done in three dimensions. This gives a basis to both compare future three-dimensional implementations of the parabolic approximation (as in Fig. 2-3 for two-dimensional version) and to investigate the errors made in reducing the three-dimensional experimental problem to a two-dimensional model; of course, this additional knowledge comes with additional computational cost. From the two fields, $\hat{\phi}_{1,2}$ the right-hand side of Eq. (2-5) is then computed, with $f(y)$ as discussed in the following subsection. Equation (2-36) is then solved at the low frequency (k becomes $\frac{\delta\omega}{c_0}$) using $f(y)$ as the source distribution that generates the incident field as in Eq. (2-19).

2.3.4 Interaction Term — Direct Calculation

The interaction term, given on the right-hand side of Eq. (2-5), generates the source $f(y)$, in Eq. (2-18); this interaction term is not particularly difficult to compute as it consists simply of the derivatives of known fields. Since the computations are performed in the temporal frequency domain, the time derivative is computed using a straightforward multiplication by $i\omega$. The spatial gradient is slightly more complicated to compute and there are several options. For the figures that follow, the (windowed) spatial Fourier transform of the field is taken and the gradient is computed by multiplication in the Fourier domain. Both this method of calculating the derivative and the propagation of the field using the parabolic approximation are computed on a rectangular grid to make use of the standard FFT.

In Fig. 2-5, the imaginary part of $f(y)$ is shown, filtered so that only contributions at the difference frequency contribute to the result. (In the actual experiment, the hydrophone performs this filtering as it is simply not sensitive to the high-frequency fields nor to the field at the sum frequency.) The focused nature of this field makes it similar to a point-source excitation at the difference frequency; however, the shape in the figure shows clearly that the field is not simply a point excitation. Thus, the simulation, provides an improvement on simply using a focused beam at $\delta\omega$; this is also an advantage of ultrasound vibroacoustography since the size of the focal area is governed by $\omega_{1,2}$ instead of by $\delta\omega$.

Although it is the goal to focus the high-frequency energy on a small circular region this is not possible and the focused field has, in general, a more complicated shape as is illustrated in Fig. 2-5. As mentioned in the introduction, in order to accurately represent this distribution, denoted $f(y)$ above, it must be discretized on the order of the high-frequency wavelength. In any method, the presence of this second length scale calls for the use of variable grid sizes in which the vicinity of the source is sampled more densely than other regions. In a surface integral method, the source is integrated to give an incident field ψ^{inc} on the surface of the obstacle as in Eq. (2-19). The normal derivative is computed from similar expressions where G and ψ^{inc} are replaced by their normal derivatives. If the source

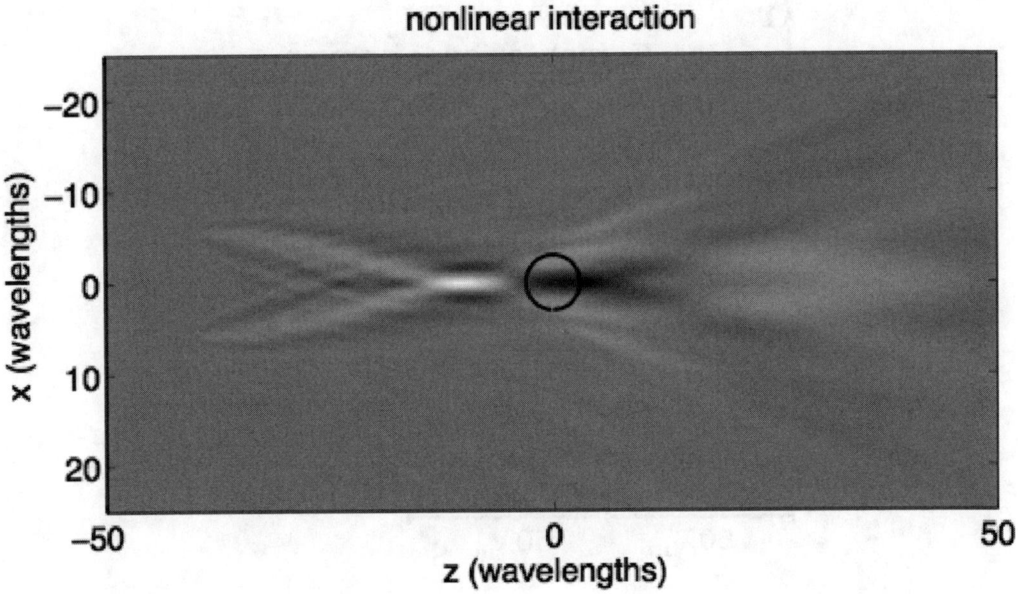

Figure 2-5. The imaginary part of the right-hand side of Eq. (2-5) computed for the same model as the fields shown in Figure 2.2

is far from the obstacle, the calculation of these integrals is straightforward and so variable grids are not needed. When the source is near the obstacle, it is more difficult to calculate because ψ^{inc} is evaluated on the boundary (i.e., $x \in \partial D$) and if $f(y)$ is also supported there (i.e., if the focal area intersects ∂D), the integrand becomes singular. Doing this computation on a regularly spaced rectangular grid is not efficient because of this singularity; an extremely dense mesh would be necessary to adequately sample the function near the singularity, making the cost of propagating the high-frequency fields untenable. Because the singularity is along the boundary of the obstacle, it is possible to compute the integral accurately and efficiently with two changes of variables. The first step is to interpolate to a coordinate system in which one coordinate is tangent to the obstacle and the other normal to it; for the case of a circular obstacle, as in the examples shown here, this is a polar grid. Following this, the normal variable is resampled in such a way as to place many more points in the vicinity of the boundary, using the polynomial resampling function suggested in [13] (p. 75) and discussed in the previous subsection. This allows the accurate calculation of the integral without requiring the computation of the field at too many points. Further improvements are possible to improve the efficiency of this method, e.g., based on the ideas in [35] to constrain the resampling to the vicinity of the boundary; their analysis and implementation, however, is left for future work.

2.4 PRELIMINARY RESULTS

This section describes preliminary, primarily two-dimensional, results using the numerical models discussed above. Two two-dimensional cases are shown in which the obstacle is relatively small and circular with a diameter of 6λ, where $\lambda = 0.5$ mm is the wavelength (in water) of the high-frequency field. The background has an acoustic velocity of 1.5 mm/μs, that of water; the obstacle has an acoustic velocity of 1.65 mm/μs in the first case and 6 mm/μs in the second case. The first case has a velocity contrast within the range of validity of the parabolic approximation, to explore the effects of different experimental configurations

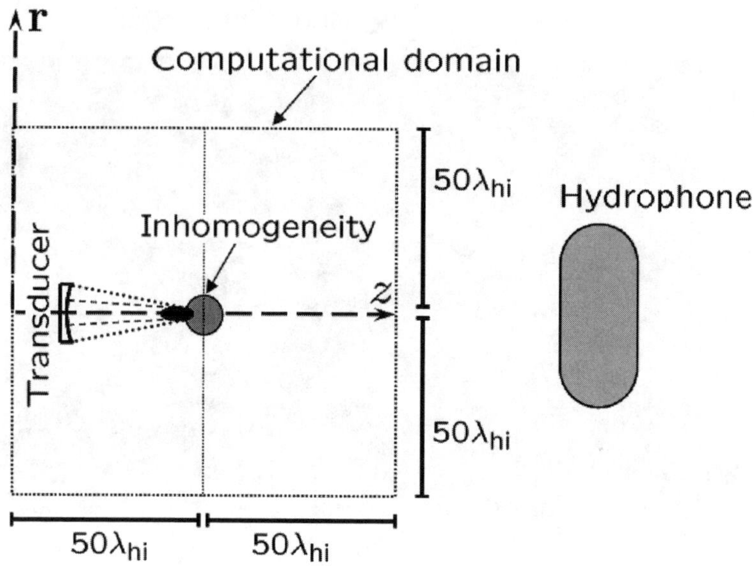

Figure 2-6. Experimental configuration for both computational experiments shown below

and to compare with three-dimensional results. The second case is compared with experimental data. For the three-dimensional result, using the volume integral equation approach outlined in Section 2.3.3 a cubic obstacle is used with dimension 3λ. High frequencies of 3 MHz and 3.05 MHz are used, and a focal length of $40\lambda = 2$ cm, with an aperture of 30°

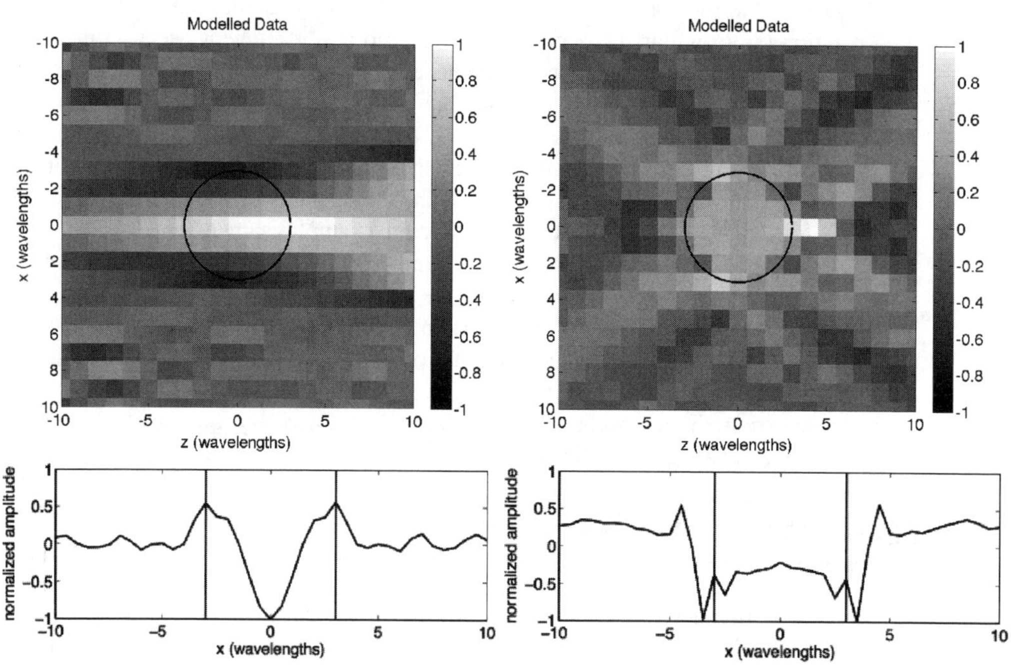

Figure 2-7. On the left is an image made with a 30° aperture as is used in the comparisons with experimental data shown in Figure 2.9. On the right is an image made with a 120° aperture. The one-dimensional images shown below are from $z = 0$, with a denser sampling in x than was used in the two-dimensional images above

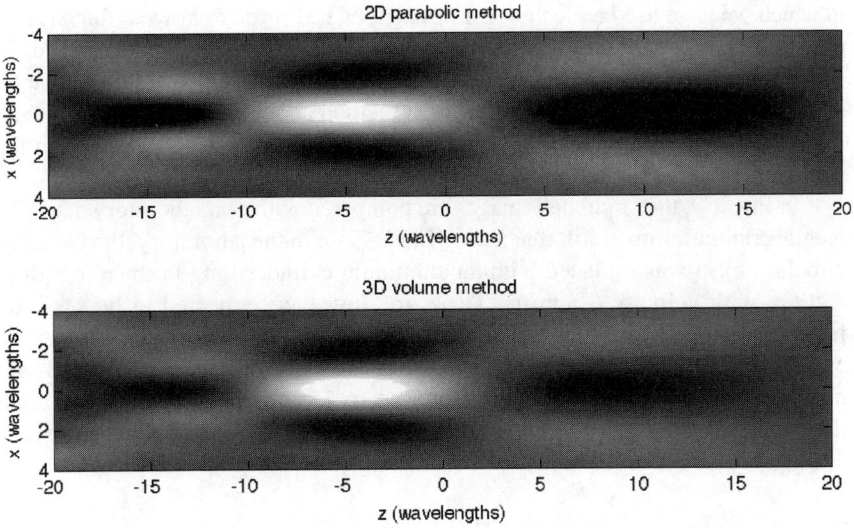

Figure 2-8. A comparison of $\phi_1\phi_2^*$ computed with two- and three-dimensional methods

unless stated otherwise. A hydrophone has been placed along the x-axis directly across from the transducer to mirror the experimental configuration. The experimental setup is shown in Fig. 2-6.

For this configuration, with a small contrast, the images shown in Fig. 2-7 are obtained. The smearing visible in the image is characteristic of the small aperture used. An image

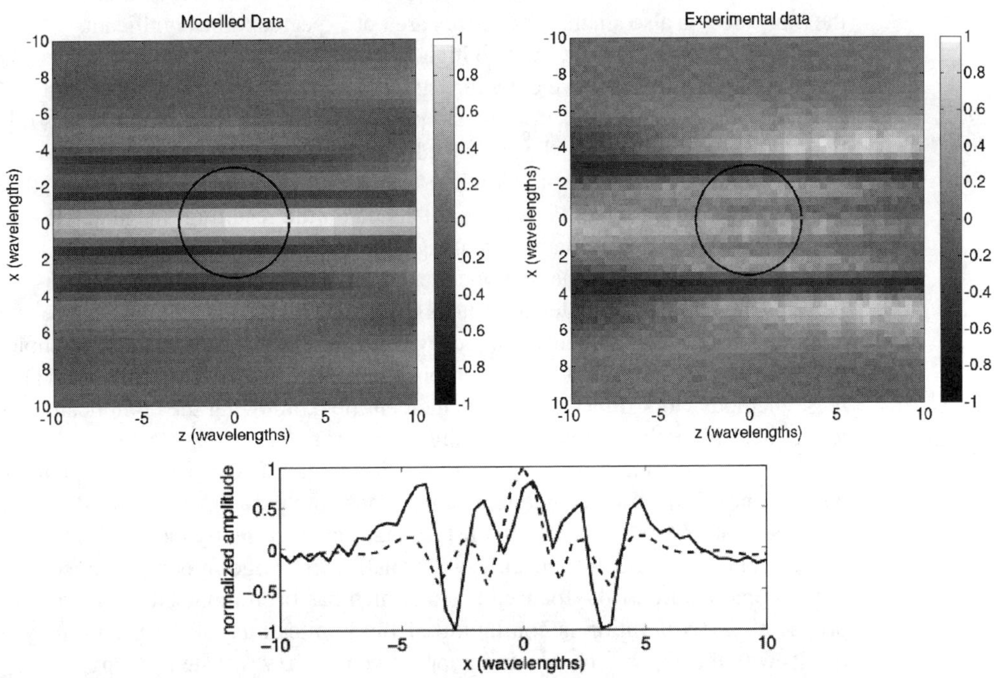

Figure 2-9. Upper left is the synthesized image, on the upper right is the experimental image; below is a comparison of a slice of the image for fixed $z = 0$

in which we have used a much larger aperture of 120° is also shown (which is not possible in the laboratory), resulting in a notable improvement in the computed image. Figure 2-8 shows the results of using the three-dimensional volume integral method; similar to Fig. 2-5, $\phi_1\phi_2^*$ computed with the three-dimensional method and with the two-dimensional method. The two results are quite similar, justifying the two-dimensional assumption for this experimental configuration.

In Fig. 2-9, the simulated images are compared with real laboratory data. The experimental configuration is still that shown in Fig. 2-6; in the laboratory, the two-dimensional circular object was replaced with an aluminum cylinder long in the z (out of plane) direction, with velocity 6 mm/μs. There are, however, expected to be some departures from a truly two-dimensional geometry because the focus of the transducer used in the experiment is three dimensional. However, a general agreement is still seen with the main structures of the image being reproduced. Aluminum has a very high contrast with the surrounding water medium, meaning that back-scattered energy is more important in this case.

2.5 SUMMARY

The material presented here is designed to explore a possible approach to the modeling of UVA, using the properties of each step in the imaging procedure to design or select methods specific to each stage. This is essential to the modeling of UVA because of the different properties of the wavefields at different stages.

For the high-frequency field excited by the transducer the parabolic approximation is found to be both appropriate and efficient for the modeling of this field. This approximation introduces errors at large propagation angles; errors that are larger with larger velocity contrasts. Because the acoustic velocity contrast in tissue is relatively small and the aperture of the transducer is also small, these errors are not expected to be significant.

Once the interaction of the two high-frequency beams is understood from the higher-order correction to the wave equation, its numerical computation is straightforward. It is essential, however, that this non-linear coupling be taken into account as it is this interaction that is responsible for the improved resolution possible from UVA as compared to imaging techniques based solely on low-frequency excitations. Furthermore, the shape of the low-frequency source has a large influence on the resulting image, as illustrated with the comparison of different aperture images.

The field recorded at the hydrophone, which is the field resulting from this interaction, presents the modeling challenge of taking into account the distributed source and yet still modeling efficiently at the low frequency. Integral methods are the natural choice for this because they are particularly computationally efficient at low frequencies and also because these methods work directly from the incident field, allowing the complicated source pattern to be dealt with in a relatively straightforward manner.

The combination of these approaches allows numerical modeling of the fields excited with this novel imaging technique. Comparisons with experimental data show recovery of the general structure of the data, while the details are incorrectly modeled. There are several sources for these errors, the most likely of which are the large velocity contrast and the three-dimensional nature of the focused beam, which has been modeled with two-dimensional propagation. Exploration is continuing of the implications of the preliminary modeling results with the hope of introducing improvements in UVA systems being developed at the Mayo Clinic. Improving the ability of the model to take into account the true structures of interest requires the extensions of the methods presented here to three dimensions (for the parabolic approximation and surface integral method) as well as to the elastic case.

Although a complete solution has been presented to the forward problem of predicting the response of a known medium, it is the inverse problem that is of most interest for medical diagnosis. In particular, the shear modulus is of most interest; this modulus can only be recovered reliably by employing shear waves in an elastic model. It is for this reason that the preliminary setup for the solution of the elastic model is discussed here, the further development of which is a subject of continuing research.

Acknowledgments

The authors are grateful for many useful discussions on this topic with Farid Mitri and Xiaoming Zhang of the Mayo Clinic and Bojan Guzina of the University of Minnesota. Thanks are also due to Randall Kinnick for collecting the experimental data.

REFERENCES

[1] Fatemi, M., and Greenleaf, J. F., 1998, "Ultrasound Stimulated Vibro-Acoustic Spectrography," *Science*, **280**, pp. 82 – 85.

[2] Fatemi, M., and Greenleaf, J. F., 1999, "Vibro-Acoustography: an Imaging Modality Based on Ultrasound-Stimulated Acoustic Emission," *Proceedings of the National Academy of Sciences of the United States of America*, **96**, pp. 6603–6608.

[3] Heikkilä, J., Karjalainen, T., Vaukonen, M., and Hynynen, K., 2004, "3D Simulations of Difference Frequency Effects on a Blood Vessel in Ultrasound Stimulation Vibro-Acoustography," 2004, *IEEE Ultrasonics Symposium*, pp. 1247–1250.

[4] Varslot, T., and Måsøy, S. -E., 2006, "Forward Propagation of Acoustic Pressure Pulses in 3D Soft Biological Tissue," *Modeling, Identification and Control*, **27**, pp. 181.

[5] Bamberger, A., Engquist, B., Halpern, L., and Joly, P., 1988, "Higher Order Paraxial Wave Equation Approximations in Heterogeneous Media," *SIAM Journal of Applied Mathematics*, **48**, pp. 129–154.

[6] Claerbout, J.F., 1985, *Imaging the Earth's Interior*, Blackwell Sci. Publ., Palo Alto, CA.

[7] Tappert, F. D., 1977, "The Parabolic Approximation Method," In Keller, J. B. and Papadakis, J. S., eds., Wave Propagation and Underwater Acoustics, Springer-Verlag, Berlin.

[8] Hasegawa, T., Kido, T., Iizuka, T., and Matsuoka, C., 2000, "A General Theory of Rayleigh and Langevin Radiation Pressures," *Journal of the Acoustical Society of Japan (E)*, **21**, pp. 145–152.

[9] Olsen, H., Romberg, W., and Wergeland, H., 1958, "Radiation Force on Bodies in a Sound Field," *Journal of the Acoustical Society of America*, **30**, pp. 69–76.

[10] Bleszynski, E., Bleszynski, M., and Jaroszewicz, T., 1996, "AIM: Adaptive Integral Method for Solving Large-Scale Electromagnetic Scattering and Radiation Problems," *Radio Science*, **31**, p. 1225.

[11] Xu, X. M., and Liu, Q. H., 2001, "Fast Spectral-Domain Method for Acoustic Scattering Problems," *IEEE Transactions on Ultrasonics, Ferroelectrics and Frequency Control*, **48**, pp. 522 – 529.

[12] Coifman, R., Rokhlin, V., and Wandzura, S., 1993, "The Fast Multiple Method for the Wave Equation: a Pedestrian Prescription," *IEEE Antennas and Propagation Magazine*, **35**, pp. 7–12.

[13] Colton, D., and Kress, R., 1998, *Inverse Acoustic and Electromagnetic Scattering*, 2nd ed., Springer, Berlin.

[14] Silva, G. T., Zeraati, M., and Fatemi, M., 2001, "Calculation of Radiation Force on Cylinders Based on Diffraction and Ray Approximation," 2001 *IEEE Ultrasonics Symposium*, pp. 1237–1240.

[15] Silva, G. T., Chen, S., and Viana, L. P., 2006, "Parametric Amplification of the Dynamic Radiation Force of Acoustic Waves in Fluids," *Physical Review Letters*, **96**, pp. 234301.

[16] Torr, G. R., 1984, "The Acoustic Radiation Force," *American Journal of Physics*, **52**, pp. 402 – 408.

[17] Ostrovsky, L., Sutin, A., Il'inkskii, Y., Rudenko, O., and Sarvazyan, A., 2007, "Radiation Force and Shear Motions in Inhomogeneous Media," *Journal of the Acoustical Society of America*, **121**, pp. 1324–1331.

[18] Callé, S., Remenieéras, J.-P., Pata, F., and Matar, O., 2003, "Application of Vibro-Acoustography to Tissue Elasticity Imaging," *Acta Acustica United with Acustica*, **89**, pp. 936–941.

[19] Stolk, C. C., 2004, "A Pseudodifferential Equation with Damping for One-Way Wave Propagation in Inhomogeneous Acoustic Media," *Wave Motion*, **40**, pp. 111–121.

[20] Jensen, F., Kuperman, W., Porter, M., and Schmidt, H., 2000, Computational Ocean Acoustics. Series in Modern Acoustics and Signal Processing. AIP.

[21] Taylor, M. E., 1981, *Pseudodifferential Operators*, Princeton University Press, Princeton, NJ.

[22] de Hoop, M. V., 1996, "Generalization of the Bremmer Coupling Series," *Journal of Mathematical Physics*, **37**, pp. 3246–3282.

[23] Treves, F., 1980, *Introduction to Pseudodifferential and Fourier Integral Operators*, Plenum Press, New York, Vol. 2.

[24] Gazdag, J., and Sguazzero, P., 1984, "Migration of Seismic Data by Phase Shift Plus Interpolation," *Geophysics*, **49**, pp. 124–131.

[25] Piwakowski, B., and Sbai, K., 1999, "A New Approach to Calculate the Field Radiated from Arbitrarily Structured Transducer Arrays," *IEEE Transactions on Ultrasonics, Ferroelectrics and Frequency Control*, **46**, pp. 422 – 440.

[26] Angelsen, B. A. J., Torp, H., Holm, S., Kristoffersen, K., and Whittingham, T. A., 1995, "Which Transducer Array is Best?," *European Journal of Ultrasound*, **2**, pp. 151–164.

[27] Colton, D., and Kress, R., 1983, *Integral Equation Methods in Scattering Theory*, Springer.

[28] Bruno, O., and Kunyansky, L., 2001, "A Fast, High-Order Algorithm for the Solution of Surface Scattering Problems: Basic Implementation, Tests and application," *Journal of Computational Physics*, **169**, pp. 80–110.

[29] Chew, W. C., Song, J. M., Cui, T. J., Velamparambil, S., Hastriter, M. L., and Hu, B., 2004, "Review of Large Scale Computing in Electromagnetics with Fast Integral Equation Solvers," *CMES Computer Modelling in Engineering and Sciences*, **5**, pp. 361–372.

[30] Abubakar, A., 2000, "Three-Dimensional Nonlinear Inversion of Electrical Conductivity," Ph.D. thesis, Delft University.

[31] Richmond, J., 1965, "Scattering by a Dielectric Cylinder of Arbitrary Cross Section Shape," *IEEE Transactions on Antennas and Propagation*, **13**, pp. 334–341.

[32] Vainikko, G., 1993, *Multidimensional Weakly Singular Integral Equations, Volume 1549 of Lecture Notes in Mathematics*, Springer-Verlag, Berlin.

[33] Greenbaum, A., 1997, *Iterative Methods for Solving Linear Systems, volume 17 of Frontiers in Applied Mathematics*, Society for Industrial and Applied Mathematics (SIAM), Philadelphia, PA.

[34] Fokkema, J. T., and van den Berg, P. M., 1993, *Seismic Applications of Acoustic Reciprocity*, Elsevier, Amsterdam.

[35] Anand, A., and Reitich, F., 2007, "An Efficient High-Order Algorithm for Acoustic Scattering from Penetrable Thin Structures in Three Dimensions," *Journal of the Acoustical Society of America*, **151**, pp. 2503–2514.

Chapter 3

STRAIN INDUCED BY DUAL ACOUSTIC RADIATION FORCE AND ITS ULTRASONIC MEASUREMENT

Hideyuki Hasegawa and Hiroshi Kanai
Graduate School of Biomedical Engineering, Tohoku University, Sendai 980-8579, Japan

ABSTRACT

There are many studies on measurement of tissue mechanical properties by applying an acoustic radiation force induced by ultrasound to an object. However, one acoustic radiation force along a direction (e.g., vertical direction) does not generate the strain of an object effectively because it also causes change in the object position, which has zero spatial gradient in displacement (= no strain). Especially, when the elastic modulus of the object is much larger than that of the surrounding tissue (such like a tumor in the breast tissue), one acoustic radiation force might generate only the change in position of the object and the strain of the object is hardly generated. In such cases, mechanical properties of the object cannot be evaluated. In this study, two cyclic acoustic radiation forces are simultaneously applied to an object along two different directions (e.g., two opposite horizontal directions) to effectively generate the strain inside the object even when the object is much harder than the surrounding tissue.

3.1 INTRODUCTION

In recent years, some remote actuation methods based on acoustic radiation force have been reported. Fatemi and coworkers proposed an imaging method called ultrasound stimulated acoustic emission (USAE) [1, 2]. Their system consists of two confocal ultrasonic transducers, and two ultrasound beams with two slightly different frequencies of f and $(f + \Delta f)$ are transmitted. Acoustic radiation pressure, $P_R(t)$, exerted on the interface between two different media is a function of the energy density, $e(t)$, and the specific acoustic impedances, Z_1 and Z_2, of the media [3]. The energy density, $e(t)$, is proportional to the square of the sum of the sound pressures, $p_1(t)$ and $p_2(t)$, generated by the two transducers. In the intersectional area of these two beams, therefore, an oscillatory radiation pressure $P_R(t)$ at the frequency difference, Δf, is applied to the interface. The radiation force produces acoustic emission that is closely related to the mechanical frequency response of the medium. By measuring the acoustic emission with a hydrophone, hard inclusions in soft material were experimentally detected. The spatial resolution in the depth direction corresponds to the size of the intersectional area.

Nightingale et al [4] proposed an alternative imaging method in which the pulsed ultrasound is employed for applying the radiation force to a soft tissue during short durations (less than 1 millisecond). The viscoelastic properties of the tissue are measured from the magnitude and the transient response of the displacement, $d(t)$, of the tissue [4-6]. In order to generate measurable displacement by several ultrasonic pulses, high-intensity pulsed ultrasound of 1000 W/cm^2 is employed. According to the safety guideline shown by the Japan Society of Ultrasonics in Medicine (JSUM), however, the intensity of ultrasound is recommended to be less than 240 mW/cm^2 and 1 W/cm^2 for the pulsed and continuous waves, respectively [7]. Therefore, the intensity of the pulsed ultrasound employed by Nightingale et al [4] is far greater than that indicated in the safety guideline.

To improve the spatial resolution in measurement of the response of the object, Michishita et al used an ultrasound correlation-based method, the ultrasonic *phased tracking method* [8-10], to measure the minute displacement, $d(t)$, caused by the acoustic radiation force [11]. In addition, to suppress the sound pressure of the employed ultrasound, a continuous wave (CW) ultrasound, which consists of the sum of frequency components at f and $f + \Delta f$, was employed to cyclically actuate the acoustic radiation pressure $P_R(t)$ at a low-frequency Δf to increase the signal-to-noise ratio in displacement estimation [11]. However, one acoustic radiation force along a direction (e.g., vertical direction) does not generate the strain of an object effectively because it also causes change in the object position, which have zero spatial gradient in displacement (= no strain). In particular, when the elastic modulus of the object is far greater than that of the surrounding media, one acoustic radiation force may generate only the change in position of the object as illustrated in Fig. 3-1a. In such cases, the mechanical properties of the object cannot be estimated. In this study, for effective generation of the strain inside the object even in such cases, two acoustic radiation forces with phases of φ_1 and φ_2 were applied to two different positions in the object from two different directions (e.g., two opposite horizontal direction) as shown in Fig. 3-1b [13]. Furthermore, the displacements of the region inside the object, where the two acoustic radiation forces were applied, were measured by the ultrasonic *phased tracking method.*

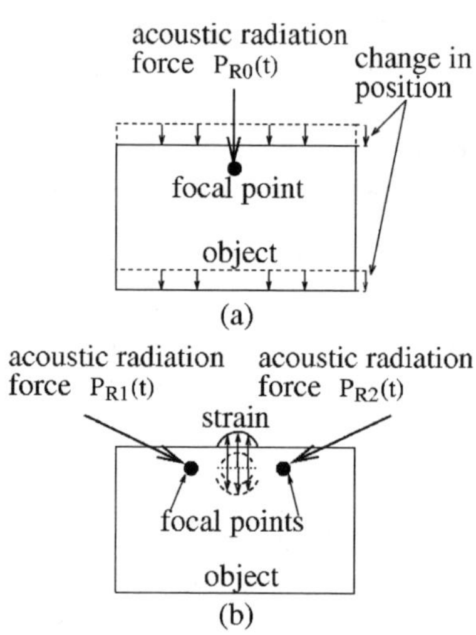

Figure 3-1. Illustrations of (a) the change in position of an object caused by only one acoustic radiation force and (b) the strain inside an object caused by two acoustic radiation forces

3.2 GENERATION OF FLUCTUATING ACOUSTIC RADIATION FORCE

When a continuous plane-wave ultrasound is incident on an interface between two different media, a constant force, which is called an acoustic radiation force, is exerted on the interface. Acoustic radiation pressure, $P_R(t)$, is defined as the acoustic radiation force per unit area of the interface as follows [1, 2]:

$$P_R(t) = (1 + R^2)\, e(t) \tag{3-1}$$

where R and $e(t)$ are the pressure reflection coefficient and the energy density at the interface, respectively. In Eq. (3-1), the transmitted wave is assumed to be perfectly absorbed in the object. Using the densities, ρ_1 and ρ_2, and sound speeds, c_1 and c_2, of the media and object, the reflection coefficient, R, and the energy density, $e(t)$, are defined by

$$R = \frac{Z_2 - Z_1}{Z_2 + Z_1} = \frac{\rho_2 c_2 - \rho_1 c_1}{\rho_2 c_2 + \rho_1 c_1} \tag{3-2}$$

$$e(t) = \frac{1}{\rho_1 c_1^2}\left\{ p(t) \right\}^2 \tag{3-3}$$

where $p(t)$ is the sound pressure at the interface. JSUM shows the safety guideline in which the intensity of the CW ultrasound should be less than 1 W/cm^2. By assuming that the density ρ_1 and sound speed c_1 of water are 10^3 kg/m^3 and 1500 m/s, respectively, the acoustic radiation pressure, $P_R(t)$, exerted on the interface is calculated to be 6.67 Pa when the ultrasound intensity is 1 W/cm^2. In this calculation, the reflection coefficient, R, is assumed to be zero. For example, sound speeds of muscle and fat are 1568 m/s and 1465 m/s, respectively [13]. By assuming densities of these tissues to be 10^3 kg/m^3, the reflection coefficient R is determined to be 0.034. Thus, the reflection coefficient can be assumed to be zero.

As shown in Eq. (3-3), the energy density $e(t)$ of the incident wave is proportional to the square of the sound pressure $p(t)$ of the ultrasound beam. When two ultrasonic beams with slightly different frequencies, f and $f + \Delta f$, are crossed each other, the sound pressure, $p_{sum}(t)$, at the intersectional area is expressed by the sum of sound pressures of two ultrasonic beams as follows:

$$p_{sum}(t) = p_0 \cos(\omega t + \varphi) + p_0 \cos(\omega + \Delta\omega)t \tag{3-4}$$

where p_0, ω, and $\Delta\omega$ are the amplitude of the sound pressure of each ultrasound beam, angular frequency of the incident wave ($\omega = 2\pi f$), and difference in angular frequency ($\Delta\omega = 2\pi\Delta f$), respectively, and φ is the initial phase of the CW at frequency f. In this case, the energy density $e(t)$ is given by

$$\begin{aligned} e(t) &= \frac{1}{\rho_1 c_1^2}\{ p_{sum}(t)\}^2 \\ &= \frac{1}{\rho_1 c_1^2}\{ p_0 \cos(\omega t + \varphi) + p_0 \cos(\omega + \Delta\omega)t\}^2 \\ &= \frac{p_0^2}{\rho_1 c_1^2}\Big[1 + \cos(\Delta\omega t - \varphi) + \cos\{(2\omega + \Delta\omega)t + \varphi\} \\ &\quad + \frac{1}{2}\cos 2(\omega t + \varphi) + \frac{1}{2}\cos 2(\omega + \Delta\omega)t\Big] \end{aligned} \tag{3-5}$$

From the second term of the right-hand side of Eq. (3-5), it is found that the energy density $e(t)$ of the incident field has a component at the frequency difference Δf. Therefore, the cyclically oscillating radiation pressure $P_R(t)$ at the frequency difference Δf is given by

$$P_R(t) = (1 + R^2)\frac{p_0^2}{\rho_1 c_1^2}\{1 + \cos(\Delta\omega t - \varphi)\} \quad (3-6)$$

Thus, using an ultrasound beam generated by the sum of signals at slightly different frequencies f and $f + \Delta f$, the oscillatory radiation force can be applied to the focal area of the beam.

In this study, in order to generate the regional strain inside the object, we use two phase-controlled acoustic radiation pressures, $P_{R1}(t)$ and $P_{R2}(t)$, given by

$$P_{R1}(t) = (1 + R^2)\frac{p_{01}^2}{\rho_1 c_1^2}\{1 + \cos(\Delta\omega t + \varphi_1)\} \quad (3-7)$$

$$P_{R2}(t) = (1 + R^2)\frac{p_{02}^2}{\rho_1 c_2^2}\{1 + \cos(\Delta\omega t + \varphi_2)\} \quad (3-8)$$

where φ_1 and φ_2 are the initial phases of CWs at frequency f for generating $P_{R1}(t)$ and $P_{R2}(t)$. These two acoustic radiation forces at Δf, whose phase difference is $\varphi_2 - \varphi_1$, are applied by setting the focal points of two transducers at two different sites inside the object, and the insonification angles are assigned to θ_1 and θ_2 for the respective transducers.

3.3 DISPLACEMENT OF OBJECT INDUCED BY FLUCTUATING ACOUSTIC RADIATION FORCE

3.3.1 Experimental Setup

An experimental setup is illustrated in Fig. 3-2. An object made of gel was placed in a water tank. For the application of radiation pressures $P_R(t)$, a concave ultrasonic transducer was employed. To generate a fluctuating acoustic radiation force, the concave ultrasonic transducer (Tokimec 1Z20I-PF50-C; center frequency: 1 MHz) was driven by a sum of two continuous waves at two slightly different frequencies of 1 MHz and 1 MHz + Δf Hz. The resultant ultrasound beam was focused at 50 mm away from the surface of the transducer, and the focal point was set at a point on the top surface of the object with beam angle $\theta = 60$ deg. In this experiment, the displacement at the focal point was measured by a laser velocimeter (ONO SOKKI, LV-1300).

In the following ultrasonic measurement, the CW ultrasound for actuation will interfere in the pulsed ultrasound for the measurement of displacements. In order to avoid this interference, it is necessary to stop actuation during transmission and reception of the ultrasonic pulse for the measurement. The application of ultrasound for actuation is controlled using a custom-made electrical switch [11]. As shown in Fig. 3-2, the electrical switch is controlled by the trigger signal from a function generator that synchronizes with the trigger signal from ultrasonic diagnostic equipment (Toshiba, SSH-160A). The time chart for controlling the application of ultrasound for actuation is shown in Fig. 3-3.

3.3.2 Experimental Results

Figure 3-4 shows the displacements $d(t)$ on the top surface of the object, which are obtained by integrating the velocities measured by the laser velocimeter. In Figs. 3-4a, b, c, and d,

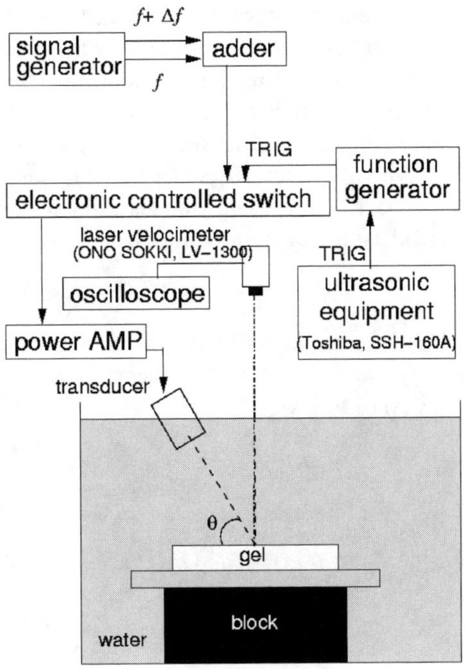

Figure 3-2. Experimental setup for measurement of the displacement on the surface of the object cyclically actuated using a concave ultrasonic transducer

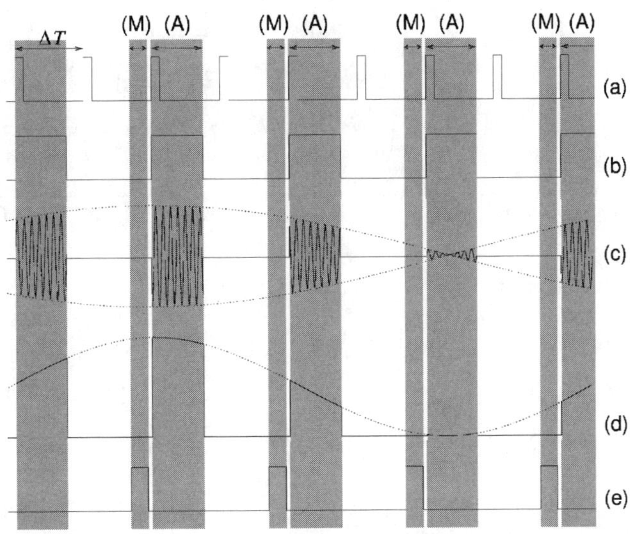

Figure 3-3. Time chart for controlling the application of ultrasound for actuation. (a) Clock signal from ultrasonic diagnostic equipment. (b) Signal for controlling the application of ultrasound for actuation. Ultrasound is applied when this signal is high. (c) Sound pressure $p_{\mathrm{sum}}(t)$ consisting of the sum of two CWs at slightly different frequencies f and $f + \Delta f$ applied intermittently only during actuation period (A). (d) Acoustic radiation pressure $P_{\mathrm{R}}(t)$ generated by the sound pressure shown in (c). (e) Sampling gate for acquiring ultrasonic pulses for measurement, which determines the period for measurement (M)

frequency difference Δf was set to 5 Hz, 10 Hz, 15 Hz, and 20 Hz, respectively. As shown in Fig. 3-4, the fluctuating displacements at the corresponding frequency difference Δf was detected. From these results, it was confirmed that the object was actuated by applying the fluctuating acoustic radiation force. In Fig. 3-5, the amplitude of displacement is plotted as a function of frequency difference Δf. As shown in Fig. 3-5, the amplitude of displacement decreases with frequency difference (= actuation frequency) Δf. When the object is assumed to be a viscoelastic material, which is explained by the Voigt model, the displacement $d(t)$ is expressed using static elastic modulus E_s and viscosity constant η as follows:

$$P_R(t) = E_s d(t) + \eta \dot{d}(t) \qquad (3-9)$$

where $\dot{d}(t)$ is the time derivative of displacement. In this experiment, acoustic radiation pressure $P_R(t)$ and resulting displacement $d(t)$ can be expressed by sinusoidal waves at frequency Δf as follows

$$P_R(t) = P_{R0} e^{j2\pi\Delta ft} \qquad (3-10)$$

$$d(t) = d_0 e^{j(2\pi\Delta ft+\phi)} \qquad (3-11)$$

By substituting Eqs. (3-10) and (3-11) into Eq. (3-9), Eq. (3-9) is modified as follows:

$$d_0 = \frac{P_{R0}}{\sqrt{E_s^2 + (2\pi\Delta f\eta)^2}} \qquad (3-12)$$

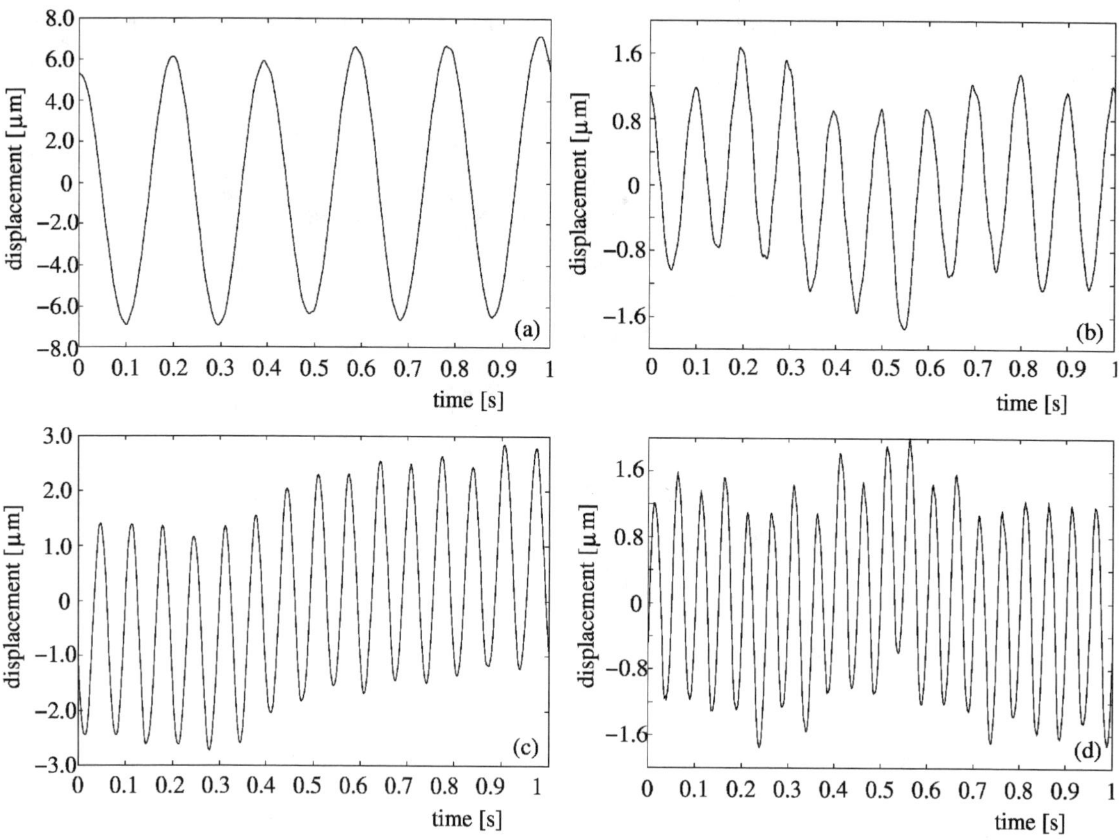

Figure 3-4. Displacement $d(t)$ on the surface of the object measured by the laser velocimeter. Object was actuated at (a) Δf = 5 Hz, (b) 10 Hz, (c) 15 Hz, and (d) 20 Hz

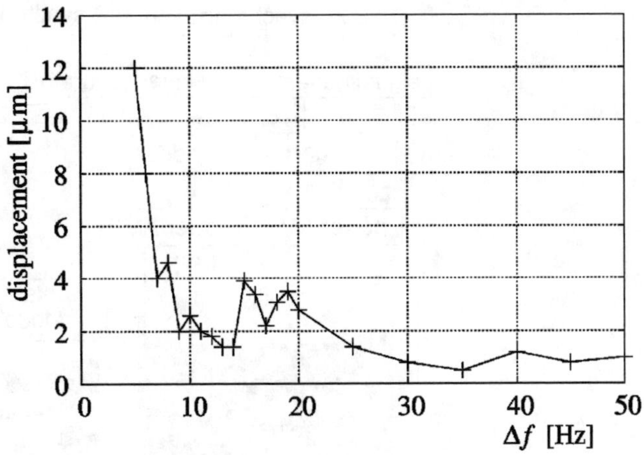

Figure 3-5. Amplitude of displacement plotted as a function of actuation frequency Δf

In this experiment, the amplitude of radiation force P_{R0} was set to be constant. The amplitude of displacement d_0 decreased with actuation frequency Δf, as shown by Eq. (3-12).

3.4 DEFORMATION INDUCED BY DUAL ACOUSTIC RADIATION FORCE CAPTURED BY VIDEO CAMERA

3.4.1 Experimental Setup

An experimental setup is illustrated in Fig. 3-6. As in the experiments described in Section 3.3, the object made of gel (45 mm × 45 mm × 17 mm) was placed in a water tank. For effective generation of the object deformation, two concave ultrasonic transducers were employed to apply two acoustic radiation pressures $P_{R1}(t)$ and $P_{R2}(t)$. A concave ultrasonic transducer (Tokimec 1Z20I-PF50-C; center frequency: 1 MHz) was driven by a sum of two CWs at two slightly different frequencies of 1 MHz and 1 MHz + 10 Hz. The resultant ultrasound beam was focused at 50 mm away from the surface of the transducer, and the focal point was set at a point, α, on the top of the object with beam angle θ_1. Another concave transducer (Tokimec 1Z20I-PF50-C; center frequency: 1 MHz) was driven by a sum of CWs at 1 MHz and 1 MHz + 10 Hz. The focal point, which is 50 mm away from the surface of the transducer, was set at a point, β, on the top of the object with beam angle θ_2. The phase difference, $\varphi_2 - \varphi_1$, between $P_{R1}(t)$ and $P_{R2}(t)$ was set to be zero.

In this experiment, the application of acoustic radiation forces was not stopped because the amplitude of displacement is reduced when acoustic radiation forces are applied intermittently. The resolution of the video camera in detection of the displacement is poorer in comparison with the laser velocimeter, and the amplitude of displacement is desired to be as large as possible. As indicated in Fig. 3-6, four different sites (a, b, c, and d) were separately measured. Therefore, the image acquisition was triggered by a rectangular signal that synchronizes with the signal applied to transducers for actuation.

3.4.2 Experimental Results

As illustrated in Fig. 3-6, four different sites (a, b, c, and d) at the surface on the side of the object were measured with a high-speed video camera at a frame rate of 100 Hz. In this

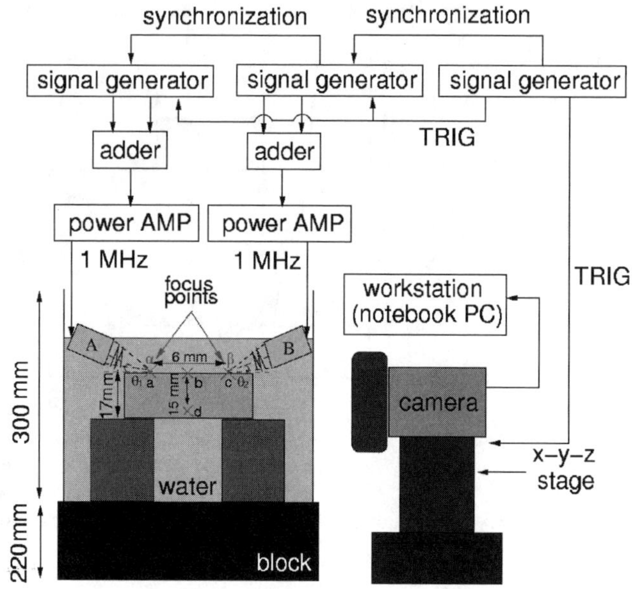

Figure 3-6. Experimental setup for measurement of deformation of the object using a video camera

experiment, beam angles θ_1 and θ_2 were set to be 22 deg. and 35 deg., respectively, and the distance between points α and β was set to be 6 mm. Focal areas of two transducers were set at the edge of the top surface of the object. The four sites, a, b, c, and d, were measured for one cycle (100 milliseconds) of acoustic radiation forces. The period for acquiring pictures in relation to acoustic radiation forces is shown in Fig. 3-7. In Figs. 3-7a to d, bright spots in measured pictures correspond to the light reflected from carbon powder, which is contained in the object, due to the irradiation of the white light using halogen lamp. At points a and c (two focal points), displacements along directions of respective ultrasonic beams were found as shown in Figs. 3-7a and c. As shown in Fig. 3-7b, point b was found to move upward when acoustic radiation forces increased. This result shows that the region between two focal points was vertically thickened (deformed) due to horizontal compression by application of dual acoustic radiation force.

3.5 ULTRASONIC MEASUREMENT OF DISPLACEMENT DISTRIBUTION INSIDE AN OBJECT INDUCED BY DUAL ACOUSTIC RADIATION FORCE

3.5.1 Experimental Setup

An experimental setup is illustrated in Fig. 3-8. In order to measure the strain using ultrasound, we employed ultrasonic diagnostic equipment (Toshiba SSH-160A) with a sector-type probe (center frequency: 5 MHz). The equipment was modified to detect the minute displacement of the object by the ultrasonic *phased tracking method*. As in Section 3.4, an object made of gel (45 mm × 45 mm × 17 mm, containing carbon powder to obtain sufficient scattering) was placed in a water tank as shown in Fig. 3-8. For the application of radiation pressures $P_{R1}(t)$ and $P_{R2}(t)$, two concave ultrasonic transducers were employed. The settings for the application of acoustic radiation forces were the same as described in Section 3.4.1.

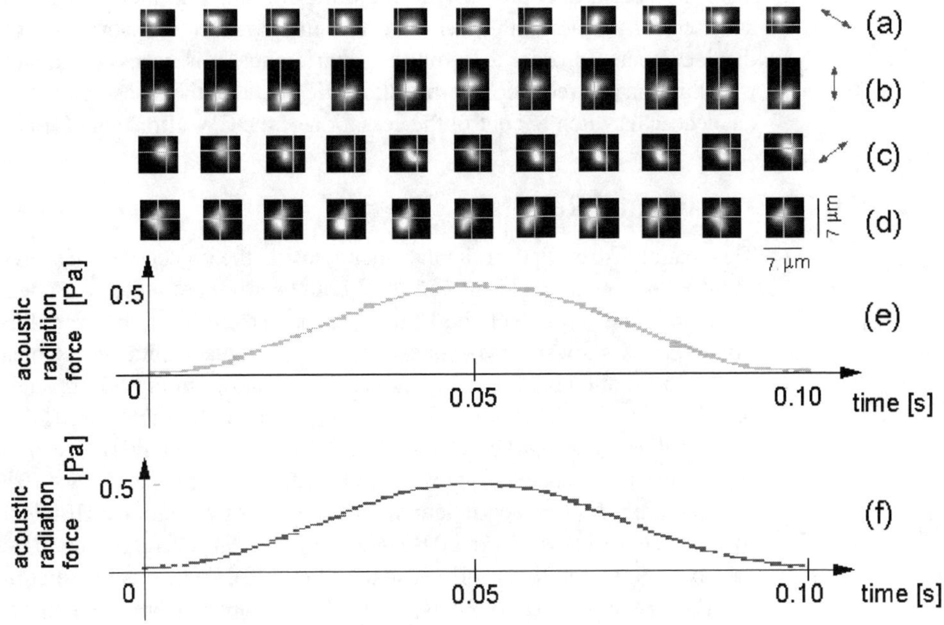

Figure 3-7. Results obtained at point a (a), b (b), c (c), and d (d) as indicated in Fig. 3-6 (e) and (f). Acoustic radiation forces radiated by two ultrasonic transducers

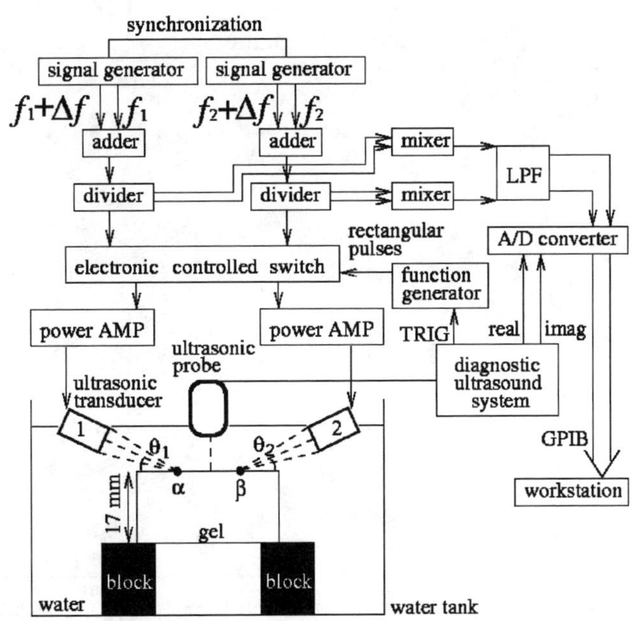

Figure 3-8. Experimental setup for measurement of the strain inside the object cyclically actuated using two concave ultrasonic transducers

As described in Section 3.3.1, in the ultrasonic measurement, when the CW ultrasounds are used for actuation, they will interfere in the pulsed ultrasound for the measurement of displacements. In order to avoid this interference, it is necessary to stop actuation during transmission and reception of the ultrasonic pulse for the measurement. Therefore, we used an electrical switch to control the cessation of the CW ultrasound for actuation [11].

3.5.2 Experimental Results

The spatial distribution of displacements inside the object was measured with ultrasound. In this experiment, angles θ_1 and θ_2 of both beams were set to be 25 deg., and the distance between α and β was set to be 15 mm. The measured positions were illustrated in Fig. 3-9. Figure 3.10a shows an M-mode image of the object obtained at measured position d. Acoustic radiation pressures $P_{R1}(t)$ and $P_{R2}(t)$, shown in Figs. 3-10b and c, were calculated based on Eq. (3-6) as follows: The density, ρ_2, and the sound speed, c_2, of the object were measured as 1.1×10^3 kg/m^3 and 1.47×10^3 m/s, respectively. By assuming the density, ρ_1, and the sound speed, c_1, of water to be 1.0×10^3 kg/m^3 and 1.5×10^3 m/s, respectively, the pressure reflection coefficient, R, and the energy reflection coefficient, R^2, were respectively calculated as 0.038 and 0.0014 using Eq. (3-2). Therefore, in this study, by assuming the object to be totally absorbing material ($R = 0$), the acoustic radiation pressures $P_{R1}(t)$ and $P_{R2}(t)$ exerted on two points, α and β, in the object were calculated based on Eq. (3-6). In this calculation, we obtained the amplitude of the focused sound pressures of p_{01} and p_{02} beforehand by measuring the acoustic fields of each ultrasound for actuation with a hydrophone (Force Institute, MH-28-10). In Fig. 3-10d, the displacement $d(t)$ at the surface (point A) of the object actuated by the acoustic radiation forces were measured by the ultra-

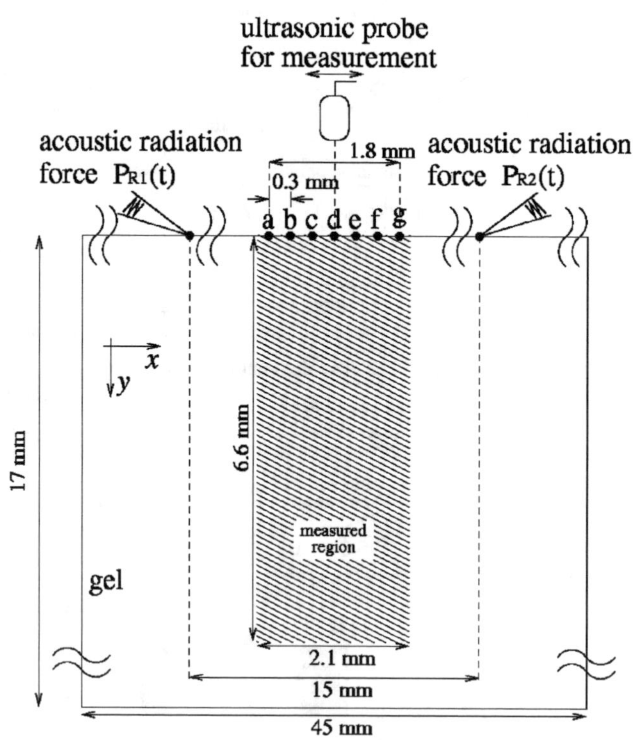

Figure 3-9. Illustration of positions measured with ultrasound

sonic *phased tracking method*. The tracking lines were superimposed by the red lines on the M-mode image in Fig. 3-10a. In Fig. 3-10d, it is found that the object was cyclically actuated with the amplitude of a few micrometers [11], which was similar to that measured with the high-speed video camera. Two acoustic radiation forces were applied by setting focal points at the upper surface of the object, and angles θ_1 and θ_2 of both ultrasonic beams for actuation were set to be 25 deg. Therefore, directions of horizontal components of these two acoustic radiation forces were opposite each other by setting $\varphi_2 - \varphi_1 = 0$. Therefore, the region between two focal points was cyclically compressed along the horizontal axis by these acoustic radiation forces, and the thickness of the region between two focal points becomes thick along the vertical axis.

Figures 3-10d, e, and f show displacements at multiple points along an ultrasonic beam. When acoustic radiation forces shown in Figs. 3-10b and c increase, the surface of the object and a point that was 1.5 mm deeper than the surface moved upward (see Figs. 3-10d and e). On the other hand, a point that was 3.0 mm deeper than the surface moved downward, as shown in Fig. 3-10f. This result shows the thickness of the region between two focal points increased along the vertical axis, which shows that the strain was generated inside the object.

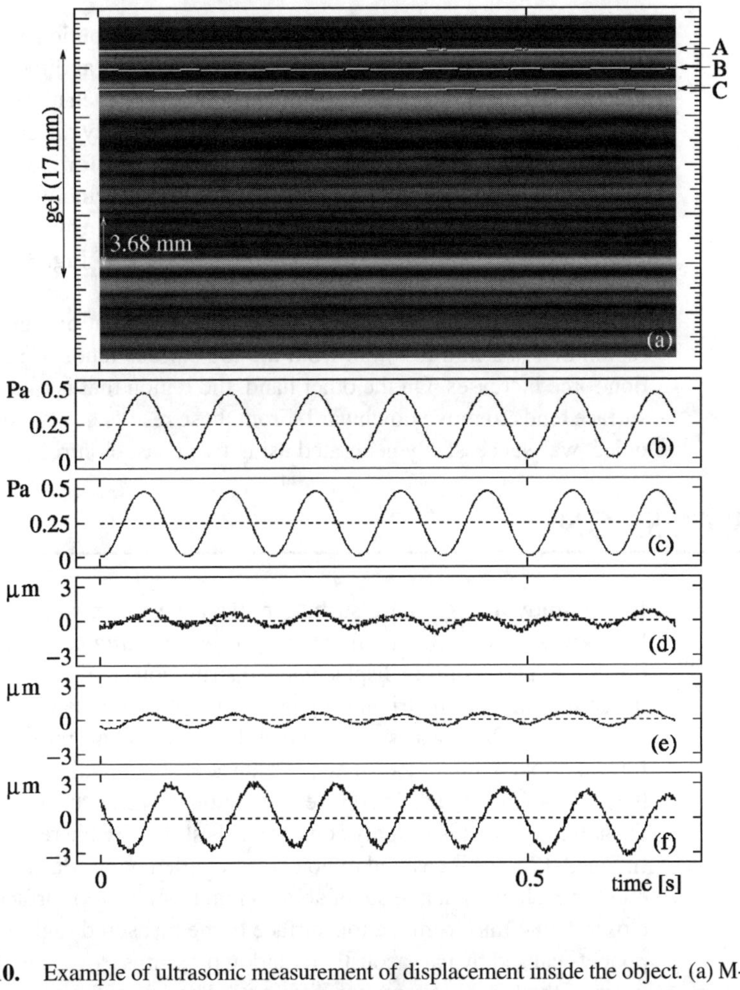

Figure 3-10. Example of ultrasonic measurement of displacement inside the object. (a) M-mode image of the object. (b) Acoustic radiation pressure, $P_{R1}(t)$. (c) Acoustic radiation pressure, $P_{R2}(t)$. (d) Displacement at point A (surface) indicated in (a). (e) Displacement at point B (1.5 mm from the surface). (f) Displacement at point C (3.0 mm from the surface)

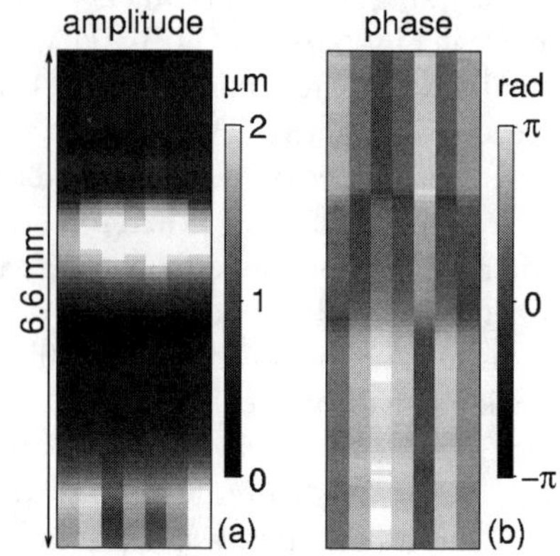

Figure 3-11. Spatial distribution of displacements inside the object. (a) Amplitudes and (b) phases of displacements

Then, the two-dimensional displacement distribution was measured by moving the ultrasonic probe for the measurement in the horizontal direction as illustrated in Fig. 3-9. The displacement $d(t)$ at every sampled point along each ultrasonic beam was obtained, and the amplitude of the displacement $d(t)$ at the frequency difference $f = 10$ Hz was estimated, as shown in Fig. 3-11a, by applying the Fourier transform to the displacement $d(t)$ and the acoustic radiation pressure $P_{R1}(t)$ with a Hanning window of 0.68 second (1024 points). In addition, the phase delay of the displacement $d(t)$ from the acoustic radiation pressure $P_{R1}(t)$ was also estimated as shown in Fig. 3-11b. In Fig. 3-11, amplitudes and phase delays of displacements $d(t)$ were shown in gray scale. The phase difference of zero shows that that point moves upward when the acoustic radiation force increases. In Fig. 3-11, the region that is 2 mm to 3 mm from the top surface tends to move upward when the radiation force increases. On the other hand, the region that is over 3.0 mm deeper than the top surface tends to move downward. From these results, it was found that the strain inside the object was successfully generated using two phase-controlled acoustic radiation forces.

3.6 DISCUSSION

In this study, the spatial distribution of displacements caused by two acoustic radiation forces was measured with ultrasonic *phased tracking method*. The applied acoustic radiation forces and resulting displacement distribution are discussed below. Figure 3-12a shows the sound pressure distribution generated by the ultrasonic transducer used for actuation. From Fig. 3-12a, it was supposed that the width at half maximum of the acoustic radiation force is about 5 mm in the x-axis, and the acoustic radiation force was almost constant along the z-axis in the object. Therefore, the region in which the acoustic radiation force was over the half maximum is roughly considered as shown by the region with shadow in Fig. 3-12b. In Fig. 3-12b, the measured region is surrounded by the dashed line. Two ultrasound beams crossed each other at a depth of about 4 mm from the top surface. At each point in the region closer than 4 mm from the top surface in the measured region, the sum of horizontal components caused by two acoustic radiation pressures $P_{R1}(t)$ and $P_{R2}(t)$ are not zero. (Magnitudes of these components are different.) Therefore, the region nearer than 4 mm from the top surface was compressed in the horizontal axis, and, as shown in Fig. 3-11, the resulting upward displacement during the increase of acoustic radiation forces was found.

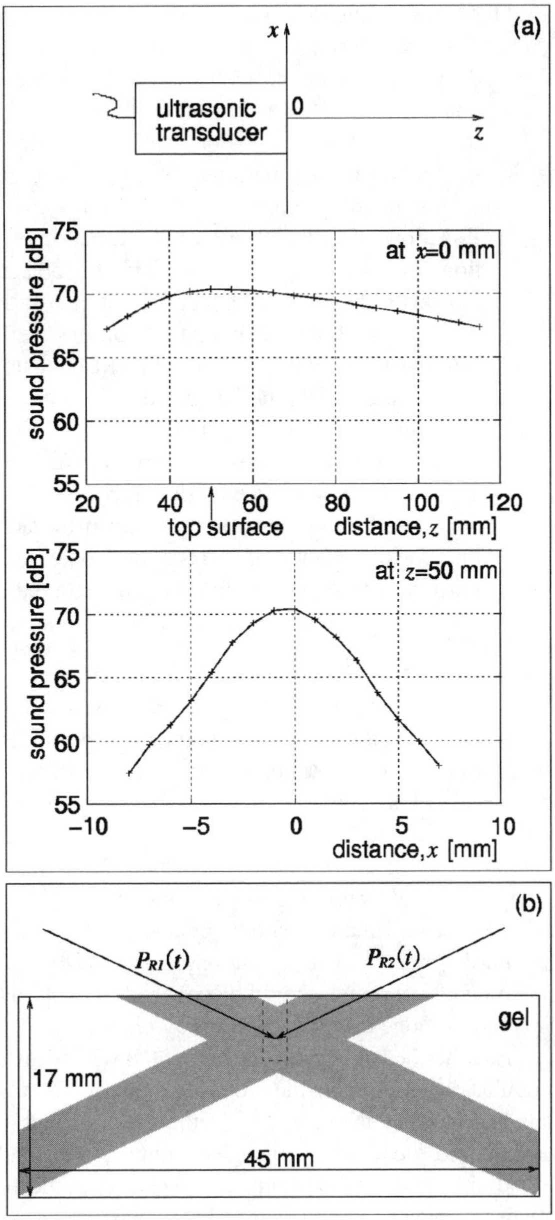

Figure 3-12. (a) Sound field generated by the ultrasonic transducer used for actuation. (b) Consideration of displacement distribution generated by two acoustic radiation forces

Around 4 mm from the top surface, horizontal components of two acoustic radiation forces were almost compensated, and only vertical components remain. Therefore, the region around 4 mm from the top surface moves downward during the increase of acoustic radiation force because of the downward displacement caused by vertical components of acoustic radiation forces and that caused by the horizontal compression in the region nearer than 4 mm from the top surface.

As shown in Fig. 3-11, the measured displacement distribution showed that the strain inside the object was successfully generated. However, at this moment, it is difficult to estimate the elastic modulus distribution precisely due to the problems described below.

(1) Measurement of Strain Distribution
The stress applied by the acoustic radiation force is not the plane stress. Therefore, the region, where the acoustic radiation force is applied, is displaced along not only one direction. However, in this study, only the beam-axis component of displacement was measured. To obtain the precise strain and elastic modulus distributions, a method for measuring three dimensional components of displacement is needed.
(2) Estimation of Magnitude of Applied Stress
Equation (3-6), which used for estimation of the magnitude of the acoustic radiation force, was derived by assuming the object to be a perfectly absorbing material. However, the attenuation coefficient of soft tissue has a finite value. For example, the mixture of muscle and fat (abdominal wall), is typically 3 dB/cm/MHz [13]. To estimate the magnitude of the applied acoustic radiation force, the attenuation coefficient of tissue should be considered.
(3) Estimation of Stress Distribution
Figure 3-12a shows the sound pressure distribution, which is directly related to the acoustic radiation force as shown by Eq. (3-6), generated by the ultrasonic transducer used for application of the acoustic radiation force. As shown in Fig. 3-12a, the acoustic radiation force is a function of the geometry, and the stress distribution must be considered in addition to the strain distribution.

In addition, the distance between two focal points, beam angles, and depth of the region of interest will be restricted by the factors described below.

When focal areas of two ultrasonic transducers overlap, the horizontal components of two acoustic radiation forces are compensated each other in the intersectional area. Therefore, for effective generation of the strain, two focal points are desired to be assigned so that sound fields generated by two transducers do not overlap within the object (= region of interest). On the other hand, when the distance between two focal points is very large in comparison with the size of the focal area, the displacement distribution is not so different from that caused by one acoustic radiation force. Therefore, the size of the region of interest (between two focal areas) should be comparable to the width of the ultrasonic beam.

Furthermore, to keep beam angles θ_1 and θ_2 similar at every depth, the distance between two focal points should be changed in relation to the depth of the region of interest. This will be done using a phased array probe.

As described above, at this moment, it will be difficult to estimate the "absolute" elastic modulus precisely using the proposed method. Therefore, the elasticity of different individuals measured by the proposed method are difficult to be compared each other. However, for example, the temporal change or frequency dependence of the strain distribution within an individual is directly related to the change in the elastic property. The proposed method can detect such a relative change in the elastic property of an object.

3.7 CONCLUSIONS

In this study, in order to generate the regional strain inside the object, we constructed an experimental setup for application of cyclic remote actuation in the object using two focused ultrasonic transducers. The phase difference between two acoustic radiation forces, which were respectively applied at two different positions in the object, was controlled to be 0 deg. The resultant displacements inside the object were successfully measured by the ultrasonic *phased tracking method.* These results show potential of the proposed method for generation of the regional strain inside the object.

Acknowledgments

The authors thank Mr. Michishita, Mr. Nishio, Mr. Takahashi, and Mr. Odagiri for their efforts in the experiments.

REFERENCES

[1] Fatemi, M., and Greenleaf, J. F., 2002, "Vibro-Acoustic Mammography," *IEEE Transactions on Medical Imaging*, **21**, pp. 1–8.

[2] Fatemi, M., and Greenleaf, J. F., 1999, "Vibro-Acoustography: An Imaging Modality Based on Ultrasound-Stimulated Acoustic Emission," *Proceedings of the National Academy of Sciences of the United States of America*, **96**, pp. 6603–6608.

[3] Torr, G. R., 1984, "The Acoustic Radiation Force," *American Journal of Physics*, **52**, pp. 402–408.

[4] Nightingale, K., Soo, M. S., Nightingale, R., and Trahey, G., 2002, "Acoustic Radiation Force Impulse Imaging: In Vivo Demonstration of Clinical Feasibility," *Ultrasound in Medicine and Biology*, **28**, pp. 227–235.

[5] Trahey, G. E., Palmeri, M. L., Bentley, R. C., and Nightingale, K. R., 2004, "Acoustic Radiation Force Impulse Imaging of the Mechanical Properties of Arteries: In Vivo and Ex Vivo Results," *Ultrasound in Medicine and Biology*, **30**, pp. 1163–1171.

[6] Fahey, B. J., Nightingale, K. R., Nelson, R. C., Palmeri, M. L., and Trahey, G. E., 2005, "Acoustic Radiation Force Impulse Imaging of the Abdomen: Demonstration of Feasibility and Utility," *Ultrasound in Medicine and Biology*, **31**, pp. 1185–1198.

[7] The Japan Society of Ultrasonics in Medicine, 1984, *Japanese Journal of Medical Ultrasonics*, **22**, pp. 41–46 [in Japanese].

[8] Kanai, H., Sato, M., Koiwa, Y., and Chubachi, N., 1996, "Transcutaneous Measurement and Spectrum Analysis of Heart Wall Vibrations," *IEEE Transactions on Ultrasonics, Ferroelectrics, and Frequency Control*, **43**, pp. 791–810.

[9] Kanai, H., Hasegawa, H., Chubachi, N., Koiwa, Y., and Tanaka, M., 1997, "Noninvasive Evaluation of Local Myocardial Thickening and Its Color-Coded Imaging," *IEEE Transactions on Ultrasonics, Ferroelectrics, and Frequency Control*, **44**, pp. 752–768.

[10] Kanai, H., Sugimura, K., Koiwa, Y., and Tsukahara, Y., 1999, "Accuracy Evaluation in Ultrasonic-Based Measurement of Microscopic Change in Thickness," *Electronics Letters*, **35**, pp. 949–950.

[11] Michishita, K., Hasegawa, H., and Kanai, H., 2003, "Ultrasonic Measurement of Minute Displacement of Object Cyclically Actuated by Acoustic Radiation Force," *Japanese Journal of Applied Physics*, **42**, pp. 4608–4612.

[12] Hasegawa, H., Takahashi, M., Nishio, Y., and Kanai, H., 2006, "Generation of Strain Inside Objects Using Dual Acoustic Radiation Force," *Japanese Journal of Applied Physics*, **45**, pp. 4706–4711.

[13] Odagiri, Y., Hasegawa, H., and Kanai, H., 2008, "Ultrasonic Measurement of Strain Distribution Inside Object Cyclically Compressed by Dual Acoustic Radiation Force," *Japanese Journal of Applied Physics*, **47** (in press).

[14] Goss, S. A., Johnston, R. L., and Dunn, F., 1978, "Comprehensive Compilation of Empirical Ultrasonic Properties of Mammalian tissues," *Journal of Acoustical Society of America*, **64**, pp. 423–457.

Chapter 4

COMPUTATIONAL INVERSE PROBLEM TECHNIQUES IN VIBROACOUSTICS

Wilkins Aquino and John C. Brigham
School of Civil and Environmental Engineering, Cornell University, Ithaca, NY 14853, USA

ABSTRACT

This chapter describes state-of-the-art numerical methods for the solution of direct and inverse vibroacoustic problems. Particular emphasis is placed on current and emerging techniques for the characterization of material properties through the use of vibroacoustic experiments. Numerical techniques such as the weak-form Galerkin finite element method and the boundary element method are presented as tools for solving the boundary-value and initial boundary-value problems occurring in coupled acoustic-structure interaction problems. Then, emerging model-updating techniques for the solution of inverse problems that use the aforementioned numerical methods coupled with classical and novel optimization and machine learning tools are presented. An example problem is presented which applies these inverse problem techniques for the characterization of viscoelastic and elastic properties of a solid using vibroacoustic testing methods. Advantages, disadvantages, and the relevance of the described methods in medical applications will be addressed throughout the chapter.

4.1 INTRODUCTION

Non-invasive methods to characterize material behavior are of particular interest in the medical field. Determining the properties of biological structures provides valuable insight into the onset and progression of various diseases, as well as the development of tools to diagnose and treat illness. Examples include the work by Baldewsing et al [1], and the work by Haemmerich et al [2], which used finite element analyses to model the behavior of biological structures. In both examples, it is crucial that accurate material properties be determined to develop modeling tools for diagnosis and treatment of diseases.

Characterization of biological soft tissues has been shown to pose particular difficulty due to the complexity of the material behavior [3]. In general, tissue mechanics can include non-linearity, anisotropy, heterogeneity, and rate dependent behaviors. Furthermore, many tissues are subjected to large strains under normal conditions, thus requiring finite strain considerations. In order to characterize these properties, it is necessary to perform tests that contain sufficient information relating to the material without disturbing the current state. For instance, characterizing rate dependent properties requires the use of dynamic testing procedures to measure features such as waveforms, wave speeds, impedances, steady-state dynamic responses, and other time- or frequency-dependent behavior. Fortunately, several methods that were initially developed for medical imaging have been shown to provide the information necessary to characterize biological material properties [4-7].

Vibroacoustography (VA) [8] is one such imaging technique, which offers the potential for the non-invasive characterization of mechanical properties of tissues [4, 7, 9, 10]. In VA, the radiation force of ultrasound is used to excite a fluid-immersed structure and the resulting acoustic response in the surrounding fluid is measured. Since the acoustic pressure in the fluid and the motion of the structure are coupled, the acoustic response of the fluid is dependent on the material properties of the structure.

This chapter describes modern techniques for approximating the solution of inverse problems related to the non-destructive and non-invasive characterization of material properties using acoustic emissions. In particular, the chapter concentrates on forward modeling methods, challenges related to the inverse characterization of viscoelastic properties, and non-linear optimization methods.

4.2 FORMULATION OF THE VIBROACOUSTIC PROBLEM

4.2.1 Solid Mechanics Formulation

Fluid-structure interaction, in the present context, can be described mathematically by a coupled system of partial differential equations (PDEs) derived from the conservation of linear momentum. For the purposes of this work, body forces are assumed to be negligible. Thus, the governing differential equation for structural behavior is given by

$$\nabla \cdot \sigma(\vec{x},t) = \rho \, \ddot{\vec{u}}(\vec{x},t) \text{ on } \Omega \tag{4-1}$$

A constitutive model describing the stress-strain relationship in a general rate dependent form can be represented as

$$\sigma(\vec{x},t) = f\left(\varepsilon(\vec{x},t), \frac{\partial \varepsilon(\vec{x},t)}{\partial t}, \frac{\partial^2 \varepsilon(\vec{x},t)}{\partial t^2}, \ldots, \frac{\partial^n \varepsilon(\vec{x},t)}{\partial t^n}\right) \tag{4-2}$$

Assuming small strains and deformations, the strains are defined as

$$\varepsilon(\vec{x},t) = \frac{1}{2}\left[\nabla \vec{u}(\vec{x},t) + \left(\nabla \vec{u}(\vec{x},t)\right)^T\right] \tag{4-3}$$

This assumption is justified since the displacements and strains resulting from vibroacoustic excitations are, in general, small. The natural boundary conditions are given by the Cauchy formula as

$$\sigma(\vec{x},t)\vec{n}(\vec{x}) = \vec{T}(\vec{x},t) \text{ on } \Gamma_T \tag{4-4}$$

and the essential boundary conditions are given as

$$\vec{u}(\vec{x},t) = \vec{u}^o(\vec{x},t) \text{ on } \Gamma_U \tag{4-5}$$

The initial displacements and velocities in the body are assumed to be zero. In Eqs. (4-1) to (4-5), $\vec{x}$ is the spatial position vector, t is time, $\sigma(\vec{x},t)$ is the stress tensor, ρ is the mass density (assumed constant in this work), $\vec{u}(\vec{x},t)$ is the displacement vector, $\varepsilon(\vec{x},t)$ is the strain tensor, $\vec{n}(\vec{x})$ is the unit vector outward normal to the surface, $\vec{T}(\vec{x},t)$ is the traction vector, Ω is the given domain of interest, Γ is the boundary of the domain, Γ_T is the portion of the boundary where external traction is specified, and Γ_U is the portion of the boundary where the displacement is specified. In this work, vectors are shown with arrows (or curved brackets for discretization), second-order tensors are shown in bold, fourth-order tensors are shown with IV as a superscript, and matrices are shown with square brackets. Each dot over a variable represents a derivative with respect to time.

A common method to define rate dependent stress-strain relationships is through the concepts of viscoelasticity, in which the constitutive model combines the behaviors of an elastic solid and a viscous fluid. An example of a linear viscoelastic model is a stress-strain relationship obeying an exponential series relaxation law (i.e., Prony series) in the time domain given by

$$\boldsymbol{\sigma}(\vec{x},t) = \mathbf{C}^{\mathrm{IV}}\boldsymbol{\varepsilon}(\vec{x},t) + \sum_{m=1}^{n}\left[\int_0^t \tilde{\mathbf{C}}_{\mathrm{m}}^{\mathrm{IV}} \exp\left(\frac{\xi - t}{\tau_{\mathrm{m}}}\right)\frac{\partial \boldsymbol{\varepsilon}(\vec{x},\xi)}{\partial \xi}\mathrm{d}\xi\right] \tag{4-6}$$

where $\mathbf{C}^{\mathrm{IV}}$ is the fourth-order tensor representing the long-term purely elastic moduli of the material, $\tilde{\mathbf{C}}_{\mathrm{m}}^{\mathrm{IV}}$ are the fourth-order tensors representing the viscous moduli, t is total time, and τ_{m} are known as the relaxation times. This viscoelastic relationship is an expansion to three dimensions of the traditional one-dimensional relationship derived from a rheological model of n parallel sets of springs (purely elastic solid behavior) and dashpots (purely viscous fluid behavior) in series (Maxwell elements), with an additional spring in parallel, and is referred to as the generalized Maxwell model. (See [11] for a complete discussion of viscoelasticity.) It is important to bear in mind that if no Maxwell units are used (i.e., $n = 0$), the representation shown in Eq. (4-6) simplifies to a purely linear elastic material model.

The time-domain relaxation model can be converted to the frequency domain by assuming harmonically varying strains and stresses given by

$$\boldsymbol{\varepsilon}(\vec{x},t) = \boldsymbol{\varepsilon}_0(\vec{x},\omega)\exp(i\omega t) \tag{4-7}$$

and

$$\boldsymbol{\sigma}(\vec{x},t) = \boldsymbol{\sigma}_0(\vec{x},\omega)\exp(i\omega t + i\delta) = \boldsymbol{\sigma}^*(\vec{x},\omega)\exp(i\omega t) \tag{4-8}$$

where

$$\boldsymbol{\sigma}^*(\vec{x},\omega) = \boldsymbol{\sigma}_0(\vec{x},\omega)\exp(i\delta) \tag{4-9}$$

In Eqs. (4-7) to (4-9), ω is the circular frequency, $\boldsymbol{\varepsilon}_0$ and $\boldsymbol{\sigma}_0$ are the strain amplitude and the stress amplitude, respectively, and δ is the phase lag angle between the stress and strain responses. Typically, δ is used as a measure of viscous damping in the material. By substituting Eqs. (4-7) to (4-9) into Eq. (4-6), the stress-strain relationship for the generalized Maxwell model can be expressed in the frequency domain as

$$\boldsymbol{\sigma}^*(\vec{x},\omega) = \mathbf{C}^{*\mathrm{IV}}(\omega)\,\boldsymbol{\varepsilon}_0(\vec{x},\omega) \tag{4-10}$$

where

$$\mathbf{C}^{*\mathrm{IV}}(\omega) = \mathbf{C}^{\mathrm{IV}} + \sum_{m=1}^{n}\tilde{\mathbf{C}}_{\mathrm{m}}^{\mathrm{IV}}\left(\frac{\omega^2\tau_{\mathrm{m}}^2 + i\omega\tau_{\mathrm{m}}}{1+\omega^2\tau_{\mathrm{m}}^2}\right) \tag{4-11}$$

$\mathbf{C}^{*\mathrm{IV}}(\omega)$ is referred to as the complex relaxation tensor of the material.

4.2.2 Acoustic Medium

For a compressible fluid with no net flow, negligible viscosity, and small pressure amplitudes, the governing PDE for conservation of linear momentum is given by [12]

$$\nabla p(\vec{x},t) + \rho_f \ddot{\vec{u}}^f(\vec{x},t) = 0 \text{ on } \Phi \tag{4-12}$$

The pressure is related to the divergence of displacements through the fluid constitutive relationship as

$$p(\vec{x},t) = -B_f \nabla \cdot \vec{u}^f(\vec{x},t) \tag{4-13}$$

and the essential and natural boundary conditions are given, respectively, as

$$p(\vec{x},t) = p_0(\vec{x},t) \text{ on } \Psi_P \tag{4-14}$$

and

$$\frac{\partial p(\vec{x},t)}{\partial \vec{n}_f(\vec{x})} = -\rho_f \ddot{u}_n^f(\vec{x},t) \text{ on } \Psi_U \tag{4-15}$$

Problems that can be represented with semi-infinite domains can be treated as finite for modeling purposes, and the boundaries that are considered to extend infinitely (i.e., non-reflecting boundaries) can be approximated using the Sommerfeld radiation condition given by

$$\frac{\partial p(\vec{x},t)}{\partial \vec{n}_f(\vec{x})} = -\beta \sqrt{\frac{\rho_f}{B_f}} \dot{p}\left(\vec{x},t\right) \text{ on } \Psi_R \tag{4-16}$$

In Eqs. (4-12) to (4-16), $p(\vec{x},t)$ is the scalar fluid pressure in excess of hydrostatic pressure, ρ_f is the mass density of the fluid, $\vec{u}^f(\vec{x},t)$ is the fluid particle displacement, B_f is the bulk modulus of the fluid, $\vec{n}_f(\vec{x})$ is the unit outward normal to the fluid surface, $\ddot{u}_n^f(\vec{x},t)$ is the acceleration of the fluid boundary in the direction of the normal, β is a geometry-specific parameters for improved non-reflecting radiation conditions [13], Φ is the fluid domain, Ψ is the boundary of the fluid domain, Ψ_P is the portion of the boundary where pressure is specified, Ψ_U is the portion of the boundary where acceleration is specified, and Ψ_R is the portion of the boundary where radiation conditions are specified.

4.2.3 Coupling Conditions

The mathematical description of a body vibrating in an acoustic fluid can be constructed by coupling the solid and acoustic problems described above. At the fluid-solid interface, Γ_{FS}, fluid pressure translates into tractions acting on the solid body, while the accelerations normal to the interface translate into gradients of pressure in the normal direction. These coupling conditions can be expressed as

$$\sigma\left(\vec{x},t\right)\vec{n}\left(\vec{x}\right) = -p\left(\vec{x}\right)\vec{n}\left(\vec{x}\right) \text{ on } \Gamma_{FS} \tag{4-17}$$

and

$$\frac{\partial p(\vec{x},t)}{\partial \vec{n}_f(\vec{x})} = -\rho_f \ddot{\vec{u}}(\vec{x},t) \cdot \vec{n}_f(\vec{x}) \text{ on } \Gamma_{FS} \tag{4-18}$$

4.3 NUMERICAL METHODS FOR SOLVING VIBROACOUSTIC PROBLEMS

The systems of coupled PDEs with corresponding boundary and initial conditions that describe vibroacoustic problems can seldom be solved analytically. Numerical methods are commonly used to obtain approximate solutions to these (initial) boundary value problems (BVP or IBVP). Some of the methods used for this purpose are the finite element, boundary element, and finite difference methods, among others. This chapter focuses on the two most common methods, namely the finite element method (FEM) and the boundary element method.

4.3.1 Finite Element Method

The FEM uses a variational or weighted integral representation of the initial boundary value problems, and the solution is represented by a linear combination of basis functions, which are compactly supported over simple domains that are called finite elements. These basis functions in most cases are taken as polynomials. Several books exist with thorough treatments of the theories and implementation of FEM. Some examples include the work by Reddy [14], Hughes [15], Cook et al [13], Bathe [16], and Ihlenburg [17], among others.

Although there are different versions of FEM, this article concentrates on the weak-form Galerkin method as this is the most commonly used approach for the problems discussed herein. The first step in obtaining a finite element representation using the weak-form Galerkin method is to construct a variational statement (i.e., weak form) of the problem at hand. For the solid mechanics problem, it can be shown that the principle of virtual work represents a variational statement. Defining a continuous virtual displacement field, $\delta\vec{u}(\vec{x})$, on Ω and Γ, such that the virtual displacement field vanishes where essential boundary conditions are specified, and expressing the stress and strain tensors in Voigt notation, the principle of virtual work can be expressed as [13]

$$\int_{\Omega} \rho\delta\vec{u}(\vec{x},t)\cdot\ddot{\vec{u}}(\vec{x},t)\mathrm{d}V + \int_{\Omega} \delta\vec{\varepsilon}(\vec{x},t)\cdot\vec{\sigma}(\vec{x},t)\,\mathrm{d}V - \int_{\Gamma_{\mathrm{T}}} \delta\vec{u}(\vec{x},t)\cdot\vec{T}(\vec{x},t)\,\mathrm{d}S = 0 \tag{4-19}$$

where $\delta\vec{\varepsilon}(\vec{x},t)$ is the virtual strain field corresponding to the virtual displacement field.

To obtain a system of ordinary differential equations in time, the domain is divided into finite elements and the displacement and virtual displacement fields are approximated within each element as

$$\vec{u}\left(\vec{x},t\right) = \left[N\left(\vec{x}\right)\right]\{u^{\mathrm{e}}(t)\} \tag{4-20}$$

and

$$\delta\vec{u}\left(\vec{x}\right) = \left[N\left(\vec{x}\right)\right]\{\delta u^{\mathrm{e}}\} \tag{4-21}$$

where $[N(\vec{x})]$ is the matrix containing the interpolation functions for an element and the superscript "e" specifies that the quantity is defined only for a single element. Substituting the field approximations into Eq. (4-19), eliminating the arbitrary virtual displacement field vector, and adding individual element contributions, the semi-discretized form of the equation of motion becomes

$$[M]\{\ddot{u}(t)\} + \{I(t)\} = \{R(t)\} \tag{4-22}$$

where $[M]$ is the global mass matrix, $\{\ddot{u}(t)\}$ is the global vector of nodal accelerations, $\{I(t)\}$ is the internal force vector, and $\{R(t)\}$ is the external force vector. These global quantities are defined as

$$[M] = \sum_{\text{elements}} \int_{\Omega^{\mathrm{e}}} \rho \left[N(\vec{x})\right]^{\mathrm{T}} \left[N(\vec{x})\right] \mathrm{d}V \tag{4-23}$$

$$\{I(t)\} = \sum_{\text{elements}} \int_{\Omega^e} \left[B(\vec{x})\right]^{\mathrm{T}} \vec{\sigma}(\vec{x},t)\mathrm{d}V \tag{4-24}$$

and

$$\{R(t)\} = \sum_{\text{elements}} \int_{\Gamma_{\mathrm{T}}^{\mathrm{e}}} \left[N(\vec{x})\right]^{\mathrm{T}} \vec{T}(\vec{x},t)\mathrm{d}S \tag{4-25}$$

where $[B(\vec{x})]$ is the matrix of spatial derivatives of the interpolation functions, and the summation over elements refers to the assembly of the element matrices. The resulting system of ordinary differential equations can then be solved using conventional time integration algorithms.

For the acoustic component of the problem, a similar approach is followed for obtaining a semi-discretized system of ordinary differential equations in time. Using a variational approach [13], an arbitrary, continuous, virtual pressure field, $\delta p(\vec{x}, t)$, is defined on Φ and Ψ, which satisfies homogeneous essential boundary conditions. Taking the divergence of Eq. (4-12) and substituting Eq. (4-13), the acoustic wave equation becomes

$$\nabla^2 p(\vec{x},t) - \frac{\rho_{\mathrm{f}}}{B_{\mathrm{f}}}\ddot{p}(\vec{x},t) = 0 \tag{4-26}$$

Then, taking the product of the virtual pressure field with Eq. (4-26), integrating over the domain, applying the divergence theorem, and substituting the natural and radiation boundary conditions, the weak form of the acoustic field problem is given by

$$\int_{\Phi} \delta p(\vec{x},t)\frac{\rho_{\mathrm{f}}}{B_{\mathrm{f}}}\ddot{p}(\vec{x},t)\mathrm{d}V + \int_{\Phi} \nabla\delta p(\vec{x},t)\cdot\nabla p(\vec{x},t)\mathrm{d}V + \cdots$$

$$\cdots + \int_{\Psi_{\mathrm{U}}} \delta p(\vec{x},t)\rho_{\mathrm{f}}\ddot{u}_n^{\mathrm{f}}(\vec{x},t)\,\mathrm{d}S + \int_{\Psi_{\mathrm{R}}} \delta p(\vec{x},t)\beta\sqrt{\frac{\rho_{\mathrm{f}}}{B_{\mathrm{f}}}}\dot{p}\left(\vec{x},t\right)\mathrm{d}S = 0 \tag{4-27}$$

Again, the domain is divided into finite elements and the pressure and virtual pressure fields are approximated within each element in an identical manner to that described for the displacement fields in the solid mechanics problem. Substituting the field approximations into Eq. (4-27) and eliminating the arbitrary virtual pressure field vector, the semi-discretized acoustic wave equation becomes

$$[M_{\mathrm{f}}]\{\ddot{p}(t)\} + [C_{\mathrm{f}}]\{\dot{p}(t)\} + [K_{\mathrm{f}}]\{p(t)\} = -\{R_{\mathrm{f}}(t)\} \tag{4-28}$$

where $\{p(t)\}$ is the vector of nodal pressures, and the global matrices are given by

$$[M_{\mathrm{f}}] = \sum_{\text{elements}} \int_{\Phi^{\mathrm{e}}} \frac{\rho_{\mathrm{f}}}{B_{\mathrm{f}}} \left[H(\vec{x})\right]^{\mathrm{T}} \left[H(\vec{x})\right] \mathrm{d}V \tag{4-29}$$

$$[K_{\mathrm{f}}] = \sum_{\text{elements}} \int_{\Phi^{\mathrm{e}}} \left[F(\vec{x})\right]^{\mathrm{T}} \left[F(\vec{x})\right] \mathrm{d}V \tag{4-30}$$

$$[C_{\mathrm{f}}] = \sum_{\text{elements}} \int_{\Psi_{\mathrm{R}}^{\mathrm{e}}} \beta\sqrt{\frac{\rho_{\mathrm{f}}}{B_{\mathrm{f}}}} \left[H\left(\vec{x}\right)\right]^{\mathrm{T}} \left[H\left(\vec{x}\right)\right] \mathrm{d}S \tag{4-31}$$

and

$$\{R_{\mathrm{f}}(t)\} = \sum_{\text{elements}} \int_{\Psi_{\mathrm{U}}^{\mathrm{e}}} \rho_{\mathrm{f}} \left[H(\vec{x})\right]^{\mathrm{T}} \ddot{u}_n(\vec{x},t)\mathrm{d}S \tag{4-32}$$

In Eqs. (4-29) to (4-32), $[H(\vec{x})]$ is a row vector of fluid interpolation functions, $[F(\vec{x})]$ is the matrix of spatial derivatives of the interpolation functions for the fluid, $[M_\mathrm{f}]$ is referred to as the fluid mass matrix, $[K_\mathrm{f}]$ is the fluid stiffness matrix, $[C_\mathrm{f}]$ is the fluid damping matrix associated with the radiation boundary condition, and $\{R_\mathrm{f}(t)\}$ is the fluid external force vector.

The structural and fluid equations are coupled through their respective external force vectors. Therefore, the semi-discretized finite element equations for the coupled fluid-structure system can be expressed as

$$[M]\{\ddot{u}(t)\} + \{I(t)\} = \{R(t)\} + [S]\{p(t)\} \tag{4-33}$$

and

$$[M_\mathrm{f}]\{\ddot{p}(t)\} + [C_\mathrm{f}]\{\dot{p}(t)\} + [K_\mathrm{f}]\{p(t)\} = -\{R_\mathrm{f}(t)\} - \rho_\mathrm{f}[S]^\mathrm{T}\{\ddot{u}(t)\} \tag{4-34}$$

where $[S]$ is the global interaction matrix defined by

$$[S] = \sum_{\text{elements}} \int_{\Gamma_\mathrm{FS}^\mathrm{e}} \left[N(\vec{x})\right]^\mathrm{T} \vec{n}_\mathrm{f}(\vec{x}) \left[H(\vec{x})\right] \mathrm{d}S \tag{4-35}$$

For steady-state dynamic analysis, structural displacement, fluid pressure, and external forces are assumed to vary harmonically as

$$\{u(t)\} = [\Re\{\bar{u}(\omega)\} + i\Im\{\bar{u}(\omega)\}]e^{i\omega t} \tag{4-36}$$

$$\{p(t)\} = [\Re\{\bar{p}(\omega)\} + i\Im\{\bar{p}(\omega)\}]e^{i\omega t} \tag{4-37}$$

$$\{R(t)\} = [\Re\{\bar{R}(\omega)\} + i\Im\{\bar{R}(\omega)\}]e^{i\omega t} \tag{4-38}$$

and

$$\{R_\mathrm{f}(t)\} = [\Re\{\bar{R}_\mathrm{f}(\omega)\} + i\Im\{\bar{R}_\mathrm{f}(\omega)\}]e^{i\omega t} \tag{4-39}$$

where $\{\bar{u}(\omega)\}$, $\{\bar{p}(\omega)\}$, $\{\bar{R}(\omega)\}$, and $\{\bar{R}_\mathrm{f}(\omega)\}$ are the complex amplitudes of the steady-state response of the displacement, pressure, and external force vectors, respectively; and $\Re$ and $\Im$ denote real and imaginary components, respectively.

Applying the generalized Maxwell model for the solid constitutive equation, and defining the elastic and viscous stiffness matrices to be

$$\left[K^\mathrm{E}\right] = \sum_{\text{elements}} \int_{\Omega^\mathrm{e}} [B]^\mathrm{T}[C][B]\,\mathrm{d}V \tag{4-40}$$

and

$$\left[K_\mathrm{m}^\mathrm{V}\right] = \sum_{\text{elements}} \int_{\Omega^\mathrm{e}} [B]^\mathrm{T}\left[\tilde{C}_\mathrm{m}\right][B]\,\mathrm{d}V \tag{4-41}$$

where $[C]$ and $[\tilde{C}_\mathrm{m}]$ are the matrix conversions of the fourth-order tensors $\mathbf{C}^\mathrm{IV}$ and $\tilde{\mathbf{C}}_\mathrm{m}^\mathrm{IV}$, respectively, to Voigt notation. Then substituting the harmonic variables, and equating real terms and imaginary terms, the coupled system of equations becomes

$$\begin{bmatrix} [A_\mathrm{D}] & -[A_\mathrm{OD}] & -[S] & 0 \\ [A_\mathrm{OD}] & [A_\mathrm{D}] & 0 & -[S] \\ -\omega^2\rho_\mathrm{f}[S]^\mathrm{T} & 0 & [B_\mathrm{D}] & -[B_\mathrm{OD}] \\ 0 & -\omega^2\rho_\mathrm{f}[S]^\mathrm{T} & [B_\mathrm{OD}] & [B_\mathrm{D}] \end{bmatrix} \begin{Bmatrix} \Re(\{\bar{u}(\omega)\}) \\ \Im(\{\bar{u}(\omega)\}) \\ \Re(\{\bar{p}(\omega)\}) \\ \Im(\{\bar{p}(\omega)\}) \end{Bmatrix} = \begin{Bmatrix} \Re(\{\bar{R}(\omega)\}) \\ \Im(\{\bar{R}(\omega)\}) \\ -\Re(\{\bar{R}_\mathrm{f}(\omega)\}) \\ -\Im(\{\bar{R}_\mathrm{f}(\omega)\}) \end{Bmatrix} \tag{4-42}$$

where

$$[A_{\mathrm{D}}] = -\omega^2 [M] + \left[K^{\mathrm{E}}\right] + \sum_{m=1}^{n} \frac{\omega^2 \tau_{\mathrm{m}}^2}{1+\omega^2 \tau_{\mathrm{m}}^2} \left[K_{\mathrm{m}}^{\mathrm{V}}\right] \quad (4-43)$$

$$[A_{\mathrm{OD}}] = \sum_{m=1}^{n} \frac{\omega \tau_{\mathrm{m}}}{1+\omega^2 \tau_{\mathrm{m}}^2} \left[K_{\mathrm{m}}^{\mathrm{V}}\right] \quad (4-44)$$

$$[B_{\mathrm{D}}] = -\omega^2 [M_{\mathrm{f}}] + [K_{\mathrm{f}}] \quad (4-45)$$

and

$$[B_{\mathrm{OD}}] = \omega [C_{\mathrm{f}}] \quad (4-46)$$

4.3.2 Boundary Element Method

Another method for the numerical solution of the BVPs and IBVPs occurring in vibro-acoustics is the boundary element method (BEM), also known as the boundary integral equation method (BIEM). The basis of BEM is to create a weighted integral relationship for the differential equation in which the field solution is obtained through integrals over the boundary of the domain of interest, rather than integrals over the entire domain as is required by FEM. Therefore, the numerical solution of the integral equations requires only discretization and approximation over the boundaries, yielding a substantially smaller system of equations than its FEM counterpart. In addition, semi-infinite domains/boundaries can be included in a natural analytic way in BEM, rather than relying on approximations such as the Sommerfeld boundary conditions shown above. Furthermore, as the solution is obtained through discretization of the boundary alone, the solution and its derivatives are continuous throughout the interior of the domain.

However, significant disadvantages also exist for BEM in comparison with FEM. First, the matrix in the system of equations developed in BEM is fully populated. Therefore, the order of operations required to solve the BEM system can quickly exceed the operations for the sparse matrices occurring in FEM, particularly in cases where one or more spatial dimensions of the domain are considerably smaller than the remaining finite dimensions. Another issue with BEM is the requirement for significant mathematical analysis of the differential equations of interest, in comparison with the relatively simple formulation and general applicability of the weak form Galerkin methods shown above. The required development of Green's functions for the adjoint operators and the solution of singular integrals can be very challenging aspects of BEM, and likely are the major deterrents in their development and use. As a result, BEM is often considered a "niche" method for the solution of BVP and IBVP, with the most common applications consisting of large (or semi-infinite) homogeneous domains, or cases in which the required solution resolution is such that discretization of the interior would be prohibitive [18-20].

A typical procedure for the BEM formulation begins with the development of the fundamental solution (a.k.a. singular solution or Green's function) for the adjoint to the differential operator. Although the development of the fundamental solution can be a mathematically daunting task, many solutions already exist in the literature for typical engineering problems. Next, a weighted integral statement is written, in much the same way as shown above for FEM. The integral statement is manipulated through integration by parts (and/or using related theorems, e.g., Gauss's Theorem and Green's Identities), to obtain a relationship in which there is a domain integral of the product between the solution variable and the adjoint operating on the weight function, and all remaining terms involving the solution variable have been transformed to integrals over the domain boundary. Then the weight function is taken to be the fundamental solution to the adjoint operator, and the limit is taken as the source point in the fundamental solution approaches the boundary of the domain. In this manner, a relationship is developed between the solution at the boundary and an integral statement where the kernels containing the solution variable and its derivatives are integrated over the domain boundary, which is known as the boundary integral equation (BIE). It is worth noting that in some cases additional integral terms over the domain interior remain, typically involving distributed body source terms such as body forces or internal heat generation. Depending on the nature of these source terms, it is often necessary to perform a separate numerical integration of these domain integrals, requiring some form of interior domain discretization. Once the BIE relationship has been obtained, the boundary of the domain is discretized into elements and the solution field and a field containing the solution derivatives, usually the natural boundary condition quantity (e.g., normal solution derivatives in potential problems, tractions in elasticity problems), are approximated. This discretization is essentially identical to the methods described above for FEM, except that the approximation of the field and its derivatives are considered separately. Finally, the BIE is transformed into a linear system of equations which can be solved for the unknown nodal values of the solution and its derivatives along the boundary. Once the boundary solution is obtained, the integral relationship, prior to taking the limit as the source point in the fundamental solution approaches the boundary, can be used to obtain the solution anywhere within the domain.

The BEM formulation will be further described through the development of the BIE for the acoustic wave equation. The governing PDE for acoustic waves, shown in Eq. (4-26), is multiplied by a weight function and integrated over the spatial and temporal domain as follows

$$\int_0^t \int_\Phi G\left(\vec{x},\zeta\right)\left(\nabla^2 p(\vec{x},\zeta) - \frac{\rho_f}{B_f}\ddot{p}(\vec{x},\zeta)\right)\mathrm{d}V\mathrm{d}\zeta = 0 \tag{4-47}$$

where $G(\vec{x},\zeta)$ is the arbitrary weighting function and ζ is the integration variable that represents time. Notice that contrary to FEM the weighted integral statement includes an integral over the temporal domain, thereby treating the time equivalently to the spatial dimensions of the problem. Applying integration by parts for both the spatial and temporal derivatives the integral equation can be shown as

$$\left[\begin{array}{l}\displaystyle\int_0^t \int_\Phi \nabla\cdot(G\nabla p) - \nabla\cdot(\nabla G p) - \frac{\rho_f}{B_f}\frac{\partial}{\partial t}(G\dot{p}) \\ \displaystyle +\frac{\rho_f}{B_f}\frac{\partial}{\partial t}(\dot{G}p) + p\left(\nabla^2 G - \frac{\rho_f}{B_f}\ddot{G}\right)\mathrm{d}V\mathrm{d}\zeta = 0\end{array}\right] \tag{4-48}$$

where for the sake of brevity the functional dependence on $\vec{x}$ and t is assumed for G and p. Using the divergence theorem and integrating with respect to time when possible, the integral equation can then be shown as

$$\int_0^t \int_\Phi -p\left(\nabla^2 G - \frac{\rho_f}{B_f}\ddot{G}\right)\mathrm{d}V\mathrm{d}\zeta = \int_0^t \int_\Psi \left(G\frac{\partial p}{\partial \vec{n}}\right) - \left(\frac{\partial G}{\partial \vec{n}}p\right)\mathrm{d}S\mathrm{d}\zeta + \cdots$$
$$\cdots + \left[\int_\Phi -\frac{\rho_f}{B_f}(G\dot{p}) + \frac{\rho_f}{B_f}(\dot{G}p)\,\mathrm{d}V\right]_{\zeta=0}^{\zeta=t} \tag{4-49}$$

Now taking the weight function to be the fundamental solution to the adjoint operator, which satisfies

$$\nabla^2 G\left(\vec{x},t;\vec{\xi},\tau\right) - \frac{\rho_f}{B_f}\ddot{G}\left(\vec{x},t;\vec{\xi},\tau\right) = -\delta\left(\vec{x},t;\vec{\xi},\tau\right) \tag{4-50}$$

where $\delta(\vec{x},t;\vec{\xi},\tau)$ is the Dirac delta function for an impulse source applied at point $\vec{\xi}$ and time τ. In addition, the fundamental solution must adhere to the causality condition (i.e., no response occurs at times prior to the application of the impulse) shown as

$$G\left(\vec{x},t;\vec{\xi},\tau\right) = 0 \text{ for } t < \tau \tag{4-51}$$

Assuming for simplicity homogeneous initial conditions, the integral equation for the solution of the acoustic pressure at any time and anywhere within the domain is given by

$$p\left(\vec{\xi},\tau\right) = \int_0^t \int_\Phi \left(G\left(\vec{x},\zeta;\vec{\xi},\tau\right) \frac{\partial p\left(\vec{x},\zeta\right)}{\partial \vec{n}} \right) - \left(\frac{\partial G\left(\vec{x},\zeta;\vec{\xi},\tau\right)}{\partial \vec{n}} p\left(\vec{x},\zeta\right) \right) \mathrm{d}S\mathrm{d}\zeta \tag{4-52}$$

Taking the limit as the point $\vec{\xi}$ approaches the boundary, the BIE is given by

$$C\left(\vec{\xi}\right) p\left(\vec{\xi},\tau\right) = \int_0^t \int_\Psi \left(G\left(\vec{x},\zeta;\vec{\xi},\tau\right) \frac{\partial p\left(\vec{x},\zeta\right)}{\partial \vec{n}} \right) - \left(\frac{\partial G\left(\vec{x},\zeta;\vec{\xi},\tau\right)}{\partial \vec{n}} p\left(\vec{x},\zeta\right) \right) \mathrm{d}S\mathrm{d}\zeta \tag{4-53}$$

where $C(\vec{\xi})$ is a function of the boundary geometry that takes into account the strong singularities in the kernel of Eq. (4-53) when $\vec{\xi}$ and $\vec{x}$ coincide, and allows the integrals to be evaluated analytically or numerically without overlooking the singularities.

All that remains is to discretize the pressure and its normal derivative in both time and space. Then a system of equations can be formed by taking the source point to be each boundary node, in turn, and evaluating the integrals in Eq. (4-53) either numerically or analytically for each element along the boundary. Once an approximate solution is found for all of the boundary unknowns, the boundary field approximations can be substituted into Eq. (4-52) to obtain the solution anywhere within the domain. In addition, Eq. (4-52) can be differentiated analytically in order to use the boundary solution to obtain the derivatives of the solution anywhere within the domain.

The BEM procedure described for the acoustic wave equation can be repeated for the solid mechanics equations to obtain a system of equations for the solution of the boundary displacements and tractions. Furthermore, again applying the continuity of the particle displacements and equivalence of the acoustic pressure with the solid traction along the interface boundary (i.e., coupling conditions), a coupled system of equations can be obtained for the acoustic structure interaction problem.

Unfortunately, the mathematical analysis of the solid mechanics equations in BEM can cause considerable difficulty. In particular, problems arise from the solid mechanics constitutive relationships in that each mechanical behavior requires a unique fundamental solution, and even more important, BEM is inadequate for the analysis of heterogeneous materials. Since heterogeneity and high variability are common traits in biological materials, BEM is not recommended in general for the analysis of biological structures. However cases may exist where certain structures or their surrounding mediums can be approximated sufficiently using simplifications such as treating a large (or semi-infinite) portion of the material as a homogeneous solid or acoustic fluid. Yet, even in these special cases it may be more efficient, if not necessary, to use similar coupling procedures as described above to combine FEM with BEM.

4.3.3 Solution of Helmholtz Equations With High Wave Number Using FEM

For steady-state dynamics, the governing acoustic and structural equations are of the Helmholtz type. Furthermore, the combination of the low wave speeds typical in soft tissues along with the relatively high excitation frequencies used in vibroacoustics results in considerably large wave numbers for the systems considered. Therefore, it is of paramount importance to discuss some numerical problems that may arise when FEM is used to approximate the solution of Helmholtz equation with high wave number. It is well documented in the literature [21, 22] that the error resulting from finite element discretization grows dramatically faster for Helmholtz equation with high wave numbers than the error predicted by asymptotic analysis. This phenomenon is commonly called "the pollution effect," and is the product of a spurious phase lag in the solution. In fact, it is impractical to use conventional low-order finite elements for Helmholtz problems with high wave number because excessive mesh refinement would be required to produce accurate solutions.

In order to alleviate the pollution effect, several methods have been proposed such as the use of spectral methods [23], least-squares finite element processes in the h-p-k framework [24], and the partition of unity method (PUM) [25], among others. It is highly recommended to the reader who plans to use FEM for vibroacoustic problems with high wave number (i.e., low wavelengths) to read the literature cited on this topic. An excellent reference on this topic is the book by Frank Ilhenburg, *Finite Element Analysis of Acoustic Scattering* [17].

4.4 INVERSE PROBLEM TECHNIQUES FOR MATERIAL CHARACTERIZATION

The inverse problems considered here consist of the evaluation of the material behavior for some part of a given fluid-structure system (e.g., elastic or viscoelastic constants of a solid) based on the measurements of the acoustic field resulting from a remote excitation of the system. As with most inverse problems in science and engineering, with any degree of complexity in the geometry, boundary conditions, or physics governing the system, analytical solutions to inverse problems of this type become intractable. The common approaches to obtain a solution therefore require the development of a mathematical/numerical representation of the forward problem of the system (e.g., FEM to obtain the acoustic field given the system geometry, boundary conditions, and material behavior), along with non-linear optimization techniques. As such, the inverse problem is cast as an optimization problem such as

$$\underset{\vec{\alpha}}{\text{Minimize}} \left(J\left(\vec{\alpha}\right) \right) \tag{4-54}$$

where $\vec{\alpha}$ is the unknown vector characterizing the material behavior and $J(\vec{\alpha})$ is an error functional which quantifies the distance (typically through some vector space norm) between the experimental response and the response computed using the system representation. This model updating approach to inverse problems has gained much popularity in recent years due to the relentless increase in computational speed [4, 26-28].

4.4.1 Ill-Posedness and Regularization

Inverse problems of this type, and most problems relating to non-destructive evaluation (NDE), are to some extent ill-posed. Ill-posed inverse problems are those in which a solution does not exist, the solution is non-unique, and/or the solution does not vary continuously with changes in the measurement data (i.e., unstable solution). In particular, non-uniqueness is a very common source of ill-posedness in inverse problems, especially those related to NDE. In some cases, this non-uniqueness is related to the formulation of the forward problem. For example, the parameters defining a rheological viscoelasticity model are typically a non-unique representation of rate dependent mechanical behavior over a finite time or frequency domain [4]. However, as long as the solution method is tolerant to this and the mechanical behavior defined by the rheological parameters is uniquely identifiable; then this form of non-uniqueness is inconsequential. Alternatively, in some cases the information content in the measurement signal is insufficient to constrain the solution to be unique due to noise or insensitivity. Often the only solution for such ill-posedness is to attempt to obtain additional experimental data, particularly in cases where the representation of the existing experiment is not strongly sensitive to the material behavior to be characterized.

Regularization techniques can in some cases be used to alleviate ill-posedness of inverse problems [29]. For example, methods such as Tikhonov regularization, which alter the inverse problem formulation, can be used to eliminate spurious oscillatory solutions and thereby reduce non-uniqueness and improve the inverse problem stability [30, 31]. Some texts also consider filtering the measurement signal to be a form of regularization, in that it can reduce the ill-posedness induced by noise in the signal. Unfortunately, in many cases manipulation of the experiment and/or inverse problem formulation (e.g., regularization) is not sufficient to guarantee a satisfactory solution, and effort must be placed to identify a suitable optimization solution method.

4.4.2 Optimization Methods

Optimization techniques are restricted to two main categories: gradient based and non-gradient based. The gradient based methods of optimization are iterative approaches that rely on the gradient of the error functional with respect to optimization parameters to locate a minimum. Examples include, among others, Newton's method and the Conjugate Gradient method [32, 33]. The main advantage of gradient based methods is that a solution can be obtained with comparatively few iterations. However, even for well-posed inverse problems, if the error surface defined by Eq. (4-54) is non-convex (i.e., multiple local solutions exist), gradient based methods will fail unless an initial solution estimate is very close to the actual global solution. In applications for NDE of biological materials the error surfaces are typically non-convex, and biological material properties can vary drastically depending on the location in the body, age, and presence of disease [34], making an accurate initial estimate difficult. Therefore, purely gradient based methods of optimization for these problems are typically considered insufficient.

Some typical non-gradient based methods include stochastic algorithms and machine learning algorithms. Examples of stochastic algorithms are the direct/random search methods [35], and evolutionary algorithms such as genetic algorithms [36, 37]. In general, the stochastic algorithms are iterative schemes that begin with generating solution estimates in some random fashion and evaluating their corresponding errors. Then simple (often heuristic) rules based on the error levels of the estimates along with some additional randomization are applied to generate the estimates for subsequent iterations. These methods typically have excellent global search capabilities (i.e., consistently unaffected by problem non-convexity), but can require a relatively large number of evaluations of the error functional. In cases where the system is particularly complex, each numerical simulation of the system can be of great computational expense, and the number of iterations required for convergence can become prohibitive.

Examples of machine learning techniques include autoprogressive and self-learning algorithms [28, 38], which rely on general mapping tools such as artificial neural networks to represent the material behavior, and successive iterations of numerical analyses to obtain and update constitutive data from which to teach the constitutive relationship to the mapping tool. Also included in the machine learning category are surrogate model methods [39, 40]. These methods apply similar general mapping tools which are trained through a data set of parameter estimates and corresponding model responses generated with the numerical representation of the system. The mapping is then used to act as a replacement (i.e., surrogate) to the numerical representation of the system, mapping the parameter estimates to the system output in the optimization problem. This much simplified numerical representation can be minimized with respect to the parameter estimates either analytically or through another optimization approach to obtain an estimate to the solution. The machine learning techniques require very few numerical analyses to obtain a solution estimate, but in general only act as local solution methods (i.e., often obtain solution estimates corresponding to a local minimum) and the results are highly dependent upon the data sets generated to train the mapping tools, and the mapping tools themselves.

In order to successfully obtain solutions to realistic material characterization problems in engineering, it is often necessary to use combinations of the above mentioned techniques. These methods range from simple approaches such as the application of a gradient based technique beginning with the initial solution estimate as the estimate obtained from a relatively short run of a stochastic global search technique, to more elegant approaches such as the surrogate model accelerated random search algorithm (SMARS) [31], which iteratively combines the global random search algorithm with a locally applied surrogate model method such that the benefits of each algorithm are enhanced, while counteracting their respective weaknesses. In all cases, these combination methods are intended to maintain global search capabilities while reducing the computational cost to find a solution.

4.5 EXAMPLE

The following example from the work shown in [4] is intended to provide further insight into the challenges and solution methods described herein for the characterization of material properties using vibroacoustic techniques. The example considers a simulated experiment based on a potential technique to identify material changes during cryoablation. Cryoablation is a process in which localized regions of undesirable tissue are destroyed through freezing, see [41, 42]. Inverse estimation of the material properties of frozen and unfrozen tissue could be used for tracking the progress of cryoablation during clinical procedures.

The simulated vibroacoustic experiment consisted of a soft tissue sphere with an ice sphere inclusion immersed in an infinite domain of water, as shown in Fig. 4-1. A harmonic pressure was applied to a small region on the surface of the ice sphere inclusion, and the acoustic pressure amplitude response was measured at a single point in the surrounding water for a series of excitation frequencies between 10 and 80 kHz. The experiment was simulated using the steady-state dynamic finite element method, as described in Section 4.3.1.

The ice was considered to be an isotropic linear elastic material, which is reasonable for the frequencies considered, and the parameters used were based on the values presented

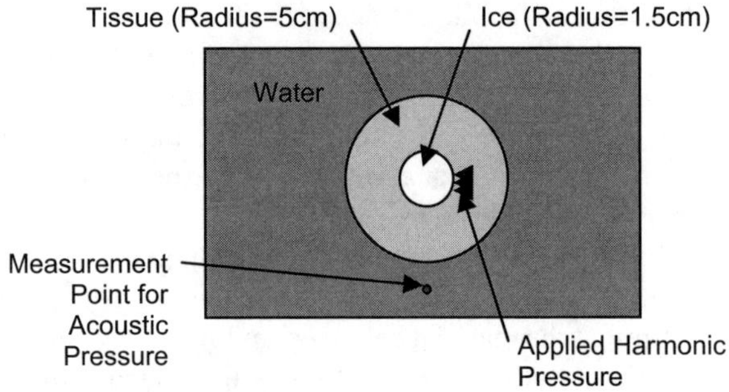

Figure 4-1. Schematic of the simulated experiment [4]

in [43]. The tissue was considered to be an isotropic, highly viscous material represented by a generalized Maxwell model with one Maxwell unit. Oftentimes, isotropic viscoelastic response is defined with two independent scalar functions of frequency. For example, in the work presented in [44], the material behavior was represented by a complex shear modulus and a purely elastic bulk modulus. In a similar manner, for this work, the material was defined by a constant Poisson's ratio and a complex Young's modulus given by

$$E^*(\omega) = E + \tilde{E}\left(\frac{\omega^2\tau^2 + i\omega\tau}{1+\omega^2\tau^2}\right) \qquad (4-55)$$

Thus, the isotropic viscoelastic constitutive model was described through the long-term elastic Young's modulus, E, the viscous Young's modulus, $\tilde{E}$, the Poisson's ratio, ν, and the relaxation time, τ. In the results presented, the viscous Young's modulus was expressed through a dimensionless viscous Young's modulus, g, given by

$$g = \frac{\tilde{E}}{E+\tilde{E}} \qquad (4-56)$$

The viscoelastic parameters used for the tissue were based on a combination of the elastic parameters reported in [34] for soft tissue and the viscoelastic parameters reported in [44] for mammalian tissue. The density of both materials was considered known for the inverse problem to be the density of water (1000 kg/m^3), which is a reasonable approximation for both soft biological tissues and ice. The tissue was considered known to be nearly incompressible, whereas the Poisson's ratio for ice was taken to be 0.33, and considered unknown in the inverse problem. Therefore, the parameters to be identified were the Young's modulus for the ice, E_1, the Poisson's ratio for the ice, ν_1, the long-term elastic Young's modulus for the surrounding tissue, E_2, the dimensionless viscous Young's modulus for the surrounding tissue, g_2, and the relaxation time for the surrounding tissue, τ_2.

It is important to realize that different combinations of viscoelastic parameters (i.e., E, $\tilde{E}$, τ), or even different viscoelastic rheological representations can result in the same material behavior over a given time period or frequency range. Alternatively, a viscoelastic material may be uniquely characterized by more physically relevant quantities, such as energy storage and dissipation over time or frequency. In the frequency domain, energy dissipation and storage can be expressed through the mechanical loss, $\tan[\delta(\omega)]$, which is associated with viscous damping, and the equivalent elastic modulus, $E^{eq}(\omega)$, respectively. The mechanical loss and equivalent elastic modulus are defined as the ratio of the imaginary to the real portion of the complex relaxation modulus, and the magnitude of the complex modulus, respectively. For the complex Young's modulus considered here these quantities are given by

$$\tan(\delta) = \left[\tilde{E}\left(\frac{\omega\tau}{1+\omega^2\tau^2}\right)\right] \Big/ \left[E + \tilde{E}\left(\frac{\omega^2\tau^2}{1+\omega^2\tau^2}\right)\right] \tag{4-57}$$

and

$$E^{\mathrm{eq}}(\omega) = \left\{\left[\tilde{E}\left(\frac{\omega\tau}{1+\omega^2\tau^2}\right)\right]^2 + \left[E + \tilde{E}\left(\frac{\omega^2\tau^2}{1+\omega^2\tau^2}\right)\right]^2\right\}^{1/2} \tag{4-58}$$

Although the generalized Maxwell model parameters defining these characteristics may be non-unique, the characteristic functions themselves uniquely define the material behavior. Therefore, in the following, the results are expressed in terms of these viscoelastic characteristic functions, $\tan[\delta(\omega)]$ and $E^{\mathrm{eq}}(\omega)$, resulting from the inversely determined viscoelastic moduli.

A model updating approach, as described in Section 4.4, was applied to solve the inverse characterization problem, again using the finite element method for the numerical representation of the system. However, for added realism and to relieve the inverse crime of using the same numerical model to simulate the experiment as to perform the model-updating optimization simulations, random Gaussian white noise was added to the simulated experimental response. The random Gaussian noise was introduced in the simulated data as

$$p_i^{\mathrm{error}} = p_i^{\mathrm{exact}}(1 + 0.1\eta) \tag{4-59}$$

where p_i^{error} is the ith response point containing the random noise, p_i^{exact} is the ith response point of the simulated test without noise, and η is a normally distributed random variable with unit variance and zero mean. It was found that this equation produced realistic values of acoustic pressure deviations. Although limitations inevitably exist when using simulated experiments, in order to represent realistic scenarios, a well developed numerical model

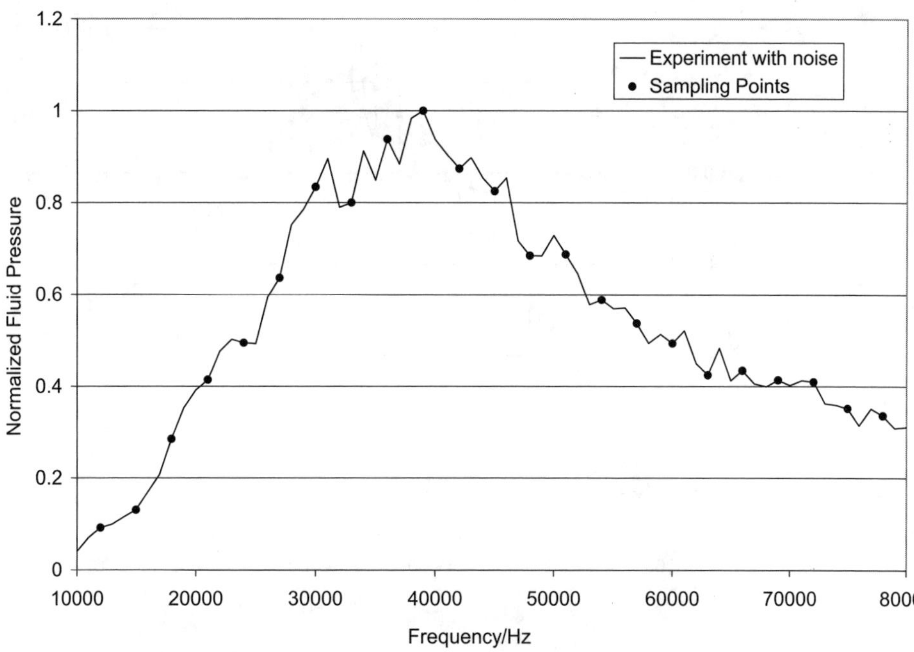

Figure 4-2. Frequency response of normalized acoustic pressure and discrete sampling points used for optimization [4]

TABLE 4-1. Experimental material properties and optimization search ranges

Optimization Parameter	Target Value	Optimization Minimum	Optimization Maximum
E_1	7.0×10^9	1.0×10^9	1.0×10^{10}
N_1	0.33	0.20	0.49
E_2	5.0×10^5	1.0×10^5	1.0×10^6
G_2	0.5	0.01	0.99
T_2	1.0×10^{-6}	1.0×10^{-7}	1.0×10^{-4}

with artificial noise is considered sufficient for the purposes of this work. Figure 4-2 shows the simulated experimental frequency response for the acoustic pressure amplitude at the chosen measurement point in the surrounding water, as well as the sampled discrete points from the spectrum used to obtain the inverse solution.

The SMARS algorithm [31] was then used to solve the model-updating optimization problem to characterize the ice and tissue properties. Table 4-1 shows the material properties used to simulate the experiment and the search ranges used for the optimization process. Although the SMARS algorithm is a method for unconstrained optimization, it is often reasonable to assume that material parameters can be bound within a certain range. It was also considered reasonable to assume that the inclusion was harder than the surrounding material for the purposes of the cryoablation application. Due to the stochastic nature of the SMARS algorithm, five independent optimization runs of the simulated example were carried out, and the mean and standard deviation of the results were calculated.

Figure 4-3 shows the mean and standard deviation of the acoustic pressure frequency responses obtained from the five optimization trials compared to the simulated experimental response. Clearly, the SMARS algorithm was able to obtain parameters that matched

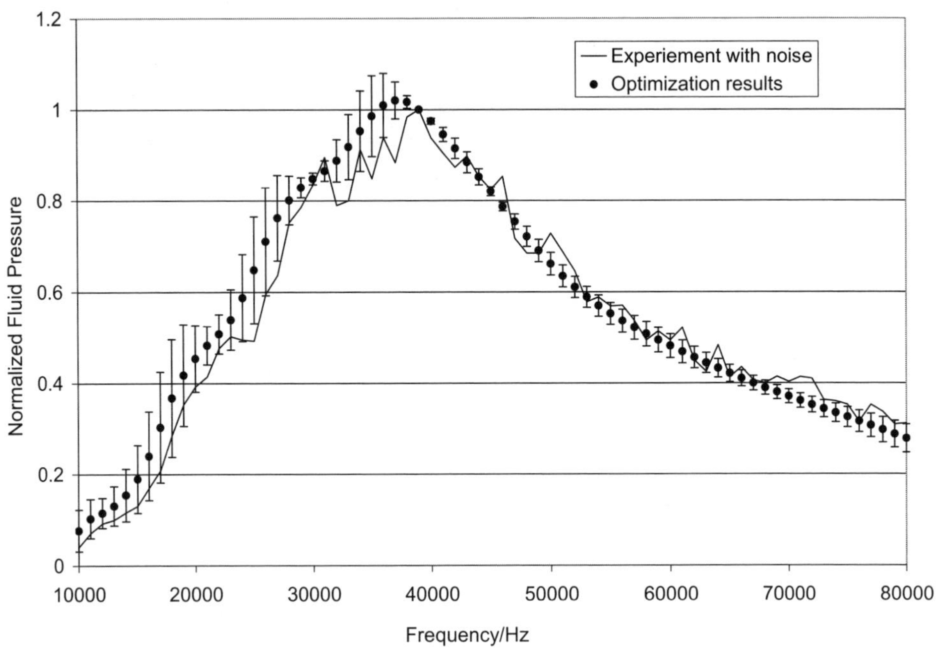

Figure 4-3. Frequency response of normalized acoustic pressure for the optimization results compared to the simulated experiment. The optimization results show the mean and standard deviation (error bars) of the five trials [4]

TABLE 4-2. Optimization results for the five optimization trials

	E_1	ν_1	E_2	g_2	τ_2
Actual	7.00×10^9	0.33	5.00×10^5	0.50	1.00×10^{-6}
Trial 1	6.96×10^9	0.37	5.10×10^5	0.51	8.90×10^{-7}
Trial 2	6.98×10^9	0.38	5.23×10^5	0.56	6.06×10^{-7}
Trial 3	6.89×10^9	0.31	3.68×10^5	0.52	1.53×10^{-6}
Trial 4	6.32×10^9	0.36	4.15×10^5	0.51	1.01×10^{-6}
Trial 5	6.61×10^9	0.29	3.80×10^5	0.50	1.39×10^{-6}
Mean	6.75×10^9	0.34	4.39×10^5	0.52	1.09×10^{-6}
Std. Dev.	2.85×10^8	0.041	7.28×10^4	0.024	3.76×10^{-7}

the target response to a satisfactory extent. Table 4-2 shows the material parameter results obtained from the optimization trials. Figures 4-4 and 4-5 show the mean and standard deviation of the mechanical loss and equivalent elastic modulus, respectively, for the resulting tissue parameters compared to the experimental values.

The elastic modulus of the ice was found with a high level of accuracy, but the results were less accurate for Poisson's ratio. Through sensitivity studies on the material parameters, it was found that the acoustic response displayed little sensitivity to changes in Poisson's ratio of the ice, while the elastic modulus of the ice showed strong sensitivity, which explains the comparative scatter and lack of accuracy for Poisson's ratio. Furthermore, the results found for the mechanical loss of the tissue showed satisfactory accuracy, but with moderate variation, whereas the equivalent elastic modulus of the tissue was found with lower accuracy and larger scatter. The sensitivity studies carried out revealed that the acoustic pressure was more affected by changes in the dimensionless viscous Young's modulus

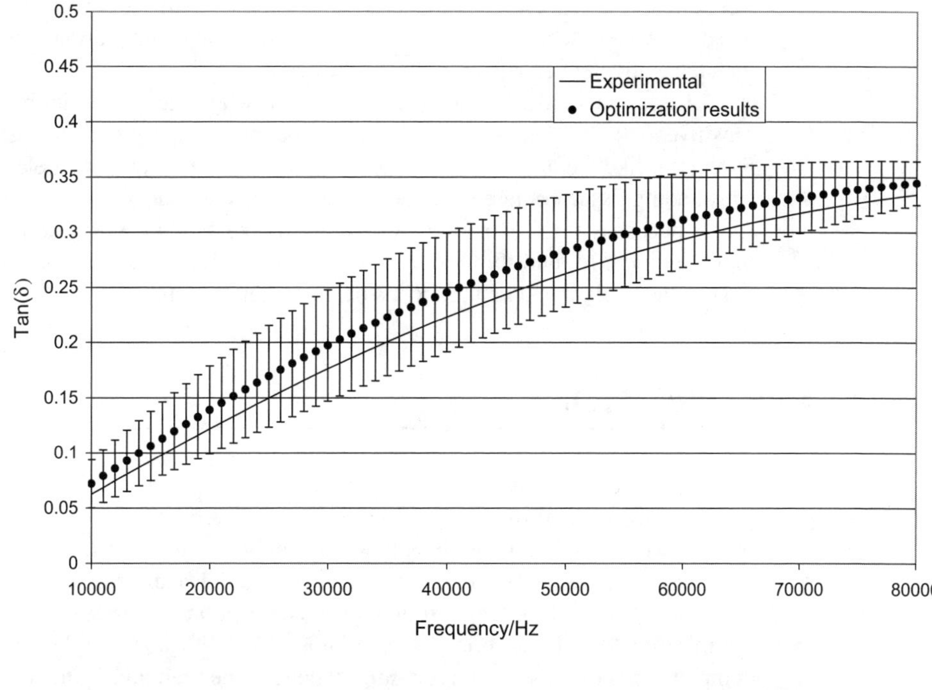

Figure 4-4. Frequency distribution of the mechanical loss (tan (δ)), for the optimization results. The optimization results show the mean and standard deviation (error bars) of the five trials [4]

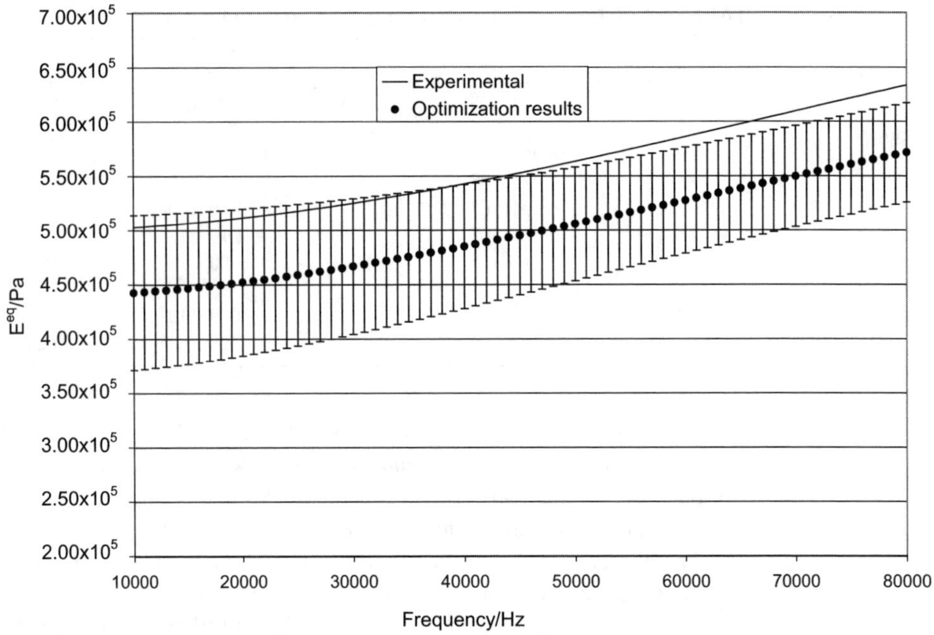

Figure 4-5. Frequency distribution of the equivalent elastic modulus (E^{eq}), for the optimization results. The optimization results show the mean and standard deviattion (error bars) of the five trials [4]

of the tissue than by changes in the long-term elastic Young's modulus of the tissue. Therefore, due to the relative dependencies of the characteristic functions on these parameters, this difference in sensitivity resulted in a more accurate identification of the mechanical loss than the equivalent elastic modulus of the tissue.

Clearly, the modes that appear in the frequency range studied in this example are very sensitive to the Young's modulus of the ice, but not to the equivalent elastic modulus of the tissue. It is likely that a more accurate identification of the equivalent elastic modulus of the surrounding tissue could be obtained if other frequency ranges were investigated. Though, for the purposes of this work and as pertains to cryoablation applications, it was considered more pertinent to identify the ice stiffness accurately, and a frequency range was chosen which was likely to excite the modes of the stiffer material.

4.6 FUTURE DIRECTIONS

The techniques presented in this chapter have shown promise for solving complex inverse problems in medical imaging and material characterization. However, further advancements are required before these methods are applicable to realistic biological problems. For instance, finite element and boundary element models can be of such high computational expense that model updating strategies become impractical, even for the most efficient optimization techniques. It is therefore imperative that techniques be devised to reduce the computational expense of the forward models, while maintaining the representation of the physics. Model reduction techniques such as proper orthogonal decomposition [45] have shown promise for improving computational speed while maintaining physics, and should be explored further from an inverse problem perspective. Furthermore, uncertainty in experimental methods and modeling techniques should be addressed quantitatively. This need calls for probability based formulations for inverse problems [46].

REFERENCES

[1] Baldewsing, R. A., De Korte, C. L., Schaar, J. A., Mastik, F., and Van Der Steen, A. F. W., 2004, "A Finite Element Model for Performing Intravascular Ultrasound Elastography of Human Atherosclerotic Coronary Arteries," *Ultrasound in Medicine and Biology,* **30,** pp. 803–813.

[2] Haemmerich, D., Tungjitkusolmun, S., Staelin, S. T., Lee, F. T., Mahvi, D. M., and Webster, J. G., 2002, "Finite-Element Analysis of Hepatic Multiple Probe Radio-Frequency Ablation," *IEEE Transactions on Biomedical Engineering*, **49,** pp. 836–842.

[3] Fung, Y. C., 1993, *Biomechanics: Mechanical Properties of Living Tissues*, 2nd ed., Springer-Verlag, New York.

[4] Brigham, J. C., Aquino, W., Mitri, F. G., Greenleaf, J. F., and Fatemi, M., 2007, "Inverse Estimation of Viscoelastic Material Properties for Solids Immersed in Fluids Using Vibroacoustic Techniques," *Journal of Applied Physics*, **101**, 23509-1-14.

[5] Doyley, M. M., Meaney, P. M., and Bamber, J. C., 2000, "Evaluation of an Iterative Reconstruction Method for Quantitative Elastography," *Physics in Medicine and Biology*, **45**, pp. 1521–1540.

[6] Raghavan, K. R., and Yagle, A. E., 1994, "Forward and Inverse Problems in Elasticity Imaging of Soft-Tissues," *IEEE Transactions on Nuclear Science*, **41,** pp. 1639–1648.

[7] Zhang, X. M., Kinnick, R. R., Fatemi, M., and Greenleaf, J. F., 2005, "Noninvasive Method for Estimation of Complex Elastic Modulus of Arterial Vessels," *IEEE Transactions on Ultrasonics Ferroelectrics and Frequency Control*, **52,** pp. 642–652.

[8] Fatemi, M., and Greenleaf, J. F., 1999, "Vibro-Acoustography: An Imaging Modality Based on Ultrasound-Stimulated Acoustic Emission," *Proceedings of the National Academy of Sciences of the United States of America*, **96**, pp. 6603–6608.

[9] Fatemi, M., and Greenleaf, J. F., 1999, "Application of Radiation Force in Noncontact Measurement of the Elastic Parameters," *Ultrasonic Imaging*, **21**, pp. 147–154.

[10] Mitri, F. G., Trompette, P., and Chapelon, J. Y., 2003, "Detection of Object Resonances by Vibro-Acoustography and Numerical Vibrational Mode Identification," *Journal of the Acoustical Society of America*, **114,** pp. 2648–2653.

[11] Findley, W. N., Lai, J. S., and Onaran, K., 1976, *Creep and Relaxation of Nonlinear Viscoelastic Materials, with an Introduction to Linear Viscoelasticity*, North-Holland Pub. Co., Amsterdam.

[12] Kinsler, L. E., Frey, A. R., Coppens, A. B., and Sanders, J. V., 2000, *Fundamentals of Acoustics*, 4th ed., Wiley, New York.

[13] Cook, R. D., Malkus, D. S., Plesha, M. E., and Witt, R. J., 2002, *Concepts and Applications of Finite Element Analysis*, 4th ed., John Wiley & Sons, New York, NY.

[14] Reddy, J. N., 2006, *An Introduction to the Finite Element Method*, 3rd ed., McGraw-Hill, New York.

[15] Hughes, T. J. R., 1987, *The Finite Element Method: Linear Static and Dynamic Finite Element Analysis*, Prentice-Hall, Englewood Cliffs, NJ.

[16] Bathe, K., Jr., 1996, *Finite Element Procedures*, Prentice-Hall, Englewood Cliffs, NJ.

[17] Ihlenburg, F., 1998, *Finite Element Analysis of Acoustic Scattering*, Applied Mathematical Sciences, Vol. 132, Springer, New York.

[18] Cheng, A. H. D., and Cheng, D. T., 2005, "Heritage and Early History of the Boundary Element Method," *Engineering Analysis with Boundary Elements*, **29**, pp. 268–302.

[19] Banerjee, P. K., 1994, *The Boundary Element Methods in Engineering*, McGraw-Hill Book Company, London.

[20] Brebbia, C. A., 1984, *The Boundary Element Method for Engineers*, Pentech Press, London.

[21] Ihlenburg, F., and Babuska, I., 1995, "Finite-Element Solution of the Helmholtz-Equation with High Wave-Number. 1. The H-Version of the FEM," *Computers & Mathematics with Applications*, **30**, pp. 9–37.

[22] Ihlenburg, F., and Babuska, I., 1997, "Finite Element Solution of the Helmholtz Equation with High Wave Number. 2. The h-p Version of the FEM," *SIAM Journal on Numerical Analysis*, **34**, pp. 315–358.

[23] Shen, J., and Wang, L. L., 2005, "Spectral Aproximation of the Helmholtz Equation with High Wave Numbers," *Siam Journal on Numerical Analysis*, **43**, pp. 623–644.

[24] Surana, K. S., Gupta, P., and Reddy, J. N., 2007, "Galerkin and Least-Squares Finite Element Processes for 2-D Helmholtz Euation in h, p, k Framework," *International Journal of Computational Methods in Engineering Science and Mechanics*, **8**, pp. 341–361.

[25] Melenk, J. M., and Babuska, I., 1996, "The Partition of Unity Finite Element Method: Basic Theory and Applications," *Computer Methods in Applied Mechanics and Engineering*, **139**, pp. 289–314.

[26] Asma, F., and Bouazzouni, A., 2005, "Finite Element Model Updating Using FRF Measurements," *Shock and Vibration*, **12**, pp. 377–388.

[27] Bagchi, A., 2005, "Updating the Mathematical Model of a Structure Using Vibration Data," *Journal of Vibration and Control*, **11**, pp. 1469–1486.

[28] Aquino, W., and Brigham, J. C., 2006, "Self-Learning Finite Elements for Inverse Estimation of Thermal Constitutive Models," *International Journal of Heat and Mass Transfer,* **49**, pp. 2466–2478.

[29] Liu, G. R. and Han, X., 2003,*Computational Inverse Techniques in Nondestructive Evaluation*, CRC Press, Boca Raton.

[30] Phillips, S. W., Aquino, W., and Chirdon, W. C., 2007, "Simultaneous Inverse Identification of Transient Thermal Properties and Heat Sources Using Sparse Sensor Information," *Journal of Engineering Mechanics-ASCE.*

[31] Brigham, J. C. and Aquino, W., 2007, "Surrogate-Model Accelerated Random Search Algorithm for Global Optimization with Applications to Inverse Material Identification," *Computer Methods in Applied Mechanics and Engineering*, **196**, pp. 4561–4576.

[32] Nocedal, J., and Wright, S. J., 2006, *Numerical Optimization*, Springer, New York.

[33] Luenberger, D. G., 1990, *Optmization by Vector Space Methods*, John Wiley & Sons, New York.

[34] Duck, F. A., 1990, *Physical Properties of Tissue: A Comprehensive Reference Book*, Academic Press, London.

[35] Brooks, S. H., 1958, "A Discussion of Random Methods for Seeking Maxima," *Operations Research*, **6**, pp. 244–251.

[36] Goldberg, D. E., 1989, *Genetic Algorithms in Search, Optimization, and Machine Learning*, Addison-Wesley, Reading, Mass.

[37] Mitchell, M., 1998, *An Introduction to Genetic Algorithm*, MIT Press, Cambridge, Mass.

[38] Ghaboussi, J., Pecknold, D. A., Zhang, M. F., and Haj-Ali, R. M., 1998, "Autoprogressive Training of Neural Network Constitutive Models," *International Journal for Numerical Methods in Engineering*, **42**, pp. 105–126.

[39] Rogers, L. L., Dowla, F. U., and Johnson, V. M., 1995, "Optimal Field-Scale Groundwater Remediation Using Neural Networks and the Genetic Algorithm," *Environmental Science & Technology*, **29**, pp. 1145–1155.

[40] Yan, S., and Minsker, B., 2004, "A Dynamic Meta-Model Approach to Genetic Algorithm Solution of a Risk-Based Groundwater Remediation Design Model," *2004 World Water and Environmental Resources Congress: Critical Transitions in Water and Environmental Resources Management*, **3**, pp. 1962–1971, American Society of Mechanical Engineers.

[41] Rubinsky, B., 2000, "Cryosurgery," *Annual Review of Biomedical Engineering*, **2**, pp. 157–187.

[42] Baust, J., Gage, A. A., Ma, H. W., and Zhang, C. M., 1997, "Minimally Invasive Cryosurgery — Technological advances," *Cryobiology*, **34**, pp. 373–384.

[43] Pounder, E. R., 1965, *The Physics of Ice*, Pergamon Press, Oxford.

[44] Urban, M. W, Kinnick, R. R., and Greenleaf, J. F., 2005, "Measuring the Phase of Vibration of Spheres in a Viscoelastic Medium as an Image Contrast Modality," *Journal of the Acoustical Society of America*, **118**, pp. 3465–3472.

[45] Aquino, W., 2007, "An Object-Oriented Framework for Reduced-Order Models Using Proper Orthogonal Decomposition (POD)," *Computer Methods in Applied Mechanics and Engineering*, **196**, pp. 4375–4390.

[46] Kaipio, J., and Somersalo, E., 2005, *Statistical and Computational Inverse Problems*, Applied Mathematical Sciences, Vol. 160, Springer, New York.

Chapter 5

ACOUSTIC RADIATION FORCE IMPULSE (ARFI) IMAGING: FUNDAMENTAL CONCEPTS AND IMAGE FORMATION

Kathryn R. Nightingale and Mark L. Palmeri
Department of Biomedical Engineering, Duke University, Durham, NC, USA

ABSTRACT

Acoustic radiation force imaging methods utilize acoustic radiation force to mechanically excite tissue, and the tissue response is monitored with conventional imaging methods. The data provide information about the stiffness of tissue. Acoustic Radiation Force Impulse (ARFI) imaging is one such method that is implemented on a modified diagnostic ultrasound scanner, using the same transducer for both the excitation and for monitoring the tissue response. ARFI images are generated of the tissue displacement magnitude within the region of excitation at a given time after radiation force application. ARFI images portray relative differences in tissue stiffness and provide interesting structural information that is well correlated with matched B-mode images. In many instances, ARFI images demonstrate improved contrast over conventional ultrasound images. This imaging mode is under investigation for many clinical applications including: breast, liver, and colorectal imaging, and monitoring RF-ablation procedures. This chapter discusses some of the fundamental concepts associated with ARFI imaging.

5.1 INTRODUCTION

For centuries, clinicians have utilized manual palpation as a method to identify pathologic tissues. For structures that are large and/or superficial, such as some breast and prostate tumors, clinical palpation remains a key approach for disease detection. However, for lesions that are smaller or that lie deep within the body, manual palpation is often unsuccessful. Palpation identifies differences in the elastic moduli, texture, and structure of diseased tissues as compared to normal tissues. Conventional imaging methods do not provide information about tissue stiffness; however, "elastography" imaging methods have been under investigation for the past two decades and have recently become commercially available [5, 51]. All elastography methods introduce mechanical excitation to tissue and then monitor the tissue response with conventional imaging methods.

A rapidly developing area of research in elastographic imaging includes methods that utilize acoustic radiation force to mechanically excite tissue [6-15]. Radiation force based methods overcome some of the challenges of more traditional elastographic approaches, as the mechanical excitation is applied directly within the region of interest as a body force, rather than using external compression or vibration sources. Acoustic Radiation Force Impulse (ARFI) imaging is one such method that uses short duration acoustic radiation force to generate localized displacements in tissue, and these displacements are tracked within the region of excitation using ultrasonic correlation based methods [6, 16]. The tissue response to these forces is monitored both spatially and temporally. A single, diagnostic ultrasonic transducer and system is typically utilized both to generate the acoustic radiation force and track the resulting tissue displacement response. This chapter summarizes some of the basic concepts that have been investigated to date in the context of ARFI imaging.

5.2 ACOUSTIC RADIATION FORCE

The spatial distribution of a focused radiation force field (i.e., the region of excitation [ROE]) is determined by both the acoustic parameters of the transmitted acoustic beam and the tissue's properties. At ultrasonic frequencies for compressional wave propagation, soft tissue is typically modeled as a fluid, and shear stresses are neglected. Additionally, in soft tissues the majority of attenuation results from absorption [17, 18], and thus in the following discussion, radiation forces generated from scattering and reflection are neglected. Under plane wave assumptions, Nyborg's approach to modeling the ultrasonic generation of acoustic streaming in a viscous fluid can be used to determine the radiation force magnitude generated by a propagating ultrasonic wave at a given location in soft tissue [19, 20]:

$$F = \frac{W_{absorbed}}{c} = \frac{2\alpha I}{c}, \qquad (5-1)$$

where F (dyn/[1000 cm]^3 or kg/sec^2 cm^2) is acoustic radiation force, $W_{absorbed}$ (W/[100 cm]^3) is the power absorbed by the medium at a given spatial location, c (m/s) is the speed of sound in the medium, α (m^{-1}) is the absorption coefficient of the medium, and I (W/cm^2) is the in situ temporal average intensity at a given spatial location. This force is associated with a transfer of momentum from the propagating acoustic wave to the tissue directed along the Poynting vector of the propagating acoustic wave. For tissues with relatively low attenuations (0.3 dB/cm/MHz to 0.5 dB/cm/MHz), or when relatively low frequencies (1 MHz to 3 MHz) and low *f*-number (1 to 2) focal configurations are utilized for radiation force excitation, the majority of the energy is absorbed within the focal region of the acoustic beam (~10 mm^3); the region of excitation can approach a spatial impulse under these conditions. For tissues with higher attenuations (0.7 dB/cm/MHz to 1.5 dB/cm/MHz), or when higher frequencies (7 MHz to 10 MHz) or higher *f*-number (3 to 5) focal configurations are utilized, the excitation energy is typically distributed throughout the geometric shadow of the active transmit aperture, and the ROE is larger with a more complex spatial distribution (~200 mm^3), as shown in Fig. 5-1.

It is interesting to note that although tissue is modeled as a fluid to quantify acoustic radiation force, the tissue response to the momentum transfer is on a much slower time scale (a few milliseconds) in contrast with a propagation time on the order of microseconds. In this longer temporal domain, tissue is typically modeled as a (visco)elastic solid.

The temporal duration of the radiation force application impacts the tissue response. In some methods, radiation force is applied for durations long enough to reach a steady state response [11, 12]. For impulsive radiation force imaging techniques [6, 10, 13], the

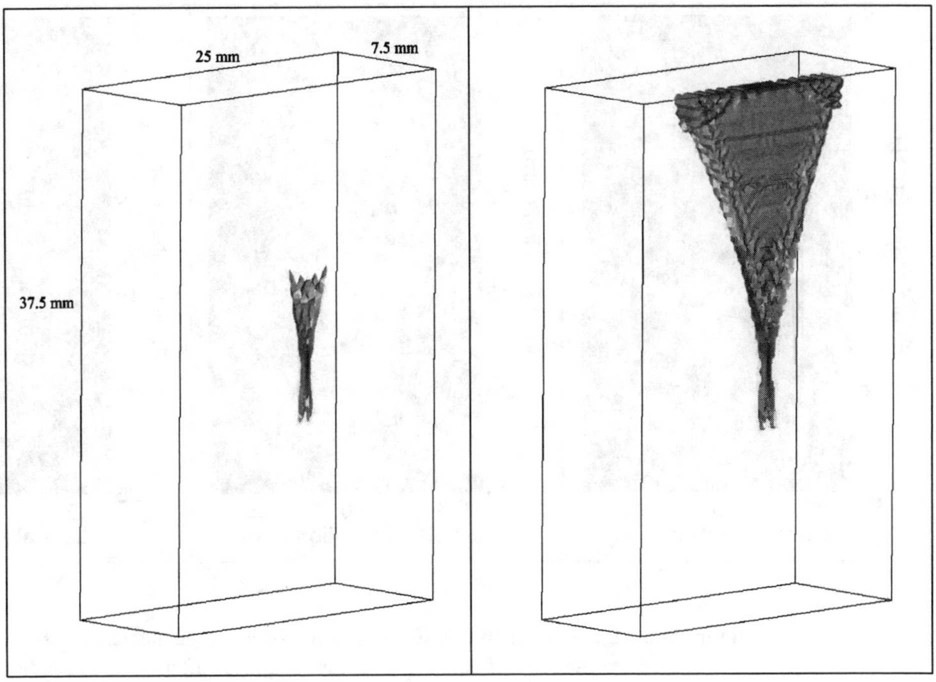

Figure 5-1. Simulated regions of excitation (volumes of appreciable radiation force magnitudes) generated by a linear array with an *f*/1.3 focal configuration at 7.2 MHz frequency. The transducer would be located at the top of the figure. The forward plane represents the axial/lateral plane, centered in elevation, in tissue with an attenuation of 0.5 dB/cm/MHz (left) and 1.0 dB/cm/MHz (right). In both cases, the peak is at the focal depth (20 mm); however, the amplitude is smaller for the higher attenuation, as more energy is distributed over a greater volume of tissue in the nearfield as the wave propagates to the focus. Images reproduced from Ref. [21]

temporal duration of each radiation force excitation is typically limited to less than 1 millisecond, generating a transient response that includes the generation of shear waves that propagate away from the ROE, as shown in Fig. 5-2. If a true spatial and temporal impulse excitation were achieved within the tissue, and the tissue response was evaluated prior to appreciable shear wave propagation, then the displacement response would reflect the local tissue mechanical properties, with minimal interference from surrounding structures. However, when the spatial distribution of the radiation force excitation occupies an appreciable volume, the dynamic response of tissue to this radiation force becomes more complicated due to spatial gradients in the field.

5.3 ULTRASONIC MONITORING OF TISSUE RESPONSE TO IMPULSIVE RADIATION FORCE

Once the tissue has been mechanically excited using acoustic radiation force, its response must be monitored to derive information about the tissue's mechanical properties. This can be done with any imaging method; however, it has typically been performed with ultrasonic imaging, which allows the same transducer to be used to generate the acoustic radiation

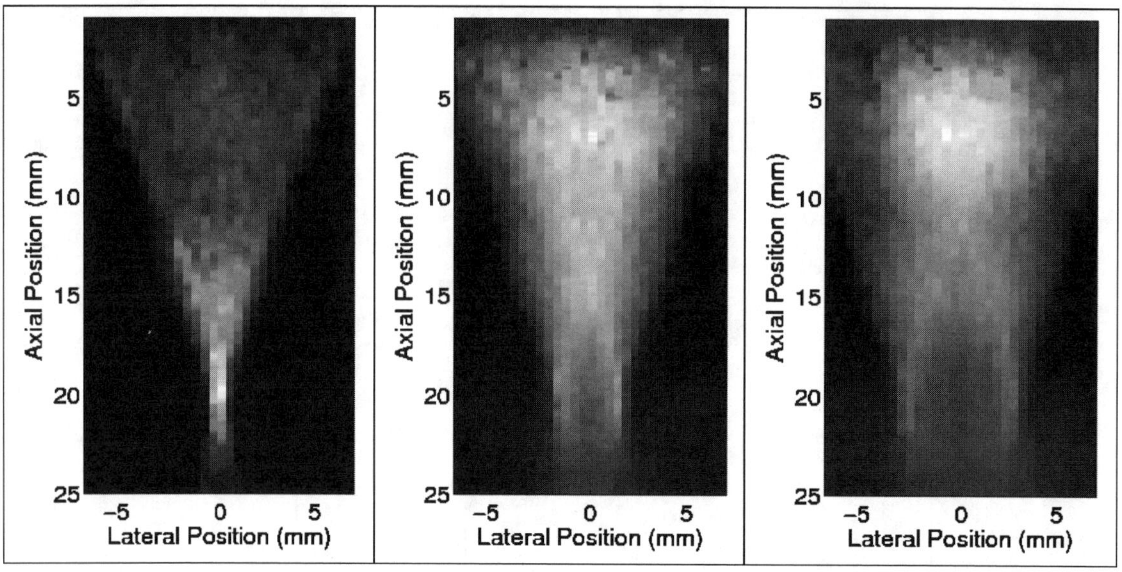

Figure 5-2. Experimentally measured axial displacements at increasing times after a radiation force excitation (0.6 msec, 1 msec, and 2.2 msec after a 30-μsec radiation force excitation, focused at 20 mm in depth, in the left, center, and right images, respectively). Note that the initial displacement response is very similar in geometry to the applied radiation force field as shown in Fig. 5-1 (left). However, with increasing time after excitation, the ROE recovers to its initial position, while shear waves propagate away from the ROE. The grayscales are the same for each image, with white indicating larger axial displacement and black indicating zero axial displacement. Images reproduced from Ref. [22]

force and track the resulting dynamic displacement fields. This concept is similar to that used in elastography imaging methods: acquire a reference line (or image), apply radiation force to locally displace the tissue, and follow this with a series of tracking lines (or images) to monitor how the tissue was perturbed by the radiation force. The tissue displacement response can then be monitored using conventional ultrasonic methods (i.e., cross correlation of radio-frequency (RF) data, or using Kasai or Loupas phase shift algorithms on IQ data) [16]. Typically, in off-line processing algorithms where speed is not of concern, normalized cross correlation methods are utilized to monitor radiation force induced tissue displacement using small kernels (~2 wavelengths) with considerable overlap of the kernel regions (70% to 95%) in RF data [16].

Ultrasonic tracking of speckle patterns has been used to estimate blood motion [23-25] and local strain estimation in solids [2, 3]. Ultrasonic displacement tracking in solid media, like fluids, can suffer from underestimation of the peak displacement when the scatterer distribution within the point spread function (PSF) of the tracking beam is distorted. This distortion is known as shearing [1, 4, 25, 26]. Analytic studies have demonstrated such an underestimation when tracking a steady-state Gaussian distribution of displacement narrower than the tracking beam PSF [27]. Ultrasonic tracking also suffers from tracking inaccuracies (jitter) that are inversely related to tracking frequency, transducer bandwidth, signal-to-noise ratio (SNR), kernel length, and the correlation coefficient between RF lines being tracked [28-30].

In elastography, the external compression of tissue provides a relatively uniform compression of the scatterer distribution within the tracking beam PSF. This compression causes minimal shearing in the lateral and elevation dimensions, but leads to decorrelation of the RF signal in the axial dimension. This problem has been addressed in elastography with companding [31, 32]. In contrast, signal decorrelation in the axial dimension is not a significant problem in radiation force methods since acoustic radiation force typically induces much smaller axial displacements (microns) than those applied in elastography (millimeters) [22]. However, depending upon the relative dimensions of the pushing and tracking beams used for radiation force imaging, the deformation of tissue in the focal zone can introduce significant shearing in the lateral and elevation dimensions that can result in signal decorrelation and introduce jitter in tissue displacement estimates. As shown in Fig. 5-3, ultrasonic tracking of radiation force induced displacement can be associated with significant underestimation of the actual displacement; under certain tracking configurations, this can be as much as 50%. The error bars in this figure represent the mean and standard deviation (i.e., jitter) over 100 independent speckle realizations using the FEM displacement as input to shift the scatterers.

Jitter increases with increasing tissue displacement and increasing lateral and elevation shearing due to decreased correlation coefficients [22]. Displacement underestimation and jitter decrease with time after an ARFI excitation as shear waves propagate away from the ROE, reducing the amount of shearing within the track PSF, as shown in Fig. 5-3. In stiffer media with higher shear wave speeds, this improvement occurs sooner than in more compliant media as the amount of shearing decreases faster. Similar trends can be achieved by using higher *f*-number focal configurations on the excitation beam compared to the tracking beams, creating less lateral shearing within the tracking beams and greater displacement tracking accuracy, as shown in Fig. 5-4. As discussed in Section 5.4.3, these biases can sometimes complicate the interpretation of ARFI parametric images.

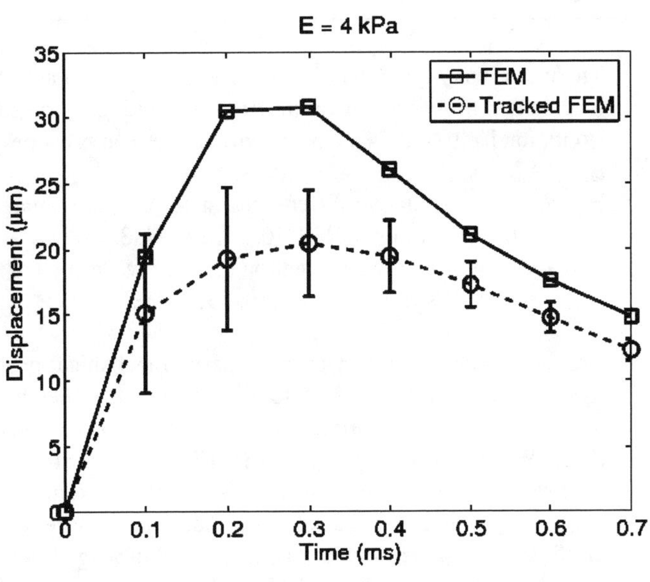

Figure 5-3. Simulated displacement data at the focal point through time in a 4-kPa homogeneous medium. The simulated data without ultrasonic tracking are shown in the solid line, while the ultrasonically tracked FEM data are shown in the dashed line. The data points correspond to an experimentally feasible PRF of 10 kHz. The underestimation of displacement magnitude decreases with increasing time, due to decreasing shearing under the track PSF. Plot reproduced from Ref. [33]

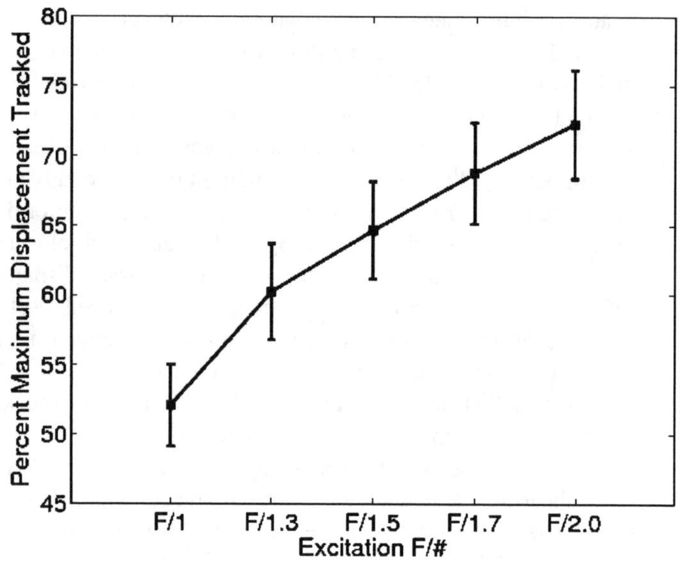

Figure 5-4. Focal point displacement underestimation is reduced when using higher excitation *f*-number focal configurations as compared to the tracking beam focal configuration due to reduced scatterer shearing in the lateral dimension of the tracking PSF. The tracking beam was held at a constant *f*/1 focal configuration. Plot reproduced from Ref. [22]

5.4 GENERATING IMAGES FROM IMPULSIVE RADIATION FORCE DATA

There are many approaches that have been investigated to derive information about the mechanical properties of tissue from evaluating its response to transient radiation force excitation [6, 10, 12-34]. One can divide these approaches into two categories:

- Imaging relative differences in displacement magnitude by monitoring tissue response within the ROE [6, 14, 15, 34, 35].
- Monitoring the propagation of shear waves that are generated by radiation force to quantify shear modulus [10, 13, 36-38].

Both approaches to reconstructing tissue mechanical properties have been widely investigated. There are advantages to each approach, depending upon implementation and the ultimate clinical application. Monitoring the propagation of shear waves has been termed Shear Wave Elasticity Imaging (SWEI) [10] and Supersonic Shear Imaging (SSI) [13]. In these approaches, impulsive radiation force excitations are applied and the resulting displacement fields are tracked as a function of space and time. These displacement data are fit to an assumed model of shear wave propagation that allows the local tissue shear modulus to be estimated [10, 13, 36-38]. Reconstructing shear moduli from propagating shear waves has the potential for modulus quantification independent of the magnitude of the applied radiation force and the potential to characterize the viscous properties of tissues, either through evaluation of the frequency dependence of shear wave speeds or the attenuation of the propagating shear wave [10, 13, 38]. In order for shear waves to propagate over appreciable spatial domains (several millimeters), relatively large excitation amplitudes, and thus large instantaneous acoustic energies, are necessary to generate measurable displacements. Shear wave reconstruction methods can be challenged if structures appreciably reflect incident shear waves, limiting the propagation of the shear waves into the structures and creating shear wave interference patterns inside and outside of structures.

5.4.1 ARFI Imaging: Generating Images of Relative Differences in Displacement Response Within the ROE

For clinical applications focused on detecting abnormal structures in otherwise healthy tissues, images of relative differences in tissue response (i.e., the displacement response within the ROE) can provide useful information. This approach has been termed "Acoustic Radiation Force Impulse (ARFI)" imaging [6]. One advantage to creating images from displacements within the ROE is the increased displacement SNR associated with monitoring the region where the acoustic energy is applied. This approach generally requires smaller instantaneous energies than the propagation approaches. However, one must apply energy throughout the entire region of interest, so the overall temporal average energy can be higher than with the propagation approaches. In addition, because the actual tissue absorption varies among patients and tissues, the force magnitude is unknown, making quantification of elastic moduli within the ROE considerably more complicated. Some clinical applications that have been investigated include tumor detection in the breast, prostate, liver, and colon [6, 39-41]; characterization of hard and soft plaques in vascular disease [8]; and monitoring tissue thermal ablation procedures [42, 43]. For these applications, a short duration, impulsive radiation force excitation beam is co-located with the tracking beams, and the response in each location can be monitored for several milliseconds. The co-located pushing/tracking beam ensemble is then translated to a new location and repeated to obtain data throughout a two-dimensional plane. Stiffer structures (e.g., tumors and ablated tissue) exhibit smaller displacements than more compliant surrounding structures, and the temporal responses differ within and outside of inclusions. Images can be generated of displacement at a given time after excitation, as well as parametric images such as time of peak displacement and recovery time, as shown in Fig. 5-5. These data can be overlaid on matched B-mode images that are generated using the reference lines before the acoustic radiation force excitations. The displacement and parametric images are speckle-free, with resolution comparable to that of B-mode images, and can exhibit higher contrast than B-mode images [6, 8, 39-41, 44].

5.4.2 ARFI Imaging of Homogeneous Tissues

When using impulsive acoustic radiation force to image homogeneous media, the dynamic displacement profiles tracked in the region of excitation are dependent on the underlying material stiffness and density, with time-to-peak displacement, displacement magnitude, and recovery time all being inversely related to tissue stiffness, as shown in Fig. 5-6.

All of these measured dynamic parameters are also dependent on the focal configuration of the excitation, where broader spatial distributions of radiation force in the lateral and elevation dimensions at a given depth lead to later times to peak displacement and recovery times (see Fig. 5-6). These simplified dynamic trends become significantly more complicated when imaging structures in otherwise homogeneous media.

5.4.3 ARFI Imaging of Tissues With Internal Structures

Spherical inclusions provide a simplified geometry that can represent a variety of pathologic masses that are targets of clinical imaging and are amenable to quantifying dynamic

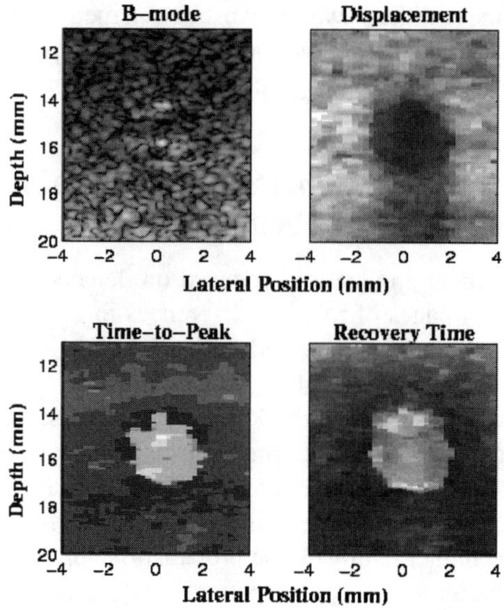

Figure 5-5. Typical ARFI parametric images from a tissue-mimicking phantom with a 3-mm diameter spherical inclusion that is four times stiffer than the background material, centered at a depth of 15 mm. The conventional B-mode image is shown in the top left and does not delineate the inclusion. The brighter pixels in the maximum displacement image (top right) represent greater displacement away from the transducer (top of the image), with displacements ranging from 0 ∝m to 12 ∝m. As expected, the more compliant background material moves farther than the stiffer inclusion material in the displacement image. The brighter pixels in the time to peak (bottom left) and recovery time (bottom right) images represent greater times from their respective references (initiation of the excitation pulse and time of peak displacement), providing additional dynamic metrics to delineate the inclusion. Image reproduced from Ref. [33]

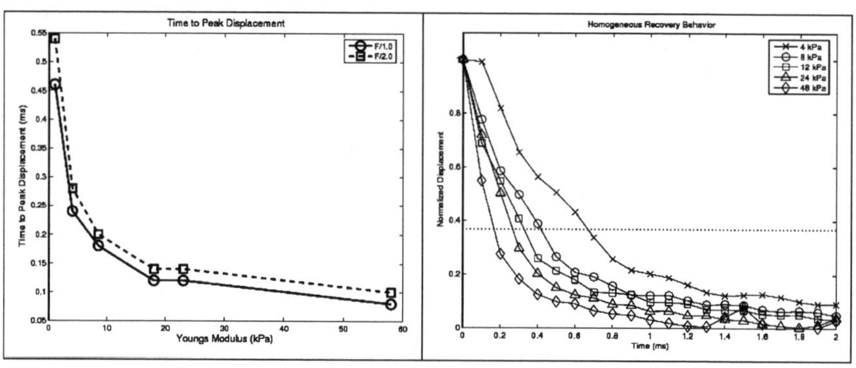

Figure 5-6. Left: Simulated trends in time to peak displacement as a function of the stiffness of the material being characterized. Two different lateral excitation focal configurations are shown. These times to peak displacement are a function of the lateral and elevation distribution of the excitation beam along with the stiffness of the material. Right: Normalized recovery behavior for different stiffness media (Young's moduli), with the time axis referring to time elapsed after peak displacement is reached in each material. The horizontal dotted line represents the recovery level (37%) at which recovery time parametric images are usually generated in ARFI imaging. Plots reproduced from Ref. [33]

responses to impulsive acoustic radiation force excitations. As shown in Fig. 5-7, after impulsive excitations, shear waves propagate away from the ROE that encounter boundaries in the lateral and elevation dimension. Differences in density and shear modulus between the inclusion and the background medium lead to reflected and transmitted shear waves that alter the displacement fields within the ROE that would exist if the inclusion were not present. Strain energy can remain confined to within the inclusion when it is stiffer than its background due to significant internal reflections of shear wave displacements. These reflected shear waves can constructively interfere with one another, leading to time to peak displacements in the inclusion that are significantly later than those that would be expected in a homogeneous material of the inclusion's stiffness, as shown in Figs. 5-5 and 5-7. The magnitude and timing of this interference are related to the stiffness contrast between the lesion and background material and the diameter of the lesion, as exemplified in Fig. 5-8.

As discussed in Section 5.3 ultrasonically tracking these displacements suffers from scatterer shearing and jitter that can lead to variations in the estimations of displacement, time to peak displacement, and recovery time parameters that complicates their representation of the underlying tissue material properties. These phenomena can also lead to an apparent expansion of spherical inclusions with increasing time after excitation, as shown in Fig. 5-9. When imaging structures using ARFI imaging, the dynamic parameters may no longer be exclusively dependent on localized material properties, but may be significantly influenced by nearby material boundaries [33].

5.4.4 Contrast in ARFI Images

While displacement is, in general, inversely related to tissue stiffness [7], the focal gain of the excitation beam leads to spatial gradients in the force magnitude, resulting in variable displacements throughout the ROE, even in a homogeneous phantom. This focal gain effect

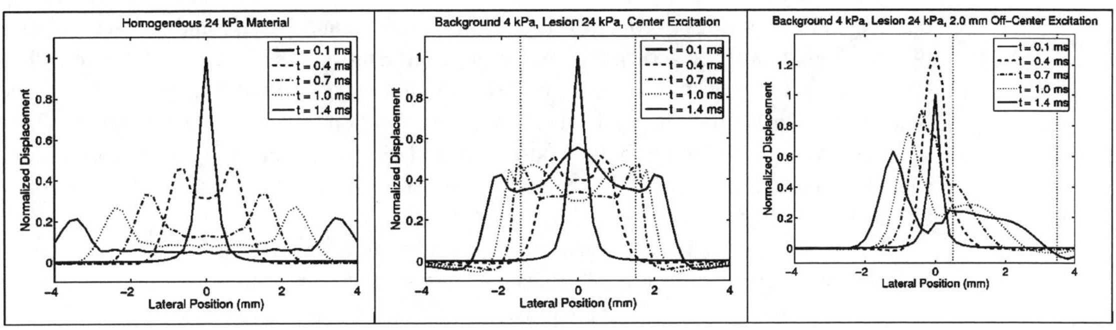

Figure 5-7. (Left) Shear waves propagating away from the region of excitation at the focal depth in a homogeneous medium ($E = 24$ kPa). (Center) Shear waves propagating away from the ROE for an excitation that is 1.5 mm from the inner inclusion boundary of a 3-mm diameter, 24-kPa inclusion, with a background stiffness of 4 kPa. (Right) Shear waves propagating away from the ROE for an excitation that is 0.5 mm outside the inclusion boundary in a 3-mm diameter, 24-kPa inclusion, with a background stiffness of 4 kPa. All displacement profiles are normalized to their respective displacement profiles at 0.1 msec. Note the constructive interference of the reflected shear waves in the center plot, and the relative null in displacement caused by the inversion of the reflected shear waves in the plot on the right. Plots reproduced from Ref. [33]

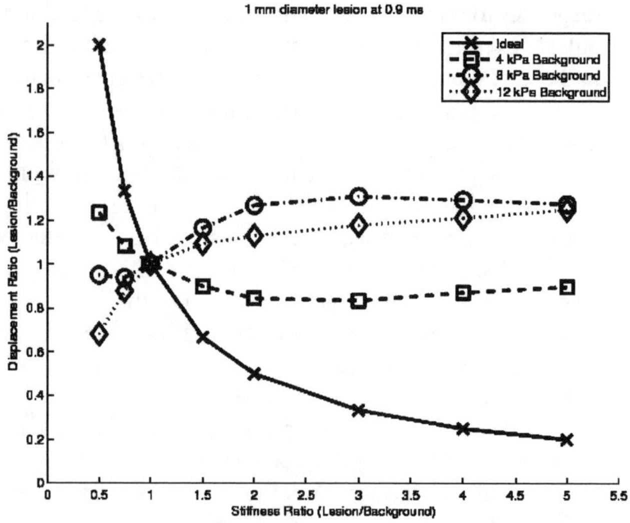

Figure 5-8. Simulated mean displacement ratios inside and outside a 1-mm diameter inclusion, centered at the focal depth of 11.5 mm, vs. the background to inclusion stiffness ratio for background stiffnesses of 4 kPa, 8 kPa, and 12 kPa. The displacement ratios are normalized to the homogeneous displacements at the focal point for each time step at 0.9 msec after excitation. The ideal displacement to stiffness ratio is shown in the solid line. Plot reproduced from Ref. [33]

is evident in the center image in the top row of Fig. 5-9, where larger displacements occur in the background material around the focal depth (near 15 mm) than above and below this region. Corrections can be applied by using either multi-focal techniques or applying normalization curves scaled by the assumed distribution of force within the ROE [43, 44].

The contrast of a lesion in an ARFI image is a function of several parameters, including both the system settings (frequency, focal configuration) and the size and relative stiffness of the lesion. It has been shown that maximum contrast in an ARFI image is achieved when the radiation force beam cross section (elevation/lateral) is smaller than the size of the structure (lesion) being imaged [44]. Thus, for applications such as breast imaging, where structures vary considerably both acoustically and elastically, many tightly focused interrogations throughout a two-dimensional region are desirable. This multi-focal approach was utilized to generate the images in Figs. 5-5 and 5-10.

For clinical applications in which the target stiffness contrast is high, tradeoffs can be made between the required number of interrogation locations and frame rate. An example is ablation monitoring, where ablation lesions have been reported to be approximately 8 times stiffer than normal tissue [42, 43, 46]. In this application, less tightly focused ARFI interrogations can be utilized because the associated decrease in contrast is not significant compared to the gain in frame rate [47]. The ultimate limit on frame rate is generally determined by tissue heating, as discussed below, and the amount of parallel receive processing available in a system.

5.5 MOTION FILTERING FOR IN VIVO APPLICATIONS

For in vivo acquisitions, some type of motion filter is generally required to isolate the tissue displacement response to ARFI excitations from other forms of motion (cardiac, respiratory, and transducer motion) [6, 39]. Although the interrogation of an entire two-dimensional

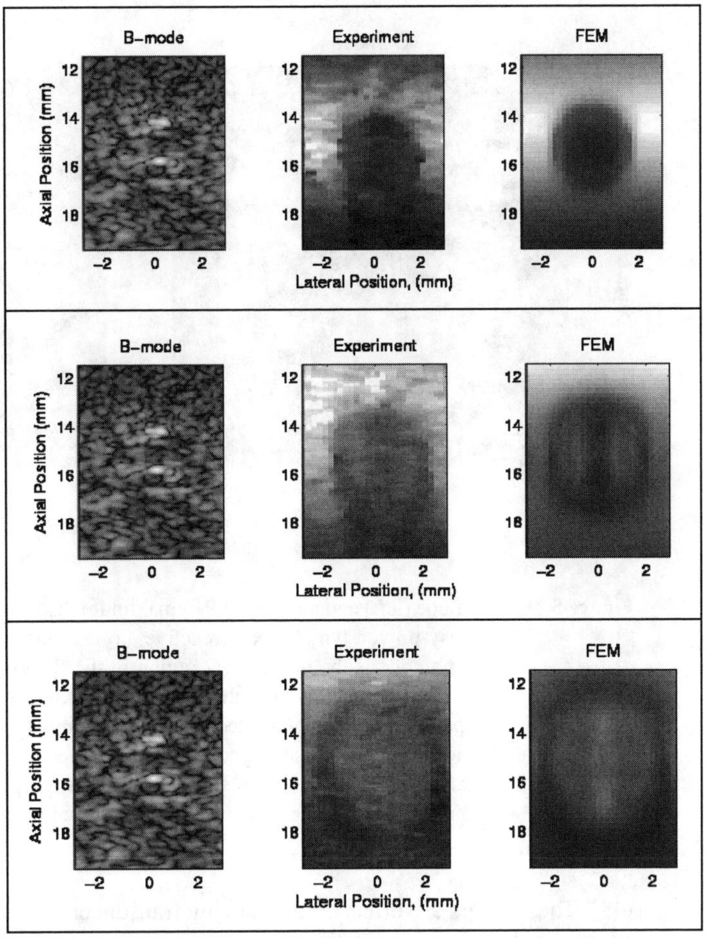

Figure 5-9. Matched simulation and experimental displacement profiles in the axial-lateral plane, centered in elevation, demonstrating similar temporal behavior for a stiff spherical inclusion in an otherwise homogeneous medium. The images at each time step are normalized by the maximum displacement value for that time step. Each row represents a different time step: 0.4 msec, 0.9 msec, and 1.8 msec for the top, middle, and bottom rows, respectively. The Young's moduli for the background and lesion in both simulation and experiment were 4.0 kPa and 15 kPa, respectively, with a lesion diameter of 3 mm. Initially, the displacements are indicative of differences in stiffness; however, later in time, wave propagation effects are portrayed, and a reversal in contrast and apparent expansion of the lesion is observed in the displacement profiles. Note that the simulation images do not include the effects of speckle tracking the ARFI-induced displacements. Images reproduced from Ref. [45]

image plane takes on the order of 100 milliseconds, each individual ARFI excitation is monitored for only a few milliseconds. Therefore, residual motion can be filtered, and the displacements generated by the radiation force excitation can be extracted. Motion artifacts can exhibit linear and sometimes higher-order behaviors over the time domains over which displacements are tracked (typically ranging from 5 milliseconds to 15 milliseconds after a radiation force excitation). Many approaches have been taken to filtering this motion, including linear fits [6], parabolic fits [43], and blind source separation (BSS) filtering

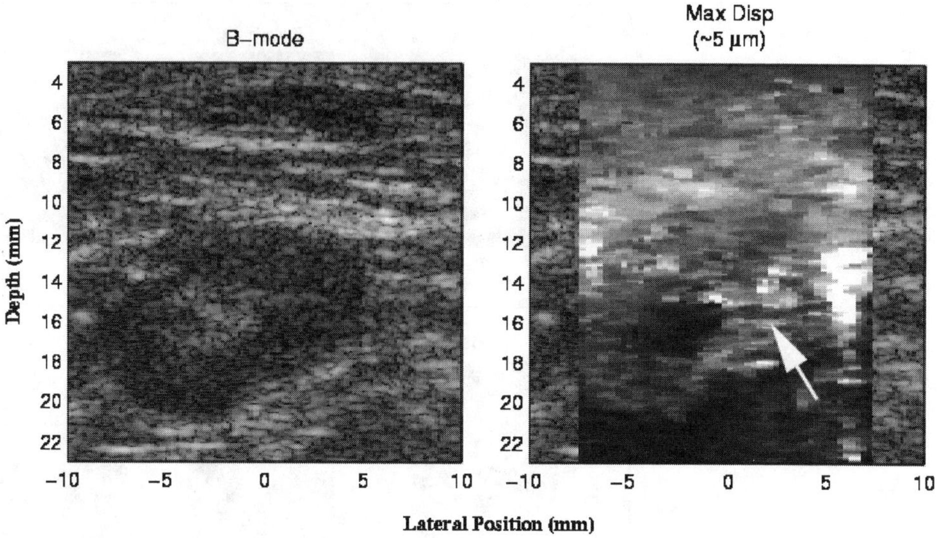

Figure 5-10. B-mode (left) and matched ARFI maximum displacement image (right) of an *in vivo*, biopsy-proven lymph node (reactive, no malignant involvement) in the breast. The lymph node is the darker oval region in the B-mode image appearing from 12 mm to 22 mm in depth. The echogenic center of the lymph node is the hilum that corresponds to the region of decreased displacement in the ARFI image (darker central region). The node itself appears slightly stiffer (darker) than the surrounding tissue, with an apparent ductal structure (arrow) (image reproduced from Ref. [40]). Note the improved contrast and structural visualization in the ARFI image

[48]. These filters are effective at reducing transducer motion and physiologic motion, in addition to low-frequency artifacts that can arise from transducer heating and power supply variations. Typical motion artifacts before filtering can reach tens of microns of displacement, several times greater than the radiation force-induced displacements being estimated, but are typically suppressed to below 1 μm after filtering operations are performed.

5.6 SAFETY CONSIDERATIONS

The acoustic energy required to displace tissue tens of microns also results in tissue heating. Typical in situ, spatial peak pulse average intensities (measured using linear extrapolation of small signal measurements) utilized for ARFI imaging range from 1000 W/cm^2 to 3000 W/cm^2. These energy levels are comparable to those used for High-Intensity Focused Ultrasound (HIFU) therapies, but they are applied for much shorter durations. For ARFI excitations, pulse durations are on the order of 50 microseconds to 200 microseconds, which are 3 orders of magnitude less than those for HIFU procedures. Typical heating associated with a single ARFI excitation is 0.05°C to 0.2°C, depending upon tissue thermal properties, acoustic attenuation, and spatial and temporal energy distribution [21, 49]. For the creation of two-dimensional images, and clinical implementations approaching real-time frame rates (0.25 fps to 2 fps) where repeated excitations can be applied in the same location, the spatial and temporal distribution of the excitation beams must be carefully controlled to maintain total tissue temperature increases within the diagnostic range (i.e., less than 6°C [50]). Because diagnostic transducers are utilized for excitations, noticeable transducer heating can occur, particularly for aggressive output sequences [49].

5.7 CONCLUSIONS

Acoustic radiation force presents a novel method for mechanically exciting tissue. As such, there are many opportunities for deriving information about tissue response to this type of excitation. The primary advantages of radiation force excitations include the localized nature of the excitation, and the ability to couple radiation force directly into the tissue of interest. In addition to the sometimes challenging interpretation of the complicated dynamic response of tissue to localized radiation force excitations, challenges for these methods can include tissue and transducer heating, which must be monitored as imaging approaches are developed.

Acknowledgments

This work was supported by NIH grants R01 EB 002132 and R01 CA 114075.

REFERENCES

[1] van der Steen et al, 1999, "Echo Decorrelation and Displacement Gradients in Elasticity and Velocity Estimation," *IEEE Ultrasonics, Ferroelectrics and Frequency Control*, **46**(4), pp. 791–801.

[2] Ophir et al, 1999, "Elastography: Ultrasonic Estimation and Imaging of the Elastic Properties of Tissue," *Proceedings of the Institution of Mechanical Engineers*, **213**, pp. 203–233.

[3] O'Donnell et al, 1994 "Internal Displacement and Strain Imaging Using Ultrasonic Speckle Tracking," *IEEE Ultrasonics, Ferroelectrics and Frequency Control*, **41**, pp. 314–325.

[4] Bilgen, M., and Insana, M., 1997, "Error Analysis in Acoustic Elastography. II. Strain Estimation and SNR Analysis," *Journal of the American Statistical Association*, **101**(2), pp. 1147–1154.

[5] Greenleaf, J. F., Fatemi, M., and Insana, M., 2003, "Selected Methods for Imaging Elastic Properties of Biological Tissues," *Annual Review of Biomedical Engineering,* **5**, pp. 57–78.

[6] Nightingale, K. R., Soo, M. S., Nightingale, R. W., and Trahey, G. E. "Acoustic Radiation Force Impulse Imaging: *In vivo* Demonstration of Clinical Feasibility," *Ultrasound in Medicine and Biology*, **28**(2), pp. 227–235.

[7] Nightingale, K. R., Nightingale, R. W., Stutz, D. L., and Trahey, G. E., 2002, "Acoustic Radiation Force Impulse Imaging of *in Vivo* Vastus Medialis Muscle Under Varying Isometric Load, *Ultrasonic Imaging*, **24**(2), pp. 100–108.

[8] Trahey, G. E., Palmeri, M. L., Bentley, R. C., and Nightingale, K. R., 2004, "Acoustic Radiation Force Impulse Imaging of the Mechanical Properties of Arteries: In Vivo and Ex Vivo Results," *Ultrasound in Medicine and Biology*, **30**(9), pp. 1163–1171.

[9] Sugimoto, T., Ueha, S., and Itoh, K., 1990, "Tissue Hardness Measurement Using the Radiation Force of Focused Ultrasound. In *Proceedings of the 1990 Ultrasonics Symposium*, pp. 1377–1380.

[10] Sarvazyan, A., Rudenko, O., Swanson, S., Fowlkes, J., and Emelianov, S., 1998, "Shear Wave Elasticity Imaging: A New Ultrasonic Technology of Medical Diagnostics." *Ultrasound in Medicine and Biology*, **24**(9), pp. 1419–1435.

[11] Walker, W., Fernandez, F., and Negro, L., 2000, "A Method of Imaging Viscoelastic Parameters with Acoustic Radiation Force," *Physics in Medicine and Biology*, **45**(6), pp. 1437–1447.

[12] Fatemi, M., and Greenleaf, J., 1998, "Ultrasound-Stimulated Vibro-Acoustic Spectrography," *Science*, **280**, pp. 82–85.

[13] Bercoff, J., Tanter, M., and Fink, M., 2004, "Supersonic Shear Imaging: A New Technique for Soft Tissue Elasticity Mapping," *IEEE Ultrasonics, Ferroelectrics and Frequency Control*, 51(4), 396–409.

[14] Lizzi et al, 2003, "Radiation-Force Technique to Monitor Lesions During Ultrasonic Therapy," *Ultrasound in Medicine and Biology*, **29**(11), pp. 1593–1605.

[15] Melodelima et al, 2006, "Elastography for Breast Cancer Diagnosis Using Radiation Force: System Development and Performance Evaluation," *Ultrasound in Medicine and Biology*, **32**, pp. 387–396.

[16] Pinton, G.F., Dahl, J.J., and Trahey, G.E., 2006, Rapid Tracking of Small Displacements with Ultrasound," *IEEE Ultrasonics, Ferroelectrics and Frequency Control*, **53**(6), pp. 1103–1117.
[17] Parker, K., 1983, "Ultrasonic Attenuation and Absorption in Liver Tissue," *Ultrasound in Medicine and Biology*, **9**(4), pp. 363–369.
[18] Christensen, D., 1988, *Ultrasonic Bioinstrumentation*. John Wiley & Sons, New York.
[19] Torr, G. R., 1984, "The Acoustic Radiation Force," *American Journal of Physics*, **52**, pp. 402–408.
[20] Nyborg, W. L. M., 1965, "Acoustic Streaming," *Physical Acoustics*, W. P. Mason, ed., Academic Press Inc, New York, Vol. IIB, chapter 11, pp. 265–331.
[21] Palmeri, M. L., and ,Nightingale, K. R., 2004, "On the Thermal Effects Associated with Radiation Force Imaging of Soft Tissue," *IEEE Ultrasonics, Ferroelectrics and Frequency Control*, **51**(5), pp. 551–565.
[22] Palmeri et al, 2006, "Ultrasonic Tracking of Acoustic Radiation Force-Induced Displacements in Homogeneous Media," *IEEE Ultrasonics, Ferroelectrics and Frequency Control*, **53**(7), pp. 1300–1313.
[23] Bohs et al, 1993, "A Real-Time System for Quantifying and Displaying Two-Dimensional Velocities Using Ultrasound," *Ultrasound in Medicine and Biology*, **19**(9), pp. 751–761.
[24] Bonnefous, O., and Pesque, A., 1986, "Time Domain Formulation of Pulse-Doppler Ultrasound and Blood Velocity Estimation by Cross Correlation," *Ultrasonic Imaging*, **8**, 73–85.
[25] Embree, A., and O'Brien, W., 1990, "Volumetric Blood Flow via Time-Domain Correlation: Experimental Verification," *IEEE Ultrasonics, Ferroelectrics and Frequency Control*, **37**, pp. 176–189.
[26] Friemel et al, 1998, "Speckle Decorrelation Due to Two-Dimensional Flow Gradients," *IEEE Ultrasonics, Ferroelectrics and Frequency Control*, **45**(2), pp. 317–327.
[27] McAleavey et al, 2003, "Estimates of Echo Correlation and Measurement Bias in Acoustic Radiation Force Impulse Imaging," *IEEE Ultrasonics, Ferroelectrics and Frequency Control*, **50**(6), pp. 631–641.
[28] Walker, W., and Trahey, G., 1995, "A Fundamental Limit on Delay Estimation Using Partially Correlated Speckle Signals," *IEEE Ultrasonics, Ferroelectrics and Frequency Control*, **42**(2), pp. 301–308.
[29] Cespedes, I., Insana, M. F., and Ophir, J., 1995, "Theoretical Bounds on Strain Estimation in Elastography," *IEEE Transactions on Ultrasonics, Ferroelectrics and Frequency Control*, **42**, pp. 969–972.
[30] Viola, F., and Walker, W. F., 2003, "Radiation Force Imaging of Viscoelastic Properties with Reduced Artifacts," *IEEE Transactions on Ultrasonics, Ferroelectrics and Frequency Control,* **50**(6), pp. 736–742.
[31] Chaturvedi, P., Insana, M. F., and Hall, T. J., 1998, "Testing the Limitations of 2-D Local Companding in Strain Imaging Using Phantoms," *IEEE Transactions on Ultrasonics, Ferroelectrics and Frequency Control*, **45**, pp. 1022–1031.
[32] Chaturvedi, P., Insana, M. F., and Hall, T. J., 1998, "2-D Local Companding for Noise Reduction in Strain Imaging," *IEEE Transactions on Ultrasonics, Ferroelectrics and Frequency Control*, **45**, pp. 179–191.
[33] Palmeri et al, 2006, "Dynamic Mechanical Response of Elastic Spherical Inclusion to Impulsive Acoustic Radiation Force Excitation," *IEEE Ultrasonics, Ferroelectrics and Frequency Control*, **53**(11), pp. 2065–2079.
[34] Konofagou, E. E., and Hynynen, K., 2003, "Localized Harmonic Motion Imaging: Theory, Simulations and Experiments," *Ultrasound in Medicine and Biology,* **29**, pp. 1405–1413.
[35] Aglyamov, S.R., and Skovoroda, A. R., 2002, "On Diagnostics of Inhomogeneities of Visco-Elastic Properties in Soft Tissues Exposed to Low Frequencies," *Biophysics, Pergamon*, **47**(3), pp. 553–558.
[36] Palmeri, M. L., Dahl, J. J., Wang, M. H., Frinkley, K. D., and Nightingale, K. R., in press, "Quantifying Hepatic Shear Modulus in Vivo Using Acoustic Radiation Force," *Ultrasound in Medicine and Biology*, **34** doi 10.1016.
[37] Chen, S., Fatemi, M., Greenleaf, J., 2004. "Quantifying elasticity and viscosity from measurement of shear wave dispersion," *Journal of the Acrustical Society of America*, **115**(6), pp. 2781–2785.
[38] McLaughlin, J., and Renzi, D., 2006, "Using Level Set Based Inversion of Arrival Times to Recover Shear Wave Speed in Transient Elastography and Supersonic Imaging," *Inverse Problems*, **22**, pp. 707–725.

[39] Fahey, B. J., Nightingale, K. R., Nelson, R. C., Palmeri, M. L., and Trahey, G. E., 2005, "Acoustic Radiation Force Impulse Imaging of the Abdomen: Demonstration of Feasibility and Utility," *Ultrasound in Medicine and Biology*, **31**(9), pp. 1185–1198.

[40] Palmeri et al, 2005, "Acoustic Radiation Force Impulse (ARFI) Imaging of the Gastrointestinal Tract," *Ultrasonic Imaging*, **27**, pp. 75–88.

[41] Nightingale, K. R., Bentley, R., and Trahey, G. E., 2002, "Observations of Tissue Response to Acoustic Radiation Force: Opportunities for Imaging," *Ultrasonic Imaging*, **24**, pp. 100–108.

[42] Fahey, B. J., Nightingale, K. R., Palmeri, M. L., McAleavey, S. A., Wolf, P., and Trahey, G. E., 2005, "Acoustic Radiation Force Impulse Imaging of Myocardial Radio-Frequency Ablation: Initial in Vivo Results," *IEEE Ultrasonics, Ferroelectrics and Frequency Control*, **52**(4), pp. 631–641.

[43] Fahey, B. J., Hsu, S. J., Wolf, P. D., Nelson, R. C., and Trahey, G.E., 2007, "Liver Ablation Guidance with Acoustic Radiation Force Impulse Imaging: Challenges and Opportunities," *Ultrasound in Medicine and Biology*, **51**, pp. 3785–3808.

[44] Nightingale, K. R., Palmeri, M.L., and Trahey, G. E., 2006, "Analysis of Contrast Generated with Transient Acoustic Radiation Force," *Ultrasound in Medicine and Biology*, **32**(1), pp. 61–72.

[45] Palmeri et al, 2005, "A Finite Element Method Model of Soft Tissue Response to Impulsive Acoustic Radiation Force," *IEEE Ultrasonics, Ferroelectrics and Frequency Control*, **52**(10), pp. 1688–1712.

[46] Varghese, T. et al, 2003, "Elastographic Measurement of the Area and Volume of Thermal Lesions Resulting from Radiofrequency Ablation: Pathologic Correlation," *AJR American Journal of Roentgenology*, **181**, pp. 701–707.

[47] Dahl et al, 2007, "A Parallel Tracking Method for Acoustic Radiation Force Impulse Imaging." *IEEE Ultrasonics, Ferroelectrics and Frequency Control*, **54**(2), pp. 301–312.

[48] Gallippi, C. M., Nightingale, K. R., and Trahey, G. E., 2003, "BSS-Based Filtering of Physiological and ARFI-Induced Tissue and Blood Motion," *Ultrasound in Medicine and Biology*, **29**(11), pp. 1583–1592.

[49] Palmeri, M. L., Frinkley, K. D., and Nightingale, K. R., 2004, "Experimental Studies of the Thermal Effects Associated with Radiation Force Imaging of Soft Tissue," *Ultrasonic Imaging*, **26**, pp. 100–114.

[50] Herman, B. A., and Harris, G. R., 2002, "Models and Regulatory Considerations for the Transient Temperature Rise During Diagnostic Ultrasound Pulses," *Ultrasound in Medicine and Biology*, **28**(9), pp. 1217–1224.

[51] Hall, T. J., Zhu, Y. N., and Spalding, C. S., 2003, "In Vivo Real-Time Freehand Palpation Imaging," *Ultrasound in Medicine and Biology*, **29**(3), pp. 427–435.

Chapter 6

CARDIOVASCULAR ELASTICITY IMAGING

E. Konofagou, W.-N. Lee, and J. Luo
Department of Biomedical Engineering, Columbia University, New York, NY, USA

ABSTRACT

The leading cause of death in men and women in the United States is cardiovascular disease. Its early detection is the key in reducing the associated death toll. The ultimate objective in cardiac diagnostic imaging of this project is to develop a highly reliable and low-cost imaging technique for the mapping of ischemia and infarction. Cardiovascular elasticity imaging can, in the myocardium, detect and quantify the extent of ischemia and infarction, at and beyond its onset due to the associated significantly altered stiffness of the muscle. Further benefiting from the real-time capability and portability of this technology, this information can ultimately be delivered in an emergency room setting to aid the early diagnosis and treatment of people suffering from myocardial infarcts. Before these benefits can be realized, however, the elastographic technique must be validated and angle-independent regional deformation estimates of the myocardium throughout a cardiac cycle obtained through the use of a theoretical, biomechanical analysis, experimental phantoms and appropriate animal models. The proposed methodology was also compared to the current gold standard for mapping of myocardial deformation, MRI tagging. In the vascular application, Pulse Wave Imaging is a technique that can detect abdominal aortic aneurysms at an early stage by utilizing the pulse wave propagation and its changed characteristics in the presence of disease. Should the results of this study indicate high reliability of the elasticity imaging findings for detection of early ischemic onset, these novel imaging techniques could be readily applied in an emergency room setting as part of a standard clinical protocol for early detection, and thus prevention, of cardiac or vascular disease as well as a reliable tool for better treatment guidance. The eventual goal of these technologies is thus to become a specific method in standard clinical practice for real-time imaging of the position and severity of cardiovascular abnormalities, improving care and outcomes at little more cost or risk than that of a clinical ultrasound.

6.1 CARDIOVASCULAR APPLICATIONS

6.1.1 Cardiac Applications

Cardiovascular diseases rank as America's no. 1 killer, claiming the lives of over 41.4% of more than 2.4 million Americans who die each year. Cancer follows, killing 23%. All other causes of death account for about 35.3%. In addition, 61.8 million Americans have some form of cardiovascular disease. This includes diseases of the heart, stroke, high blood pressure, congestive heart failure, congenital heart defects, hardening of the arteries, and other diseases of the circulatory system. Americans will pay about $286.5 billion for cardiovascular disease-related medical costs and disability.

Early detection of abnormality is thus the key in treating cardiovascular disease early and reducing the enormous death toll. The diagnosis of myocardial ischemia is often difficult to establish in its early stages when treatment is most effective. Patients suffering from myocardial ischemia may present to an emergency room or acute care facility with typical cardiac symptoms such as chest pain, described as tightness, pressure, or squeezing, but some patients may have other symptoms such as arm or chin pain, nausea, sweating, or abdominal pain. Standard techniques such as the electrocardiogram (ECG) often provide inconclusive findings regarding ischemia, and sometimes may even be unable to identify situations, in which ischemia has progressed to cell damage and myocardial infarction (MI). More reliable techniques are available for diagnosing infarction relative to its predecessor, ischemia. For example, a blood test to measure the creatine kinase-MB (CK-MB) enzyme level is used for detection of myocardial cell damage. Other useful serum markers include troponin I and, to a lesser extent, myoglobin. However, the blood levels of these compounds may take several hours to rise, so that diagnosis of MI may be delayed. Reliance on blood tests alone may result in a significant loss of time when early aggressive therapy is warranted. A viable alternative has been shown to be the assessment of the change in mechanical properties of the heart as a result of MI. In this section, the normal cardiac function is briefly described, and the myocardial, mechanical properties are then discussed regarding their variation in the presence of disease.

The Normal Cardiac Function During a Cardiac Cycle. The role of the left ventricle is to pump the oxygenated blood, which flows through the pulmonary artery and the left atrium into the ventricle, through the aorta and into the vital organs of the body (see Fig. 6-1a) [1]. The pumping-like function of the heart is generated through a cyclic contraction, i.e., successive shortening and lengthening of the individual heart cells, or myocytes, and thus of the muscle itself. A myocyte undergoes shortening (or lengthening) through a flux of calcium ions within the cell, which in turn results in a shape change of the myosin proteins under a force change of about 3 to 4 pN [2]. The regional function of the myocardium can be defined as the regional ability of the muscle to develop a force during the ion flux occurring throughout each cardiac cycle and, thus, actively and adequately contribute to the normal pumping function of the heart. The cyclic mechanical function can be appreciated by following the change of the blood pressure versus time (see Fig. 6-1b). During each cardiac cycle, the following events occur: 1) *End of filling phase (point A):* The left ventricle is first filled with blood from the left atrium through the mitral valve (see Fig. 6-1a). The flow stops when the pressure gradient across the valve leaflets is reversed through blood deceleration; causing, thus, the valve to close suddenly (point A in Fig. 6-1b). 2) *Isovolumic contraction (point A to B):* At the same time as the valve closes, the contraction of the ventricular muscle causes the pressure inside the chamber to increase rapidly. 3) *Ejection phase (point B to C):* At point B, the left-ventricular pressure exceeds the aortic pressure, the aortic valve opens and the blood volume is ejected into the systemic circulation. The aortic valve closes at point C after the ventricular pressure falls below the aortic pressure due to blood deceleration. 4) *Isovolumic relaxation phase (point C to D):* The ventricular pressure drops following aortic valve closure and until the mitral valve reopens (at point D) due to the pressure gradient resulting from the elastic recoil of the ventricular wall. 5) *Fast-filling and diastasis phases (point D to A)*: The chamber starts to refill, rapidly at first (fast-filling) and then much slower (diastasis) until the end of the filling phase is reached (point A) and the cycle continues.

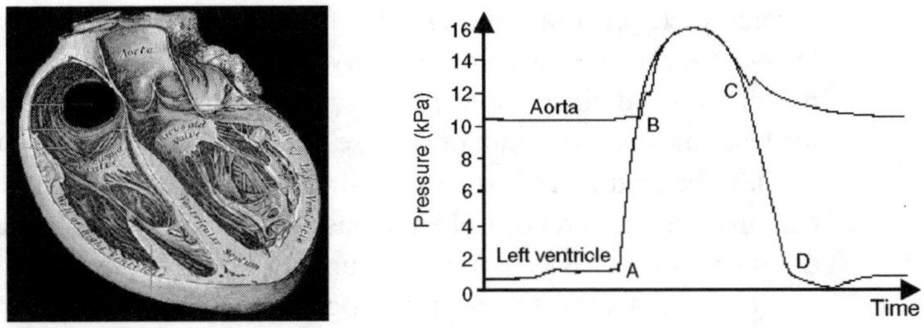

(a) (b)

Figure 6-1. (a) Anatomical section of a human heart showing the four main chambers (©The Bartleby.com edition of Gray's Anatomy of the Human Body). (b) Left-ventricular pressure and aortic pressure temporal variations during a single cardiac cycle (see Fig. 5-1 from Glass et al [1])

Mechanical Deformation of Normal and Ischemic or Infarcted Myocardium. Detection of cardiac dysfunction through assessment of the mechanical properties of the heart, and more specifically, the left ventricular, muscle has thus been a long-term goal in diagnostic cardiology. This is because both ischemia [3], i.e., the reduced oxygenation of the muscle necessary for its contraction, and infarction [4], i.e., the complete loss of blood supply inducing myocyte death, alter the mechanical properties and contractility of the myocardium. Acute myocardial infarction caused by partial or total blockage of one or more coronary arteries can cause complex structural alterations of the left-ventricular muscle [5]. These alterations may lead to collagen synthesis and scar formation, which can cause the myocardium to irreversibly change its mechanical properties. Holmes et al [6] reported that this myocardial stiffening can be measured within the first 5 minutes after ischemic onset. The most comprehensive study on experimental measurement of the stiffness of an acutely infarcted muscle with age was reported by Gupta et al [5]. In that study, in vitro mechanical testing, i.e., biaxial stretching, of a healing myocardial infarct corresponding to various times lapsing (0 hours to 6 weeks) after the induced infarction showed that the stiffness of the infarcted region increases within the first 4 hours, continues to rise by up to 20 times, peaking 1 to 2 weeks following the infarct and decreasing 4 weeks later (down to 1 to 10 times the non-infarcted value). The non-infarcted, or remote, myocardium was shown to follow a similar time course but to a much lesser extent. Some of the causes of these mechanical changes have also been reported [7, 8]. Within the first week, deposition of immature collagen as well as increases in fibroblast formation and resorption of necrotic cells occur [7]. By the second week, total loss of normal collagen matrix occurs, followed by abnormal replacement with mature collagen while over time necrotic tissue continues to be resorbed and replaced with scar tissue and some viable muscle cells [9]; hence, the partial decrease of the stiffness after 6 weeks [5]. The aforementioned complex cellular and histochemical changes directly dictate the mechanical changes of the myocardium. The fact that the mechanical properties induced by the infarct change right at the onset, continue changing thereafter and peak 2 weeks later, indicate the potential for a mechanically-based imaging technique to detect the infarct extent early.

Monitoring and Imaging Mechanical Deformation of the Myocardium. Due to the changes in local myocardial stiffness as a result of disease, monitoring of these mechanical changes could successfully identify pathological regions. In order to assess the local deformation, implantable markers in the form of metallic beads (for X-rays [10]) or sonomicrometers (ultrasound crystals [11]) have been traditionally used. In cardiology, the left-ventricular pressure-volume curve has been used extensively for the assessment of the myocardial and chamber compliances in order to detect the acute and chronic changes on the heart muscle as a result of disease [12]. However, apart from the invasive nature of these studies, the results from several of them have been either conflicting or representative of a certain type of infarct or ischemia onset [13], mainly due to the fact that the whole myocardium or chamber was assumed to undergo uniform motion and deformation; an assumption that is not valid under physiological conditions. In clinical practice, these techniques obviously cannot be applied for diagnostic purposes, and existent imaging modalities have been widely utilized in an effort to non-invasively highlight the mechanical properties of the myocardium. The principal techniques are discussed below and their advantages and pitfalls briefly described. Nuclear imaging, despite its associated quality, typically requires injection of a radioisotope into the patient in order to assess myocardial perfusion for the detection of coronary artery disease [14]. MR cardiac tagging is the only technique that has been shown capable of estimating all principal strain components of the strain tensor by utilizing "tags," i.e., deposition of planes of presaturation intersecting the myocardium prior to the playout of the MRI imaging sequence itself [15-17]. Despite the fact that the MR cardiac tagging technique has permitted important new insights with regard to normal physiology and myocardial disease, it has been slowed down by the limitations of availability, speed, and effort required.

Echocardiography remains the predominant imaging modality in diagnostic cardiology due to its real-time feedback, portability, and high-temporal resolution. In the echocardiography study rooms, cardiologists study ciné-loops of echocardiograms, or B-mode ultrasound images of the heart at various sonographic views. The main purpose is to identify regions of dyskinetic (limited contraction) or akinetic (absence of contraction) myocardium and, thus, detect regional ischemia and infarction. The most identifiable feature of a dyskinetic or akinetic myocardium is the limited "wall motion," or endocardial wall excursion during a cardiac cycle. Automation of this qualitative process through appropriate processing of the ultrasonic signals and/or images has been attempted at length over the past 20 years but without significant clinical impact. The two main areas of such investigations lie in tissue characterization and motion estimation. The field of ultrasound tissue characterization measures acoustic parameters, such as attenuation [18], speed of sound [19], and integrated backscatter (IB) [20] in order to determine such myocardial attributes as thickening and thinning (cyclic variation) [21], and anisotropy [22]. The main difficulty associated with these methods is that they can be highly dependent on the ultrasound system or scanning view used, thus restricting their repeatability across different applications.

In order to avoid dependence on the ultrasound system parameters, differential measurements, such as motion estimation through time-shift or phase-shift measurements can be made. Phase-shift techniques employ autocorrelation techniques to track the phase shift on the signal that results from cardiac motion. Phase-shift estimators are based on the fact that the phase of a sinusoidal signal can be fully characterized by two samples quarter-of-a-wavelength apart. Therefore, a phase shift, or change in the phase of the same signal, will directly indicate the velocity of the tissue scanned [23]. For higher computational efficiency, the autocorrelation function is employed in order to directly estimate the phase of the signal, and thereby the phase shift. Phase-shift motion estimation techniques include Doppler myocardial imaging (DMI) [24-26] and strain rate imaging (SRI) [27], which apply Doppler-based techniques to obtain regional velocity estimates and velocity gradients (or, strain rates) of the myocardium, respectively. Despite their associated low computation cost and therefore, featuring on certain commercially available ultrasound systems (for Doppler, DMI, and SRI), there have been several disadvantages associated with phase shift-based methods discussed in the literature [23, 28, 29]. The main pitfalls are the limitation to small bandwidth signals, i.e., low axial resolution [23, 28-30], aliasing (occurring at half the center frequency), the increase in ambiguity of the estimation with the center frequency [28], and attenuation effects [31]. It is of no surprise that DMI and SRI methods inherit the shortcomings associated with the phase shift (or, autocorrelation) methods [25]. In the case of the time-shift techniques, the time shift, or delay, between successively received echo signals is directly equivalent to the amount of displacement that the tissue traveled between two consecutive acquisitions. The speed of sound of the myocardium is assumed constant throughout the cardiac cycle, which is generally a safe assumption, so that distance and time are directly proportional. In general, time-shift-based methods, such as cross correlation and Sum of Absolute Difference (SAD) estimators, have been repeatedly shown to overcome the main limitations of low resolution and aliasing associated with the phase-shift methods while at the same time providing higher precision [23, 28, 29, 32]. However, time-shift estimation techniques in cardiac applications in vivo have mainly been reported on M-mode data [33], narrow B-mode region of interest (ROI) envelope-detected signals [34] due to the higher frame rate and the previous unavailability of the radio frequency signals, respectively. Time-shift techniques for elasticity imaging are discussed in the next section.

Elasticity Imaging and Elastography. In the previous section, the motion estimation techniques that emerged in echocardiography from the Doppler-based principle were discussed. This section focuses on the time-shift techniques that have been developed in the field of elasticity imaging, which will be used in the proposed study. Elasticity imaging is a relatively new field that has dealt with the estimation and imaging of mechanically-related responses and properties for detection of pathological diseases, most notably cancer [35-37] and has thus emerged as an important field complementary to ultrasonic imaging. Krouskop et al [38] estimated vibration motion amplitudes and related them to estimates of elastic modulus by measuring the Doppler shifts of ultrasonic tone bursts. Lerner and Parker [39] and Parker et al [37] measured the tissue response to mechanical vibrations for "sonoelasticity imaging" and have mainly concentrated on the applications on in vitro muscle [40] and in vitro prostate [41]. Ophir et al [35] developed the method of elastography that applies a small external static compression (on the order of 1%) and uses cross correlation techniques on RF signals in order to estimate the displacement and strain in the tissue resulting from the compression [42-44]. O'Donnell et al [36] and Skovoroda et al [45] developed quasi-static, phase-tracking imaging techniques using the baseband RF signal for displacement, strain, and modulus reconstruction using the RF ultrasonic signal phase information. Elastography has also had an important impact on several other applications, such as normal and shear tissue strain [44, 46] and poroelasticity imaging [47].

Myocardial Elastography. More recently the focus of the elasticity imaging field has been steered towards cardiac applications [48-52]. The authors are part of a group that has been working on the development of myocardial elastography for more than 8 years [48, 50, 53-55]. This technique encompasses imaging of any kind of mechanical parameter that would highlight the mechanical property of the myocardium, such as displacement, strain, strain rate, velocity, shear strain, rotation angle, etc. In the same fashion as in standard elastography, the parameter that can relate directly to the underlying stiffness, and thereby to onset of disease, is the strain and the image that depicts the strain is called an elastogram. Therefore, since assessment of the myocardial mechanical parameters has proven to be a crucial step in the detection of cardiac abnormalities, myocardial elastography can help make a significant impact in this field by measuring the mechanical response of the cardiac muscle at the various phases of the cardiac cycle [50]. Myocardial elastography benefits from the development of techniques that can be used for high-precision two-dimensional time-shift-based strain estimation techniques [44] and high frame rate currently available in ultrasound scanners [50] to obtain a detailed map of the transmural strain in normal [48, 50] and pathological cases [54] over different phases and over several cardiac cycles. Other groups have also been able to apply similar techniques; these are mainly time-shift based techniques for strain rate imaging, which have been successful in temporal imaging, i.e., in an M-mode fashion, of the strain or strain rate [27, 51] with several applications ranging from theoretical elasticity reconstruction models [52] and phantoms [49, 56] to resynchronized heart monitoring [57] and stress echocardiography [58]. The elastographic techniques that have been developed by our group over the past years have been highly precise mainly due to correction techniques [44] and customized cross correlation methods [59] and, therefore, successful in spatially mapping the two-dimensional principal strain components [60, 61]. In this chapter, for the purpose of cardiac diagnostics, methods for obtaining myocardial strain estimates in two dimensions at high precision, high spatial, and high temporal resolution in a full echocardiographic view are described [50, 60].

6.1.2 Vascular and Abdominal Aortic Aneurysm (AAA) Applications

Abdominal Aortic Aneurysm. An abdominal aortic aneurysm (see Fig. 6-2) is a condition whereby the terminal aorta permanently dilates to dangerous proportions, risking rupture [62]. The biomechanics of AAA has been studied with great interest since aneurysm rupture is a mechanical failure of the degenerated aortic wall and is a significant cause of death in developed countries. The development of AAAs is linked to degradation of the elastin and collagen in the three main layers of the aortic wall: intima, media, and adventitia [63]. Most AAAs with a diameter large enough to indicate surgical repair contain intraluminal thrombus (ILT). The elastin and collagen changes are most pronounced in the aneurysm wall covered by the thrombus, which shows higher inflammation and lower thickness than the part of the aneurysm exposed to blood flow. The rate of AAA diameter increase is proportional to the thrombus growth and rupture.

Aortic Stiffness and Pulse Wave Velocity (PWV). Aortic stiffness has been indicated as an early predictor of all-cause and cardiovascular mortality, primary coronary events, and fatal stroke [64-69]. In cases of aortic aneurysms, disruption of the medial elastic fibers may produce aortic stiffening, which may amplify the aneurysmal process [70]. At every heartbeat, blood is pumped from the left ventricle to the rest of the circulation. Upon left-ventricular ejection and opening of the aortic valve, pressure, and diameter waves throughout the entire circulation are produced from the blood traveling through the aortic cavity and branching out to smaller vessels thereafter. Topical reflection waves are also generated due to localized impedance mismatches as the vessels branch out. Pulse wave velocity is typically measured by considering the propagation velocity of the blood flow wave. Due to

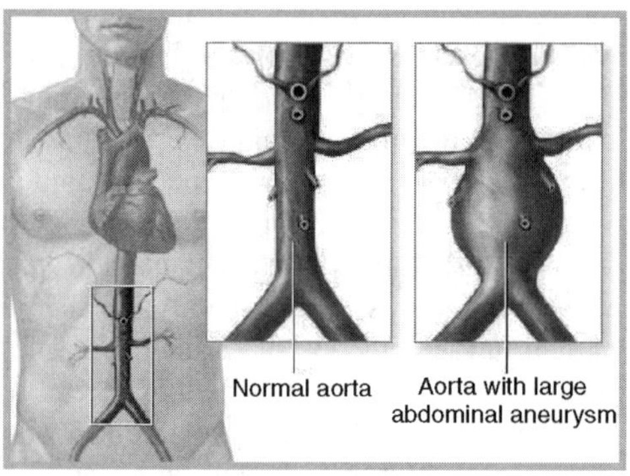

Figure 6-2. Location of the abdominal aorta and its normal and AAA state (from http://www.nlm.nih.gov/medlineplus, 2006)

important information on the distensibility of arteries, PWV is widely used for estimating the stiffness of arteries [71-73], and has been shown to be an early indicator of atherosclerosis [74]. Sumner et al [75] found that aneurysmal portions of a vessel were stiffer and contained less collagen and elastin than the adjacent non-aneurysmal segment. These findings were, in part, corroborated by a later study by He and Roach [76], who obtained uniaxial sub-failure tensile data from human AAA and non-aneurysmal abdominal aortic specimens. Correlations of the aortic PWV and the morphological changes associated with atherosclerosis have been previously reported [77, 78]. Aging of the arterial system is accompanied by structural changes in the aortic wall, which leads to stiffening of the vasculature and increased velocity of the pressure wave [79]. The PWV in brachial arteries has also been proposed as a powerful independent indicator of all cause mortality in diabetes [80] and end-stage renal disease [81]. As such, PWV measured along the aortic and aorto-femoral pathways is most clinically relevant because the aorta and its first branches are responsible for most of the pathophysiological effects of arterial stiffness, and it has been indicated to be a strong predictor of future all-cause mortality [66].

Ultrasound allows non-invasive measurement of PWV, and has been an area of recent interest [82-86]. The PWV is defined as the distance between two measurement points divided by the PW transit time-delay from these two points. Despite the simple definition of PWV, some problems still remain, which limit the interpretation of the available findings and the general applicability of the PWV measurement [87-89]. As pointed out recently, it has been very difficult to compare the values of PWV measurements across different research groups worldwide because of the lack of measurement standardization [88]. The accuracy of the PWV measured from two points suffers from errors of distance measurements and/or time-delay measurements [87]. In addition, the reflected waves and noise from blood echoes may also affect these time-delay measurements. It is difficult to measure small time changes, because the temporal resolution is low. Typically, a large time change over a relative larger distance (a few cm) is needed for reliable measurements. Therefore, the PWV measured represents the average value between two measurement sites. In addition, the measurement of distance is based on the assumption that the arterial wall is relatively straight between two measurement points, which may not always hold [89]. A method with higher temporal and spatial resolution is needed in order to calculate the regional PWV. The PWI method can be such a method and is expected to be useful to establish standards for comparisons of data from different laboratories.

Pulse Wave Imaging. Mice and humans have pulse wave velocities that vary within the same range in the abdominal aorta, i.e., 2.2 to 8.5 m/sec [84] and 5 to 5.5 m/sec [90], respectively. Due to the high velocities involved, measurements are typically taken at high frame rates and at sufficiently large distances to allow for reliable velocity measurements. Until now, the characteristics of the pulse wave could be quantified, but not regionally measured or imaged. Pulse wave imaging, recently introduced by the group [91-94], currently constitutes the only elasticity imaging technique for mapping of the pulse wave in arteries during its propagation. High-frequency and ECG-gated, ultra-high frame-rate, ultrasound systems have recently become commercially available for small animal use. To overcome limitations of frame rate in high-frequency systems for cardiovascular applications in mice and rats, retrospective ECG gating or prospective ECG triggering has been applied to facilitate imaging of fast-traveling, physiologically induced waves [94-97]. Williams et al [98] evaluated the feasibility of measuring regional PWV using high-frequency ultrasound methodology by showing the similarity of PWVs derived by two image-guided approaches: a regional transit-time method and a local flow-area method. Our group recently developed an ultrafast RF data acquisition system based on a Vevo 770 system (VisualSonics Inc., Toronto, ON, Canada) and the ECG gating technique [94, 97]. RF data and ECG signals were acquired simultaneously, and RF data corresponding to a complete cardiac cycle were then gated using the ECG to reconstruct the full-view two-dimensional images. The RF signals of the murine aorta could be obtained at an extremely high frame rate (8000 frames/sec or 8 kHz), and used to visually color-map the small motion across the murine myocardium and abdominal aorta [91, 92, 94, 99]. The two-dimensional wall displacement maps at different times indicated wave propagation in the myocardium and aorta over a cardiac cycle [91-94, 97].

6.2 ELASTICITY IMAGING METHODS AND FINDINGS

6.2.1 Two-dimensional Myocardial Elastography

The Finite-Element (FE) Canine Left-Ventricular Model. Three-dimensional FE models of a normal (Control) and left circumflex ischemic (LCx) left ventricle (LV) of an open-chest mongrel dog were employed for quality assessment of myocardial elastography (see Fig. 6-3). The models have been previously reported [100, 101]. Figures 6-3a and b show anterior views of the normal (Control) and ischemic (LCx) models, respectively, at ES. The LCx model has been designed to simulate an ischemic region 2 minutes after occlusion of the left circumflex coronary artery. Abnormal contractile function is a result of inhibited activation in the posterior-lateral region, which is perfused by the LCx coronary artery. The ischemic myocardium behaves as a passive tissue and lacks the ability to contract during the entire cardiac cycle. The ischemic muscle occupies approximately 40% of the left-ventricular wall (see Fig. 6-3b). The employed model uses the so-called "time-varying elastance" scheme to simulate the end-systolic configuration of the left ventricle.

The three-dimensional Control and LCx ischemic models were initially described with 48 nodes and 24 bicubic elements [101]. The LV wall was two elements thick, with endocardial, midwall, and epicardial surfaces defined in Cartesian coordinates (x, y, z). Both the normal and ischemic canine models have wall thicknesses of 10.92 mm and 17.8 mm in the septal and lateral regions, respectively, at the end-diastolic phase. In the two-dimensional ultrasound simulation, a two-dimensional, equatorial short-axis slice at the model (i.e., $z = 0$), which intersected the ischemic region of the LCx model, was considered to represent the two-dimensional ultrasound image plane (see Fig. 6-3). Using cubic Hermite polynomials,

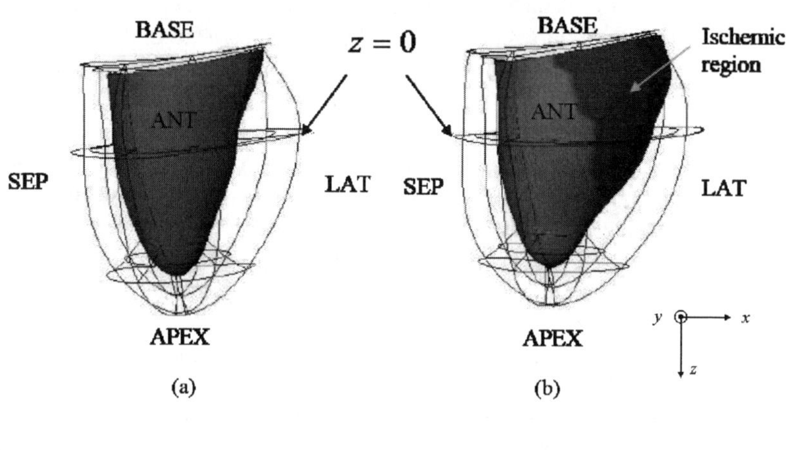

Figure 6-3. Anterior views of three-dimensional FE left-ventricular meshes at end-systole: (a) normal (Control) and (b) left-circumflex ischemic (LCx) models shown in the Cartesian coordinates (x, y, z), where positive x, y, and z point toward lateral wall, anterior wall, and apex, respectively. Endocardium and epicardium are represented by (light grey) shading and wire frame, respectively. The dark grey shaded surface depicts the ischemic region on the endocardium. The base, apex, anterior (ANT), lateral (LAT), and septal (SEP) regions are indicated [Figure courtesy of Susan Herz and Jeffrey Holmes, Department of Biomedical Engineering, Columbia University]

which allowed continuity of both the displacements and strains across all element boundaries and assured continuous representation of myocardial deformation, any arbitrary number of locations could be extracted within each element throughout the entire model. For this study, 840 evenly distributed material points (21 radially × 40 circumferentially) were used to track the regional deformation of the equatorial model slice. Considering a two-dimensional short-axis slice in a two-dimensional ultrasound image formation model, only the in-plane displacement components, i.e., the x and y components in this case, were taken into account. The out-of-plane motion was first ignored in order to examine the performance of the two-dimensional myocardial elastography (2DME) under optimal conditions, i.e., in a simplified scheme with pure two-dimensional motion. To represent the two-dimensional displacement field throughout the short-axis slice of the myocardium, bicubic interpolation of the extracted material points was used.

The two-dimensional motion vectors of the equatorial short-axis slice for the first incremental step during passive filling are shown in Fig. 6-4g, where the inner and outer contours indicate the endocardial (inner) and epicardial (outer) boundaries, respectively, and where the endocardium experiences larger motion than the epicardium. A two-dimensional motion vector can be obtained by the vector sum of its horizontal (x) and vertical (y) components. The horizontal and vertical displacement fields are shown in Figs. 6-4a and d, respectively. The three-dimensional finite-element model used has been validated in an open-chest canine left ventricle [1]. Therefore, a linear array configuration is suitable for the experiment simulated. Since a linear array image formation model is employed (see "Image Information Model: Two-Dimensional Simulation" and "Image Information Model: Three-Dimensional Simulation"),

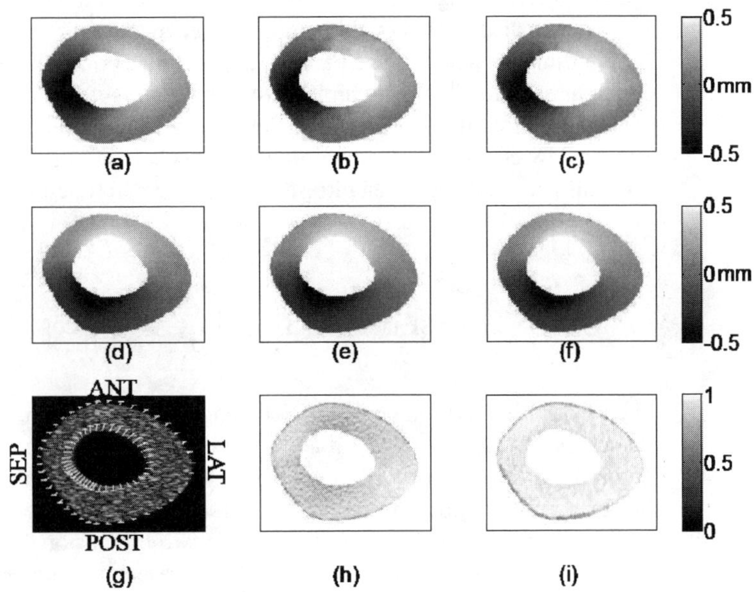

Figure 6-4. An example of the effect of the recorrelation method for the first incremental step from the unloaded state to ED at the short-axis slice at the equator of the left ventricle: (a and d) FE results; (b and e) first 2DME lateral and axial (respectively) displacement estimation (without recorrelation); (c and f) second 2DME lateral and axial (respectively) displacement estimation (with recorrelation); (g) motion vectors of the inner (endocardium) and outer (epicardium) walls, and anterior (ANT), lateral (LAT), septal (SEP), and posterior (POST) regions are indicated for the simulated echocardiographic image; (h and i) cross-correlation coefficient (CC) maps for the first estimation (without recorrelation) and the second elastographic displacement estimation (with recorrelation), respectively. 2DME stands for 2D myocardial elastography technique

horizontal (x) and vertical (y) axes coincide with the lateral and axial ultrasound axes, respectively.

In the three-dimensional ultrasound simulation model, five short-axis slices located from the sub-basal to the sub-apical levels were considered. The purpose of the three-dimensional ultrasound simulation was to study the effect of the out-of-plane motion on the in-plane deformation estimation according to the fact that out-of-plane motion varied at elevational (or, longitudinal in cardiac coordinates) levels of the left ventricle. Different from the previous two-dimensional simulation scheme with 840 material points used, therefore, only the solutions (three-dimensional motion and nodal coordinates) at the endo- and epicardium were extracted from the finite-element model and considered so that the transmural strain variation was constant in the three-dimensional simulation scheme. The entire in-plane (lateral and axial) displacement fields in the myocardium between two consecutive configurations were calculated using bilinear interpolation with the known in-plane motion of each node at the element boundaries, while the out-of-plane (elevational) displacement field for each short-axis level within a certain slice thickness was considered as rigid motion. Unlike the physiological case in vivo, the apex of the left ventricle moved towards the base, which was constrained not to move in the finite-element model, during systole. The out-of-plane motion thereby decreased from the apex to the base.

Image Formation Model: Two-Dimensional Simulation. In the simulated two-dimensional ultrasound image corresponding to the selected equatorial short-axis slice, scatterers were assumed to be two-dimensionally distributed at 96 scatterers per square wavelength in order to simulate fully developed speckle. Note that the background was assumed to have null scattering. According to the displacement field shown in Figs. 6-4a and d, the corresponding scatterer distribution could be used to generate two successive RF images. Similarly, the two-dimensional, in-plane FE displacement field could also be employed to generate RF signals at each loading configuration during the passive filling and time-varying elastance steps. Radio-frequency (RF) signals (128 in total) with an aperture size of 80 mm were generated by convolving the two-dimensional Gaussian-distributed scatterer function with a two-dimensional simulated ultrasound point spread function (PSF):

$$I(x,y) = \mathrm{PSF}(x,y) \otimes T(x,y) \qquad (6-1)$$

$$\mathrm{PSF}(x,y) = \exp\left(\frac{-1}{2}\left(\frac{x^2}{\sigma_x^2} + \frac{y^2}{\sigma_y^2}\right)\right)\cos\left(2\pi\frac{y}{\lambda_W}\right) \qquad (6-2)$$

where $I(x,y)$ is a two-dimensional RF echographic function, $T(x,y)$ is the tissue scattering function, x and y are lateral and axial coordinates, respectively, λ_W is half the wavelength (taking into account the round-trip of the wave in pulse-echo), and σ_x and σ_y are the standard deviations of the Gaussian function in the x- and y-directions at full width half maximum (FWHM), respectively. The two-dimensional PSF had a center frequency of 2 MHz and 60% fractional bandwidth, representative of real ultrasound systems. The beamwidth was 2 mm, and the pitch was 0.625 mm (128 beams with a lateral dimension of 80 mm). The sonographic signal-to-noise ratio (SNR_S) was equal to 60 dB, similar to what was previously considered in prior literature on tissue strain estimation[102]. Despite the fact that this SNR_S may be higher than what was typically measured in vivo, preliminary studies in myocardial elastography dependence on SNR_S showed that the quality (i.e., elastographic SNR, SNR_e) of the technique remains identical when the SNR_S was higher than 20 dB. These parameters were within a reasonable range. Each simulated two-dimensional B-mode image was 80 mm × 80 mm in size.

Image Formation Model: Three-Dimensional Simulation. Similarly, a three-dimensional ultrasonic image was generated by convolving a three-dimensional scatterer distribution with a three-dimensional point spread function, formularized as Eq. (6-3).

$$\mathrm{PSF}\,(x,y,z) = \exp\left(\frac{-1}{2}\left(\frac{x^2}{\sigma_x^{\,2}} + \frac{y^2}{\sigma_y^{\,2}} + \frac{z^2}{\sigma_z^{\,2}}\right)\right)\cos\left(2\pi\frac{y}{\lambda_W}\right) \tag{6-3}$$

where z is the elevational coordinate and σ_z is the standard deviation of the Gaussian function in the z-direction at FWHM. Each simulated three-dimensional B-mode image had an $80 \times 80 \times 6$ mm^3 field of view. Scatterers were assumed to be three-dimensionally distributed at 96 scatterers per cubic wavelength in order to simulate fully developed speckle. The three-dimensional FE displacement field was employed to generate RF signals at each loading configuration during passive filling and time-varying elastance steps. The defined two-dimensional linear array consisted of 128 and 6 elements in the lateral and elevational directions, respectively, with a center frequency of 2 MHz and 60% fractional bandwidth. The lateral beamwidth was 2 mm, the elevational beamwidth 2 mm, the lateral pitch 0.625 mm, and the elevational pitch 1 mm. The $\mathrm{SNR_S}$ was equal to 60 dB.

Displacement Estimation. Only the two-dimensional, in-plane (i.e., lateral and axial) motion and deformation were estimated using the proposed myocardial elastography in both two- and three-dimensional simulations. The corresponding RF signals for each load increment could be generated using the image formation model described in "Image Formation Model." Typically, myocardial function is evaluated during systole, i.e., from end-diastole to end-systole. In the finite-element analysis used here, active contraction and passive filling were modeled separately as described in "The Finite-Element (FE) Canine Left-Ventricular Model." Finite strains from ED to ES could be computed directly from the two end-point configurations, taking advantage of the history-independent assumption of the time-varying elastance model of activation. Since myocardial elastography aimed at assessing the total systolic deformation, which was associated with contraction and independent of frame rate, by first accumulating *incremental* displacements. The incremental displacements were estimated and compiled along the path from ED to ES to obtain *cumulative* displacements. These estimates were then compared to the solution of the model with ED as the reference and ES as the deformed state.

Several sets of consecutive RF frames from ED to ES were thus used to estimate the incremental two-dimensional displacements. The reference and comparison frames respectively contained the RF signals before and after the displacements occurred. The comparison frame from the previous RF frame set thus became the reference frame in the next RF frame set. Alternatively, the incremental two-dimensional displacements were accumulated to obtain the cumulative two-dimensional displacement that occurred from ED to ES. Note that appropriate registration for each pixel on two consecutive displacement images needed to be performed in order to guarantee that the cumulative displacement represented the motion of material points; thus, the displacement information was used both for correction and accumulation.

In order to estimate the two-dimensional displacement, cross-correlation and recorrelation techniques similar to those described in Konofagou and Ophir [44] were employed. For the findings shown in this chapter, the crosscorrelation technique employed a one-dimensional matching kernel of 3 mm and an 80% overlap, and performed a search in two dimensions, i.e., across the entire image plane. A one-dimensional, instead of a two-dimensional, matching kernel was used since the lateral motion stayed within one pitch (0.625 mm) for each incremental step, in which case a larger kernel size in the lateral direction would decrease the lateral resolution of the estimation.

Recorrelation. The lateral decorrelation due to axial motion reduces the accuracy of the lateral displacement estimation [103]. Therefore, the correction in axial displacement estimation, or recorrelation, has to be performed to reduce this decorrelation noise. In the two-dimensional myocardial elastography, an axial displacement correction method was instead implemented by shifting RF signal segments according to the estimated axial displacement in the comparison frame, prior to the second lateral displacement estimation (see Fig. 6-4c). In order to precisely correct for the axial displacement of the RF signals, i.e., to correct for sub-precision displacements, an interpolation factor of 10 was used to interpolate the RF signals of the comparison frame and then correct for axial decorrelation by rearranging interpolated RF signal segments based on the estimated axial displacements. The second lateral displacement estimation was thus performed after correction of the last estimated axial displacement. Similarly, the lateral displacement correction prior to axial displacement estimation was also performed by shifting RF signal segments in the lateral direction by the amount previously estimated. An example of the recorrelation results is shown in Fig. 6-4. Figures 6-4a and d show the FE calculated lateral and axial displacements within the plane at $z = 0$ (i.e., equatorial model slice). Positive displacements indicate motion rightwards in the lateral direction or upwards in the axial direction. Figures 6-4b and e respectively show the first-iteration, lateral and axial displacement images (without correction), and Figs. 6-4c and f respectively show the second-iteration, lateral and the third-iteration axial displacement images (with correction).

The crosscorrelation coefficient maps shown in Figs. 6-4h and i indicate that the recorrelation method improves the displacement estimation. The mean values and standard deviations of the cross correlation coefficients within the myocardium, namely, in the spatial distribution, for the first (see Fig. 6-4h) and the second (see Fig. 6-4i) estimations are 0.90 ± 0.06 and 0.98 ± 0.01, respectively. The cross correlation coefficient was, on average, improved by 9% through the use of the proposed recorrelation technique.

The two-dimensional (i.e., lateral and axial) displacement images of the first incremental step during passive filling from the unloaded state to ED for the normal model are shown respectively in Figs. 6-4c and f. Note that these displacement images are mapped onto the unloaded geometry. The incremental displacements (see Figs. 6-4c and f) are, on the average, within ±0.5 mm. The estimated incremental two-dimensional displacements are in excellent agreement with the FE results qualitatively (see Figs. 6-4a, c, d, and f).

Two-Dimensional, In-Plane Strain Estimation. Strain can be defined in terms of the gradient of the displacement. The two-dimensional, in-plane displacement can be written as $\underline{\mathbf{u}} = u_x \underline{\mathbf{e}}_x + u_y \underline{\mathbf{e}}_y$, where u_x and u_y are lateral (x-direction) and axial (y-direction) displacements, respectively. $\underline{\mathbf{e}}_x$ and $\underline{\mathbf{e}}_y$ are unit coordinate base vectors in lateral and axial directions, respectively. The two-dimensional displacement gradient tensor, $\nabla \underline{\mathbf{u}}$, is defined as [104]

$$\nabla \underline{\mathbf{u}} = \begin{bmatrix} \frac{\partial u_x}{\partial x} & \frac{\partial u_x}{\partial y} \\ \frac{\partial u_y}{\partial x} & \frac{\partial u_y}{\partial y} \end{bmatrix} \qquad (6-4)$$

The two-dimensional Lagrangian finite strain tensor, $\mathbf{E}$, is defined as [104]

$$\mathbf{E} = \frac{1}{2}(\nabla \underline{\mathbf{u}} + (\nabla \underline{\mathbf{u}})^{T} + (\nabla \underline{\mathbf{u}})^{T} \nabla \underline{\mathbf{u}}) \qquad (6-5)$$

where $(\nabla \underline{\mathbf{u}})^T$ is the transpose of $\nabla \underline{\mathbf{u}}$. Lateral and axial strains are the diagonal components of $\mathbf{E}$, i.e., E_{xx} and E_{yy}, respectively. In order to improve the signal-to-noise ratio in an elastogram (SNR_e), a least-squares strain estimator (LSQSE) [105] with a kernel of 7.5 mm in the lateral direction and a kernel of 4 mm in the axial direction on an 80 mm × 80 mm field of view was used.

Radial and Circumferential Strain Calculation. The two-dimensional strain tensor (i.e., $\mathbf{E}$ in the previous section), including normal and shear strain components, is highly dependent on the orientation of the imaging beam relative to the ventricular wall, since axial and lateral directions are both defined with respect to the transducer's imaging plane. This might complicate the interpretation of the direction of in-plane myocardial deformation. In order to render them independent of the imaging plane and link them to the left-ventricular geometry, the radial and circumferential strains were first computed. A rotation matrix, **R**, for each material point within the myocardium in a two-dimensional short-axis view was written as

$$\mathbf{R} = \begin{bmatrix} \cos\theta & \sin\theta \\ -\sin\theta & \cos\theta \end{bmatrix} \qquad (6-6)$$

where θ is the angle relative to the origin of the Cartesian coordinates in the FE models. Strains in cardiac coordinates were therefore obtained by

$$\hat{\mathbf{E}} = \mathbf{R}\mathbf{E}\mathbf{R}^{\mathrm{T}} \qquad (6-7)$$

where $\hat{\mathbf{E}}$ is the two-dimensional radial-circumferential strain tensor. The diagonal components of $\hat{\mathbf{E}}$ were radial ($\mathbf{E}_{rr}$) and circumferential ($\mathbf{E}_{cc}$) strains. Positive and negative radial strains indicated myocardial thickening and thinning, respectively, while myocardial stretching and shortening were represented by positive and negative circumferential strains.

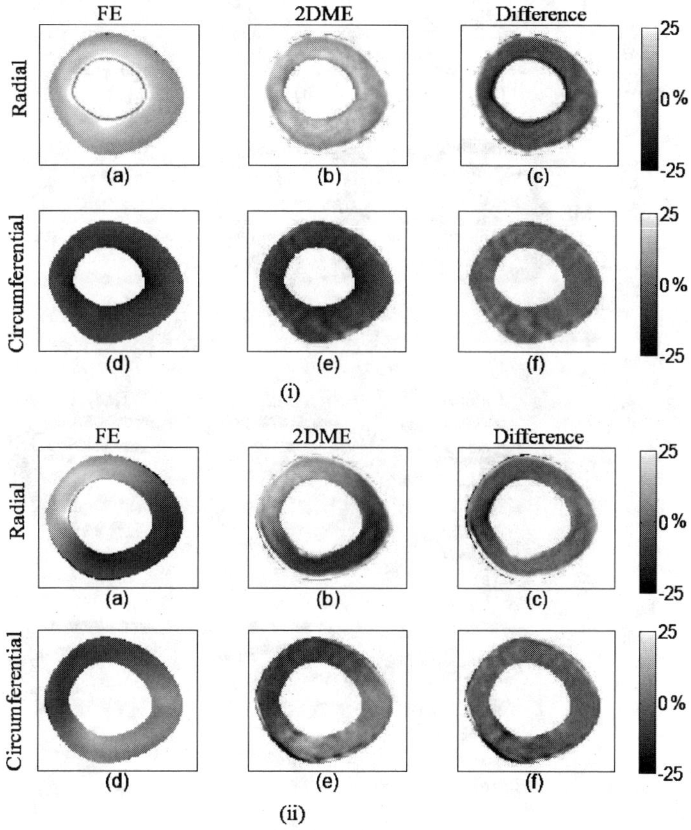

Figure 6-5. Radial and circumferential strains for (i) the normal (Control) model and (ii) the LCx (ischemic) model between ED and ES: (a and d) FE radial and circumferential strains, respectively; (b and e) estimated radial and circumferential strains, respectively; (c and f) difference images between the first and the second columns. The ischemic region is in the lower right quadrant in all images shown

Figure 6-5i shows radial thickening (a and b) as well as circumferential systolic shortening (d and e) of the contracting normal myocardium, i.e., from ED to ES. The differences between FE and estimated radial and circumferential strains are shown in Fig. 6-5i (c and f, respectively). In contrast, LCx model shows radial thinning (see Fig. 6-5ii: a and b) and circumferential stretching (see Fig. 6-5ii: d and e), instead of radial thickening and circumferential shortening, in the posterior-lateral (lower right) portion, i.e., the ischemic region, during contraction. This is consistent with the fact that the ischemic muscle behaves as a passively tethered region undergoing bulging during systole (see Fig. 6-3b) [6].

The FE and estimated Cartesian (i.e., lateral and axial) and polar (i.e., radial and circumferential) strains for the Control model at multiple slices in the three-dimensional simulation scheme are shown in Fig. 6-6i. The elevational displacement increases, while the in-plane motion and strains are smaller, near the apical level. The larger the elevational motion is, the larger the in-plane strain estimation errors become. Furthermore, Fig. 6-6i again shows radial thickening (c and g) and circumferential shortening (d and h) through-

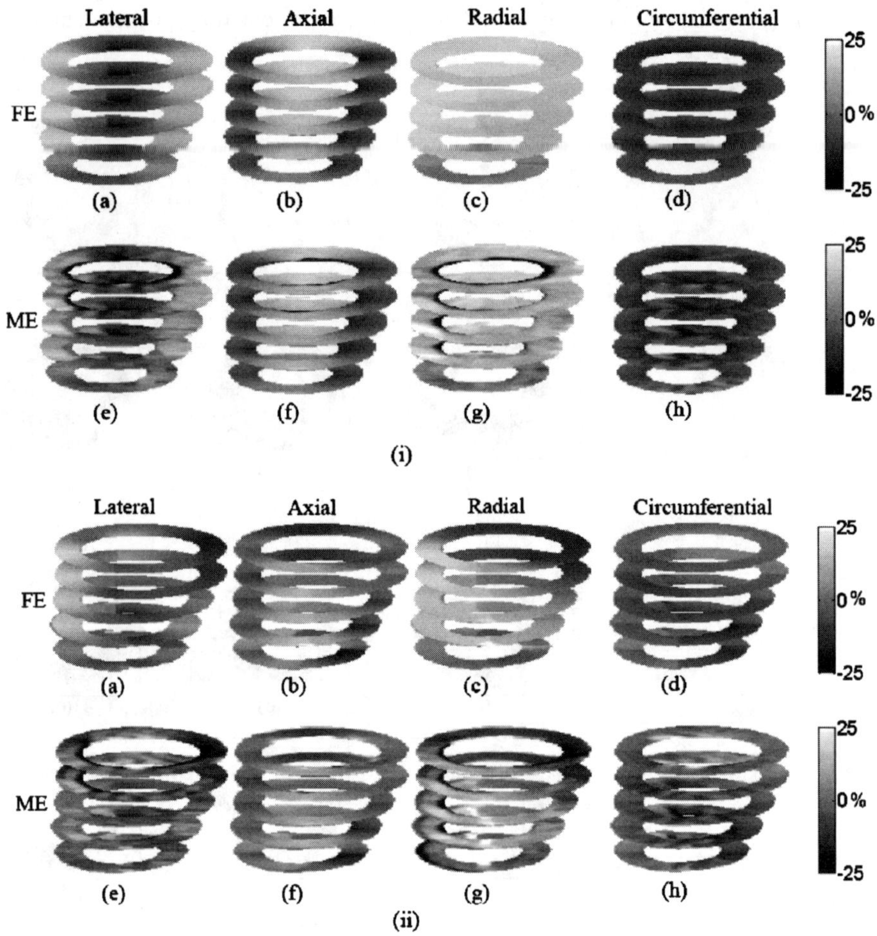

Figure 6-6. Anterior view of cumulative strains for the (i) the normal (Control) model and (ii) the LCx (ischemic) model at multiple slices in three-dimensional simulation; (a–d) FE lateral, axial, radial, and circumferential strains, respectively; (e–h) are estimated lateral, axial, radial, and circumferential strains, respectively. The images are mapped onto the end-systolic geometry. Note that the orientation is the same as Fig. 6-3

out the normal left ventricle (i.e., Control model). The FE and estimated Cartesian and polar strains for the LCx model at multiple slices in the three-dimensional simulation scheme are shown in Fig. 6-6ii. Unlike the normal case, the polar strains show radial thinning (see Fig. 6-6ii: c and g) and circumferential stretching (see Fig. 6-6ii: d and h) in the ischemic region (i.e., posterior-lateral wall) and confirm the reduced size of the ischemic region toward the apical level (see Fig. 6-3). Note that all images shown in Fig. 6-6 are mapped on the geometry at the end-systolic phase, and their orientation is the same as in Fig. 6-3.

Even though the polar strains are independent of the transducer orientation, the selection of the centroid is critical. Principal strains have been proven to be angle-independent as well as less centroid-independent than the polar strains [60, 61]. Considering a two-dimensional short-axis slice of the myocardium, solving the eigenvalue/eigenvector problem using Eq. (6-8) for the two-dimensional finite strain tensor could yield two principal strains (eigenvalues) corresponding to two principal axes (eigenvectors), which closely approximated strains in the polar directions:

$$\mathbf{E}^{\mathrm{p}} = \begin{bmatrix} \varepsilon_1 & 0 \\ 0 & \varepsilon_2 \end{bmatrix} \qquad (6-8)$$

where $\mathbf{E}^{\mathrm{p}}$ is principal strain tensor, and ε_1 and ε_2 are the first and second principal components, respectively. Note that the two principal strains were then classified according to their angles between the principal (i.e., eigenvectors) and polar directions [61]. The first and second principal strains closely approximated radial and circumferential strains, respectively [61]. Once feasibility in the theoretical framework was shown [60, 61], principal strains were further employed in a clinical setting and shown in the following section.

6.2.2 Ultrasound and Tagged MRI Clinical Data Acquisition

For the clinical validation, a clinical echocardiography ultrasound scanner (General Electric Vivid FiVe, GE Vingmed Ultrasound, Horten, Norway) with a phased array probe (FPA 2.5 MHz 1C) was used to acquire cardiac ultrasound in-phase and quadrature (I/Q) data in two-dimensional SA views at the papillary level from normal (female, 28 years old, 80 bpm) and abnormal (male, 69 years old, 73 bpm) subjects ($N = 2$) at a frame rate of 136 frames/sec using ECG-gated composite imaging [109]. The I/Q data were further up sampled to retrieve the RF signals. The lateral and axial resolutions are approximately 1.92 mm and 0.77 mm, respectively. The human subject study protocol was approved by the Institutional Review Board of Columbia University, and informed consent was obtained from all human subjects prior to scanning.

Tagged MR images were obtained on a Philips Intera 1.5-T scanner (Philips Medical Systems, Best, the Netherlands) equipped with a five-channel SENSE cardiac coil and master gradients of strength 30 mT/m and slew rate 150 T/m/sec. A multi-slice and multi-phase true SA tagged image was acquired under free-breathing with a combination of fast-field echo excitation and a multi-shot echo-planar readout (EPI-FFE) technique [106] (field-of-view [FOV] = 350 mm, TE = 4 msec, TR = 30, NSA = 4, resolution acquired/reconstructed = 192/256, flip angle = 13 deg., EPI factor = 3, and full ECG gating scan duration = 4.77 minutes). Two-dimensional grid tagging was performed yielding a 9-mm, in-plane tag resolution. The SA orientation was also acquired at the papillary muscle level from the same subjects. The nominal frame rate was 33 frames/sec.

In order to track the tagging grids and estimate the localized myocardial displacement and strain values, a template-based tracking algorithm on a two-dimensional grid-shaped mesh was implemented to obtain the displacement vectors of the crossing points (or, nodes) on the tagging grids [107, 108].

The cumulative systolic principal strains estimated from both imaging modalities at the papillary muscle level in a SA view of a normal left ventricle are shown in Fig. 6-7i, depicting positive radial (i.e., myocardial thickening in a) and b) and negative circumferential (i.e., myocardial shortening in c and d) strains. The anterior, lateral, posterior, and septal walls are in the upper right, lower right, lower left, and upper left regions, respectively.

An example of a reperfused left ventricular deformation is shown in Fig. 6-7ii. This human subject suffered a myocardial infarction caused by partial occlusion of the distal left anterior descending (LAD) coronary artery and subsequent motion abnormalities in both the septal and anterior walls. The cumulative systolic first principal (see Fig. 6-7ii, b) 2DME strain estimate for the reperfused left ventricle shows myocardial thickening in the posterior and anterior-septal walls but not in the septum or anterior regions. On the contrary, the first principal (see Fig. 6-7ii, a) tMRI strain shows thickening throughout the entire myocardium with augmented thickening in the posterior wall but with reduced

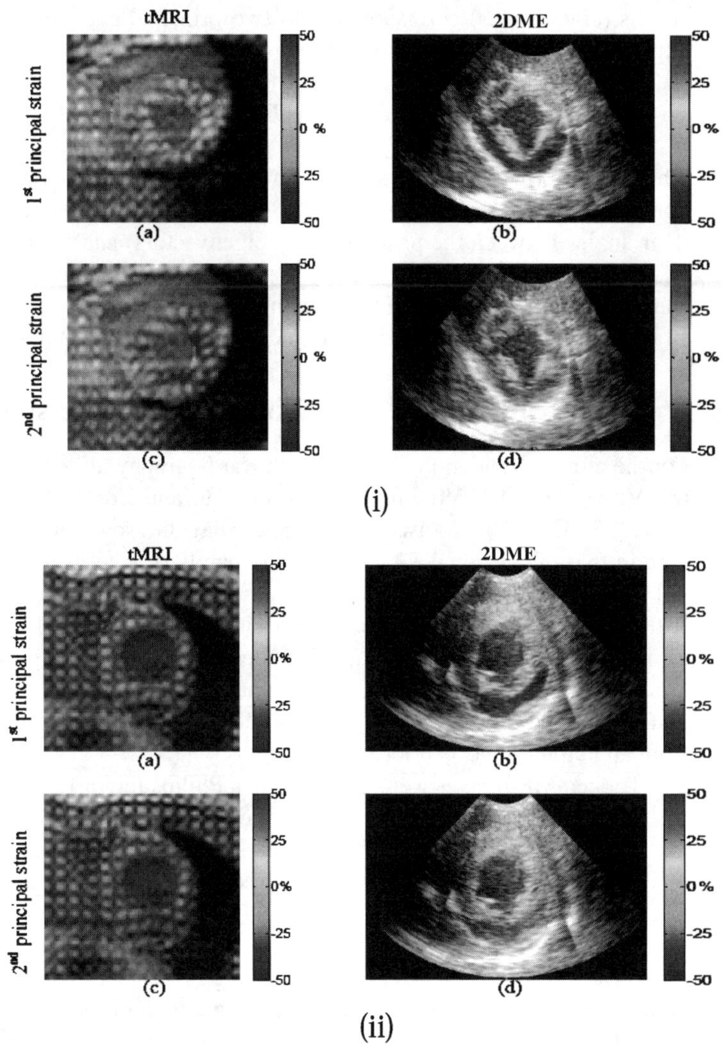

Figure 6-7. (i) A normal human left ventricle and (ii) a reperfused human left ventricle: (a and c) cumulative first and second systolic principal strains from tMRI between ED and ES, respectively; (b and d) cumulative first and second systolic principal strains from 2DME, respectively. All the images were acquired approximately at the papillary left-ventricular level and shown at end-systolic configuration. Color versions of this figure can be found in [107] and [110]

thickening in the other walls. The second principal (see Fig. 6-7ii, d) strain estimate from 2DME shows myocardial shortening in the posterior wall and slight stretching in the other regions, while the tMRI estimate (see Fig. 6-7ii, c) indicates slight stretching in the lateral, anterior, and anterior-septal walls.

Although the strains obtained from ME are overall higher compared with those from tMRI in both normal and reperfused left ventricles, these preliminary results show that the two imaging modalities are in good agreement and that ME is capable of differentiating abnormal from normal myocardium even in a post-infarction, treated left ventricle. Correct co-registration and study of the role of inherent resolution and SNR differences across the two modalities are currently ongoing.

6.3 PWI FOR VASCULAR DISEASE DETECTION

6.3.1 AAA Animal Model

The incremental/instantaneous displacements of the wall during pulse wave propagation were estimated and imaged offline using a one-dimensional normalized crosscorrelation

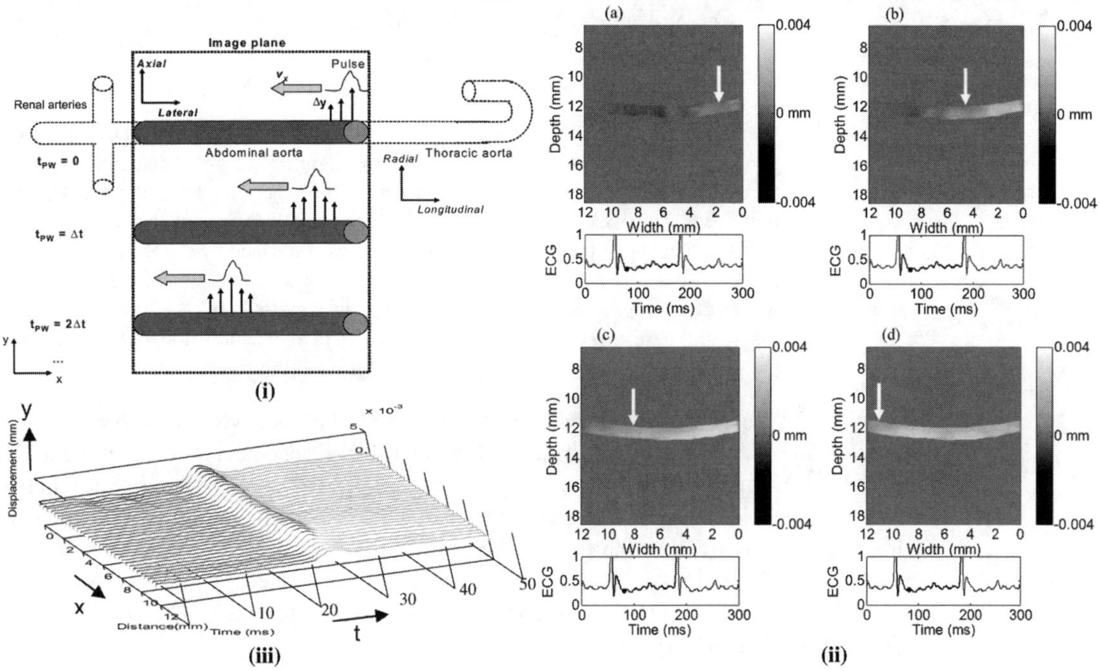

Figure 6-8. (i) Illustration of the PW imaging concept indicating the area of scanning, direction of displacements estimated (in the axial or *y* direction) and direction of pulse wave propagation (in the lateral, or *x*-direction). (ii) Visualization of the propagation of the PW along the wall in a sham murine abdominal aorta, from the heart (right side of the images) to the renal arteries (left side of the images) at (a) 21.75 msec, (b) 23 msec, (c) 26.25 msec, and (d) 25.5 msec after the R-wave. (iii) Waterfall plot of the peak displacement propagation in the longitudinal direction. Axial wall displacement maps (color maps) in mm were overlaid onto the B-mode images (not shown here to avoid confusion from overlaying two grayscale images). Only axial displacements of the anterior wall (i.e., near wall) of the aorta and some surrounding tissues were shown for better visualization. The thin white arrow denotes the approximate front of the pulse wave in (ii). The ECG (with a dot indicating the cardiac phase) is shown below each PWI image in (ii). The color version of this figure can be found in [93]

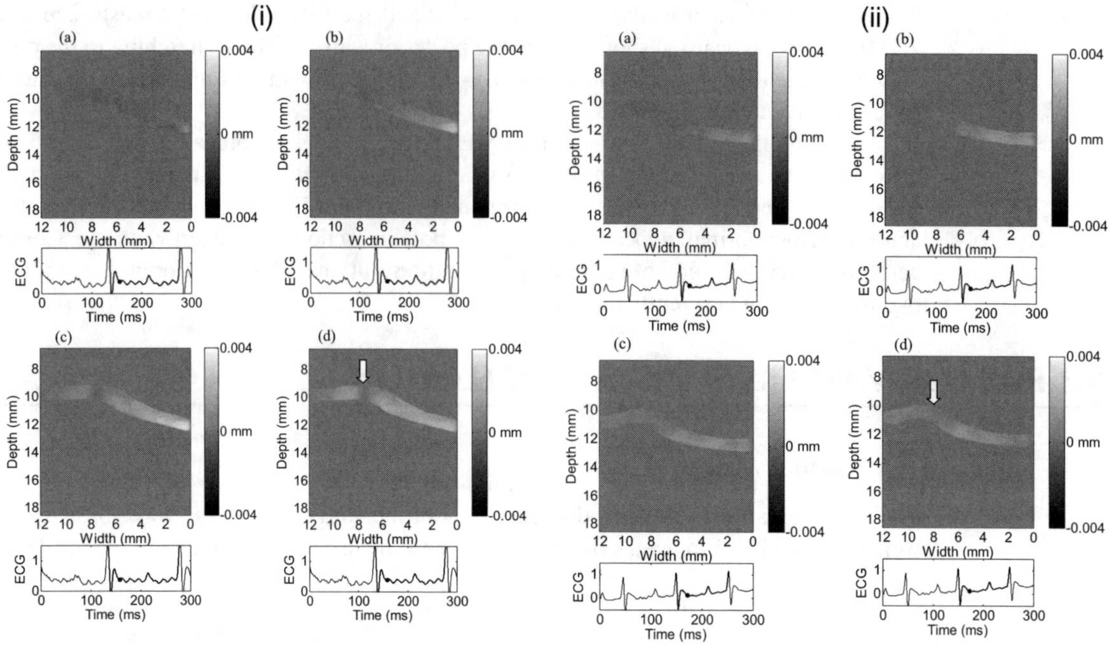

Figure 6-9. Visualization of the PW propagation along the aortic wall in the 4-week AngII-treated aorta (i) without (~30%) and (ii) with (~100%) significant diameter changes at the same time intervals as in Fig. 6-8ii. Note that the PWI images can detect the abnormality in both cases. The thick white arrow denotes the largest PW-induced displacement discontinuity occurring at the level of the aneurysm, as confirmed by the B-mode (not shown here to avoid confusion from overlaying two grayscale images), and absent in the normal case (see Fig. 6-8ii). Similar to Fig. 6-8ii, the wave in both cases travels from proximal (right) to distal (left) although not indicated to avoid confusion with localization of the displacement discontinuity. The color version of this figure can be found in [93]

technique on the RF signals obtained (window size = 0.240 mm, window shift = 0.024 mm) [35, 94] (Fig. 6-8 (ii)). Through the use of such RF-based speckle tracking techniques, displacements as small as 2 μm can be estimated at 5 MHz [102], which can be even smaller at higher frequencies due to associated decrease in the wavelength. Therefore, fundamentally, displacements as small as 0.3 μm can be estimated in soft tissues using cross correlation techniques of RF signals at 30 MHz. Both normal (wildtype) (see Fig. 6-8) and AAA-model aortas (see Fig. 6-9) were scanned. For the AAA-model aortas, the $CaCl_2$ [91, 92] and Angiotensin II (AngII) [93] models were used.

6.3.2 Human Applications

The PWI technique was recently developed for integration and application in conjunction with any clinical ultrasound scanner. The PWI system acquires ultrasound RF signals from the abdominal aorta using seven small regions-of-interest (ROIs) that, after ECG gating and signal processing offline (similar to the methods used in the mice), were combined to produce a composite image that will automatically show the amount of aortic wall displacement and the underlying B-mode (not shown) during pulse wave propagation in a full field-of-view and at a high frame rate (see Fig. 6-10). An ultrasound scanner with open architecture and control over both the acquisition and RF data was used in order to develop the first PWI human application. The scanner was a Sonix RP system from Ultrasonix Medical Corporation (Vancouver, BC, Canada). A user interface was built in Visual C++

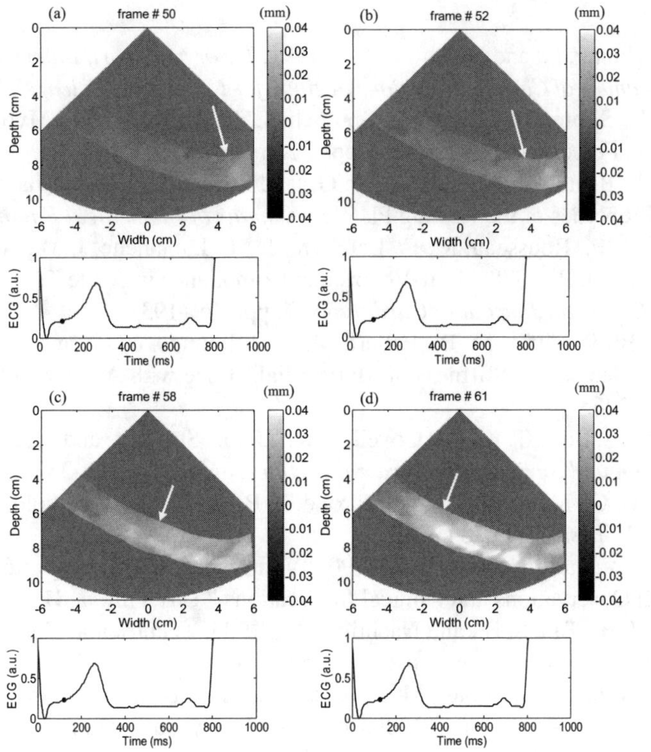

Figure 6-10. (a–d) Long-axis view of a human normal aorta obtained from a healthy 28-year-old female subject — The PWI imaging sequence showing the propagation of the pulse wave along the anterior aortic wall. The white arrow denotes the pulse wavefront. The color version of this figure can be found in [109]

that can acquire real-time ECG and RF signals and store samples automatically at 8-bit resolution. The full-view PWI ciné-loops and frames (see Fig. 6-10) can be generated at a frame rate of up to 700 frames/sec through ECG gating, allowing us to visualize the pulse wave propagation in human aortas with speeds up to 100 m/sec within an imaging width of 15 cm [109].

The frame rate required in this case is lower than that of mice since the human heart rate is typically 8 to 10 times lower than the murine one. This customized ultrasound system is currently the only system that can achieve high frame rates and visualize the pulse wave in a human aorta (see Fig. 6-10) similar to what has been observed in mice. The feasibility of the human application in normal aortas has therefore been shown, and PWI is intended for use in detecting and characterizing aneurysms in human subjects with and without AAA or other vascular disease.

6.4 CONCLUSION

The cardiovascular elasticity imaging techniques of myocardial elastography and pulse wave imaging were described, both with the potential of clinical applications for myocardial abnormalities and vascular disease detection. Both techniques can be applied in conjunction with currently used echocardiography systems and can offer important, complementary information for the early and reliable diagnosis of cardiovascular disease as well as treatment guidance.

REFERENCES

[1] Glass, L., Hunter, P., and McCulloch, A., 1991, *Theory of Heart, Biomechanics, Biophysics, and Nonlinear Dynamics of Cardiac Function (Institute for Non-Linear Science)*, Springer-Verlag, New York.

[2] Finer, J. T., Simmons, R. M., and Spudich, J. A., 1994, "Single Myosin Molecule Mechanics — Piconewton Forces and Nanometer Steps," *Nature*, **368**, pp. 113–119.

[3] Haga, J. H., Beaudoin, A. J., White, J. G., and Strony, J., 1998, "Quantification of the Passive Mechanical Properties of the Resting Platelet," *Annals of Biomedical Engineering,* **26**, pp. 268–277.

[4] Bertrand, M. E., Rousseau, M. F., Lefebvre, J. M., Lablanche, J. M., Asseman, P. H., Carre, A. G., and Lekieffre, J. P., 1978, "Left-Ventricular Compliance in Acute Transmural Myocardial Infarction in Man," *European Journal of Cardiology*, **7**, pp. 179–193.

[5] Gupta, K. B., Ratcliffe, M. B., Fallert, M. A., Edmunds, L. H., and Bogen, D. K., 1994, "Changes in Passive Mechanical Stiffness of Myocardial Tissue with Aneurysm Formation," *Circulation*, **89**, pp. 2315–2326.

[6] Holmes, J. W., Borg, T. K., and Covell, J. W., 2005, "Structure and Mechanics of Healing Myocardial Infarcts," *Annual Review in Biomedical Engineering,* **7**, pp. 223–253.

[7] Fishbein, M. C., Maclean, D., and Maroko, P. R., 1978, "Histopathologic Evolution of Myocardial-Infarction," *Chest,* **73**, pp. 843–849.

[8] Pfeffer, M. A., and Braunwald, E., 1990, "Ventricular Remodeling after Myocardial Infarction: Experimental Observations and Clinical Implications," *Circulation,* **81**, pp. 1161–1172.

[9] Caulfield, J. B., Tao, S. B., and Nachtigal, M., 1985, "Ventricular Collagen Matrix and Alterations," *Advances in Myocardiology*, **5**, pp. 257–269.

[10] Liu, K. J., Rubin, J. M., Potel, M. J., Aisen, A., Mackay, S. A., Sayre, R. E., and Anagnostopoulos, C. E., 1984, "Left-Ventricular Wall Motion — Its Dynamic Transmural Characteristics," *Journal of Surgical Research,* **36**, pp. 25–34.

[11] Lewinter, M. M., Kent, R. S., Kroener, J. M., Carew, T. E., and Covell, J. W., 1975, "Regional Differences in Myocardial Performance in Left-Ventricle of Dog," *Circulation Research,* **37**, pp. 191–199.

[12] Gaasch, W. H., Levine, H. J., Quinones, M. A., and Alexander, J. K., 1976, "Left-Ventricular Compliance — Mechanisms and Clinical Implications," *American Journal of Cardiology,* **38**, pp. 645–653.

[13] Raya, T. E., Gay, R. G., Lancaster, L., Aguirre, M., Moffett, C., and Goldman, S., 1988, "Serial Changes in Left-Ventricular Relaxation and Chamber Stiffness after Large Myocardial-Infarction in Rats," *Circulation,* **77**, pp. 1424–1431.

[14] Sabharwal, N. K., and Lahiri, A., 2003, "Role of Myocardial Perfusion Imaging for Risk Stratification in Suspected or Known Coronary Artery Disease," *Heart,* **89**, pp. 1291–1297.

[15] O'Dell, W. G., Moore, C. C., Hunter, W. C., Zerhouni, E. A., and McVeigh, E. R., 1995, "3-Dimensional Myocardial Deformations — Calculation with Displacement Field Fitting to Tagged MR Images," *Radiology*, **195**, pp. 829–835.

[16] Reichek, N., 1999, "MRI Myocardial Tagging," *Journal of Magnetic Resonance Imaging,* **10**, pp. 609–616.

[17] Osman, N. F., and Prince, J. L., 2000, "Visualizing Myocardial Function Using HARP MRI," *Physics in Medicine and Biology,* **45**, pp. 1665–1682.

[18] O'Donnell, M., Mimbs, J. W., and Miller, J. G., 1979, "Relationship between Collagen and Ultrasonic-Attenuation in Myocardial Tissue," *Journal of the Acoustical Society of America,* **65**, pp. 512–517.

[19] Auld, B. A., 1990, *Acoustic Fields and Waves in Solids*, Robert E. Krieger Publishing, Malabar, FL, 2nd ed., Vol. I.

[20] Miller, J. G., Perez, J. E., and Sobel, B. E., 1985, "Ultrasonic Characterization of Myocardium," *Progress in Cardiovascular Disease,* **28**, pp. 85–110.

[21] Madaras, E. I., Barzilai, B., Perez, J. E., Sobel, B. E., and Miller, J. G., 1983, "Changes in Myocardial Backscatter Throughout the Cardiac Cycle," *Ultrasonic Imaging,* **5**, pp. 229–239.

[22] Hoffmeister, B. K., Handley, S. M., Wickline, S. A., and Miller, J. G., 1996, "Ultrasonic Determination of the Anisotropy of Young's Modulus of Fixed Tendon and Fixed Myocardium," *Journal of the Acoustical Society of America,* **100**, pp. 3933–3940.

[23] Jensen, J. A., 1996, *Estimation of Blood Velocities Using Ultrasound*, Cambridge University Press, New York, NY.

[24] McDicken, W. N., Sutherland, G. R., Moran, C. M., and Gordon, L. N., 1992, "Colour Doppler Velocity Imaging of the Myocardium," *Ultrasound in Medicine and Biology,* **18**, pp. 651–654.

[25] Sutherland, G. R., Stewart, M. J., Groundstroem, K. W., Moran, C. M., Fleming, A., Guell-Peris, F. J., Riemersma, R. A., Fenn, L. N., Fox, K. A., and McDicken, W. N., 1994, "Color Doppler Myocardial Imaging: A New Technique for the Assessment Of Myocardial Function," *Acta Paediatrica*, **7**, pp. 441–58.

[26] Zamorano, J., Wallbridge, D. R., Ge, J., Drozd, J., Nesser, J., and Erbel, R., 1997, "Non-Invasive Assessment of Cardiac Physiology by Tissue Doppler Echocardiography — A Comparison with Invasive Haemodynamics," *European Heart Journal*, pp. 330–339.

[27] Heimdal, A., Stoylen, A., Torp, H., and Skjaerpe, T., 1998, "Real-Time Strain Rate Imaging of the Left Ventricle by Ultrasound," *Journal of the American Society of Echocardiography,* **11**, pp. 1013–1019.

[28] Bonnefous, O., and Pesque, P., 1986, "Time Domain Formulation of Pulse-Doppler Ultrasound and Blood Velocity Estimation by Cross-Correlation," *Ultrasonic Imaging,* **8**, pp. 73–85.

[29] Hein, I. A., and O'brien, W. D., 1993, "Current Time-Domain Methods for Assessing Tissue Motion by Analysis from Reflected Ultrasound Echoes — A Review," *IEEE Transactions on Ultrasonics, Ferroelectrics and Frequency Control,* **40**, pp. 84–102.

[30] de Jong, P. G. M., Arts, T., Hoeks, A. P. G., and Reneman, R. S., 1990, "Determination of Tissue Motion Velocity by Correlation Interpolation of Pulsed Ultrasonic Echo Signals," *Ultrasonic Imaging,* **12**, pp. 84–98.

[31] Brands, P. J., Hoeks, A. P. G., Ledoux, L. A. F., and Reneman, R. S., 1997, "A radio frequency domain complex cross-correlation model to estimate blood flow velocity and tissue motion by means of ultrasound," *Ultrasound in Medicine and Biology,* **23**, pp. 911–920.

[32] Bohs, L. N. and Trahey, G. E., 1991, "A Novel Method for Angle Independent Ultrasonic-Imaging of Blood-Flow and Tissue Motion," *IEEE Transactions on Biomedical Engineering,* **38**, pp. 280–286.

[33] Kanai, H., Hasegawa, H., Chubachi, N., Koiwa, Y., and Tanaka, M., 1997, "Noninvasive evaluation of local myocardial thickening and its color coded imaging," *IEEE Transactions on Ultrasonics, Ferroelectrics and Frequency Control,* **44**, pp. 752–768.

[34] Meunier, J., Bertrand, M., and Mailloux, G., 1987, "A Model for Dynamic Texture Analysis in Two-Dimensional Echocardiograms of the Myocardium," *Proceedings of SPIE,* **768**, pp. 193–200.

[35] Ophir, J., Céspedes, I., Ponnekanti, H., Yazdi, Y., and Li, X., 1991, "Elastography: A Quantitative Method for Imaging the Elasticity of Biological Tissues," *Ultrasonic Imaging,* **13**, pp. 111–134.

[36] O'Donnell, M., Skovoroda, A. R., Shapo, B. M., and Emelianov, S. Y., 1994, "Internal Displacement and Strain Imaging Using Ultrasonic Speckle Tracking," *IEEE Transactions on Ultrasonics, Ferroelectrics and Frequency Control,* **41**, pp. 314–325.

[37] Parker, K. J., Huang, S. R., Musulin, R. A., and Lerner, R. M., 1990, "Tissue-Response to Mechanical Vibrations for Sonoelasticity Imaging," *Ultrasound in Medicine and Biology,* **16**, pp. 241–246.

[38] Krouskop, T. A., Dougherty, D. R., and Vinson, F. S., 1987, "A Pulsed Doppler Ultrasonic System for Making Noninvasive Measurements of the Mechanical Properties of Soft Tissue," *Journal of Rehabilitation Research and Development,* **24**, pp. 1–8.

[39] Lerner, R. M. and Parker, K. J., 1987, "Sono-Elasticity Images for Cancer-Detection," *Ultrasonic Imaging,* **9**, pp. 63–63.

[40] Levinson, S. F., Shinagawa, M., and Sato, T., 1995, "Sonoelastic Determination of Human Skeletal-Muscle Elasticity," *Journal of Biomechanics,* **28**, pp. 1145–1154.

[41] Taylor, L. S., Porter, B. C., Rubens, D. J., and Parker, K. J., 2000, "Three-Dimensional Sonoelastography: Principles and Practices," *Physics in Medicine and Biology,* **45**, pp. 1477–1494.

[42] Céspedes, I., Ophir, J., Ponnekanti, H., and Maklad, N., 1993, "Elastography — Elasticity Imaging Using Ultrasound with Application to Muscle and Breast in-Vivo," *Ultrasonic Imaging,* **15**, pp. 73–88.

[43] Garra, B. S., Cespedes, E. I., Ophir, J., Spratt, S. R., Zuurbier, R. A., Magnant, C. M., and Pennanen, M. F., 1997, "Elastography of Breast Lesions: Initial Clinical Results," *Radiology,* **202**, pp. 79–86.

[44] Konofagou, E., and Ophir, J., 1998, "A New Elastographic Method for Estimation and Imaging of Lateral Displacements, Lateral Strains, Corrected Axial Strains and Poisson's Ratios in Tissues," *Ultrasound in Medicine and Biology,* **24**, pp. 1183–1199.

[45] Skovoroda, A. R., Emelianov, S. Y., Lubinski, M. A., Sarvazyan, A. P., and O'Donnell, M., 1994, "Theoretical-Analysis and Verification of Ultrasound Displacement and Strain Imaging," *IEEE Transactions on Ultrasonics, Ferroelectrics and Frequency Control,* **41**, pp. 302–313.

[46] Konofagou, E. E., and Ophir, J., 2000, "Precision Estimation and Imaging of Normal and Shear Components of the 3D Strain Tensor in Elastography," *Physics in Medicine and Biology,* **45**, pp. 1553–1563.

[47] Konofagou, E. E., Harrigan, T. P., Ophir, J., and Krouskop, T. A., 2001, "Poroelastography: Imaging the Poroelastic Properties of Tissues," *Ultrasound in Medicine and Biology,* **27**, pp. 1387–1397.

[48] Konofagou, E., D'hooge, J., and Ophir, J., 2000, "Cardiac Elastography — A Feasibility Study," *Proceedings IEEE Ultrasonics Symposium,* pp. 1273–1276.

[49] Kaluzynski, K., Chen, X. C., Emelianov, S. Y., Skovoroda, A. R., and O'Donnell, M., 2001, "Strain Rate Imaging Using Two-Dimensional Speckle Tracking," *IEEE Transactions on Ultrasonics, Ferroelectrics and Frequency Control,* **48**, pp. 1111–1123.

[50] Konofagou, E. E., D'hooge, J., and Ophir, J., 2002, "Myocardial Elastography — A Feasibility Study in Vivo," *Ultrasound in Medicine and Biology,* **28**, pp. 475–482.

[51] D'hooge, J., Konofagou, E., Jamal, F., Heimdal, A., Barrios, L., Bijnens, B., Thoen, J., Van de Werf, F., Sutherland, G., and Suetens, P., 2002, "Two-Dimensional Ultrasonic Strain Rate Measurement of the Human Heart in Vivo," *IEEE Transactions on Ultrasonics, Ferroelectrics and Frequency Control,* **49**, pp. 281–286.

[52] O'Donnell, M. and Skovoroda, A. R., 2004, "Prospects for Elasticity Reconstruction in the Heart," *IEEE Transactions on Ultrasonics, Ferroelectrics and Frequency Control,* **51**, pp. 322–328.

[53] Konofagou, E. E., Fung-Kee-Fung, S., Luo, J., and Pernot, M., 2006, "Imaging the Mechanics and Electromechanics of the Heart," *Proceedings of IEEE EMBS Conference,* pp. 6648–6651.

[54] Konofagou, E. E., Harrigan, T., and Solomon, S., 2001, "Assessment of Regional Myocardial Strain Using Cardiac Elastography: Distinguishing Infarcted from Non-Infarcted Myocardium," *Proceedings IEEE Ultrasonics Symposium,* pp. 1589–1592.

[55] Konofagou, E. E., Manning, W., Kissinger, K., and Solomon, S. D., 2003, "Myocardial elastography — Comparison to Results Using MR Cardiac Tagging," *Proceedings IEEE Ultrasonics Symposium,* pp. 130–133.

[56] Langeland, S., D'hooge, J., Claessens, T., Claus, P., Verdonck, P., Suetens, P., Sutherland, G. R., and Bijnens, B., 2004, "RF-Based Two-Dimensional Cardiac Strain Estimation: A Validation Study in a Tissue-Mimicking Phantom," *IEEE Transactions on Ultrasonics, Ferroelectrics and Frequency Control,* **51**, pp. 1537–1546.

[57] Breithardt, O. A., Stellbrink, C., Herbots, L., Claus, P., Sinha, A. M., Bijnens, B., Hanrath, P., and Sutherland, G. R., 2003, "Cardiac Resynchronization Therapy Can Reverse Abnormal Myocardial Strain Distribution in Patients with Heart Failure and Left Bundle Branch Block," *Journal of the American College of Cardiology,* **42**, pp. 486–494.

[58] Kowalski, M., Herregods, M. C., Herbots, L., Weidemann, F., Simmons, L., Strotmann, J., Dommke, C., D'Hooge, J., Claus, P., Bijnens, B., Hatle, L., and Sutherland, G. R., 2003, "The Feasibility of Ultrasonic Regional Strain and Strain Rate Imaging in Quantifying Dobutamine Stress Echocardiography," *European Journal of Echocardiography*, **4**, pp. 81–91.

[59] Konofagou, E. E., 1999, "Estimation and Imaging of Three-dimensional Motion and Poisson's Ratio in Elastography," Ph.D. thesis, University of Houston.

[60] Fung-Kee-Fung, S. D., Lee, W.-N., Ingrassia, C. M., Costa, K. D., and Konofagou, E. E., 2005, "Angle-Independent Strain Mapping in Myocardial Elastography," *Proceedings IEEE Ultrasonics Symposium,* pp. 516–519.

[61] Zervantonakis, I. K., Fung-Kee-Fung, S. D., Lee, W.-N., and Konofagou, E. E., 2007, "A Novel, View-Independent Method for Strain Mapping in Myocardial Elastography: Eliminating Angle and Centroid Dependence," *Physics in Medicine and Biology,* **52**, pp. 4063–4080.

[62] Vorp, D. A., 2007, "Biomechanics of Abdominal Aortic Aneurysm," *Journal of Biomechanics,* **40**, pp. 1887–1902.

[63] Tilson, M. D., 1992, "Aortic-Aneurysms and Atherosclerosis," *Circulation,* **85**, pp. 378–379.

[64] Aggoun, Y., Szezepanski, I., and Bonnet, D., 2005, "Noninvasive Assessment of Arterial Stiffness and Risk of Atherosclerotic Events in Children," *Pediatric Research,* **58**, pp. 173–178.

[65] Boutouyrie, P., Tropeano, A. I., Asmar, R., Gautier, I., Benetos, A., Lacolley, P., and Laurent, S., 2002, "Aortic Stiffness is an Independent Predictor of Primary Coronary Events in Hypertensive Patients — A Longitudinal Study," *Hypertension,* **39**, pp. 10–15.

[66] Laurent, S., Boutouyrie, P., Asmar, R., Gautier, I., Laloux, B., Guize, L., Ducimetiere, P., and Benetos, A., 2001, "Aortic Stiffness is an Independent Predictor of All-Cause and Cardiovascular Mortality in Hypertensive Patients," *Hypertension,* **37**, pp. 1236–1241.

[67] Lehmann, E. D., Hopkins, K. D., Rawesh, A., Joseph, R. C., Kongola, K., Coppack, S. W., and Gosling, R. G., 1998, "Relation Between Number of Cardiovascular Risk Factors/Events and Noninvasive Doppler Ultrasound Assessments of Aortic Compliance," *Hypertension,* **32**, pp. 565–569.

[68] Sato, H., Hayashi, J., Harashima, K., Shimazu, H., and Kitamoto, K., 2005, "A Population-Based Study of Arterial Stiffness Index in Relation to Cardiovascular Risk Factors," *Journal of Atherosclerosis and Thrombosis,* **12**, pp. 175–180.

[69] Sutton-Tyrrell, K., Najjar, S. S., Boudreau, R. M., Venkitachalam, L., Kupelian, V., Simonsick, E. M., Havlik, R., Lakatta, E. G., Spurgeon, H., Kritchevsky, S., Pahor, M., Bauer, D., and Newman, A., 2005, "Elevated Aortic Pulse Wave Velocity, a Marker of Arterial Stiffness, Predicts Cardiovascular Events in Well-Functioning Older Adults," *Circulation,* **111**, pp. 3384–3390.

[70] Marque, V., Kieffer, P., Gayraud, B., Lartaud-Idjouadiene, I., Ramirez, F., and Atkinson, J., 2001, "Aortic Wall Mechanics and Composition in a Transgenic Mouse Model of Marfan Syndrome," *Atherosclerosis, Thrombosis, and Vascular Biology,* **21**, pp. 1184–1189.

[71] Nagai, Y., Fleg, J. L., Kemper, M. K., Rywik, T. M., Earley, C. J., and Metter, E. J., 1999, "Carotid Arterial Stiffness as a Surrogate for Aortic Stiffness: Relationship Between Carotid Artery Pressure-Strain Elastic Modulus and Aortic Pulse Wave Velocity," *Ultrasound in Medicine and Biology,* **25**, pp. 181–188.

[72] Pruett, J. D., Bourland, J. D., and Geddes, L. A., 1988, "Measurement of Pulse-Wave Velocity Using a Beat-Sampling Technique," *Annals of Biomedical Engineering,* **16**, pp. 341–347.

[73] Steptoe, A., Smulyan, H., and Gribbin, B., 1976, "Pulse-Wave Velocity and Blood-Pressure Change — Calibration and Applications," *Psychophysiology,* **13**, pp. 488–493.

[74] Woolam, G. L., Schnur, P. L., Vallbona, C., and Hoff, H. E., 1962, "Pulse Wave Velocity as an Early Indicator of Atherosclerosis in Diabetic Subjects," *Circulation,* **25**, pp. 533–539.

[75] Sumner, D. S., Hokanson, D. E., and Strandness, D. E., Jr., 1970, "Stress-Strain Characteristics and Collagen-Elastin Content of Abdominal Aortic Aneurysms," *Surgery, Gynecology & Obstetrics,* **130**, pp. 459–466.

[76] He, C. M., and Roach, M. R., 1994, "The Composition and Mechanical-Properties of Abdominal Aortic-Aneurysms," *Journal of Vascular Surgery,* **20**, pp. 6–13.

[77] Farrar, D. J., Bond, M. G., Riley, W. A., and Sawyer, J. K., 1991, "Anatomic Correlates of Aortic Pulse-Wave Velocity and Carotid-Artery Elasticity During Atherosclerosis Progression and Regression in Monkeys," *Circulation,* **83**, pp. 1754–1763.

[78] Farrar, D. J., Bond, M. G., Sawyer, J. K., and Green, H. D., 1984, "Pulse-Wave Velocity and Morphological-Changes Associated with Early Atherosclerosis Progression in the Aortas of Cynomolgus Monkeys," *Cardiovascular Research,* **18**, pp. 107–118.

[79] O'Rourke, M. F., Staessen, J. A., Vlachopoulos, C., Duprez, D., and Plante, G. E., 2002, "Clinical Applications of Arterial Stiffness; Definitions and Reference Values," *American Journal of Hypertension,* **15**, pp. 426–444.

[80] Cruickshank, K., Riste, L., Anderson, S. G., Wright, J. S., Dunn, G., and Gosling, R. G., 2002, "Aortic Pulse-Wave Velocity and Its Relationship to Mortality in Diabetes and Glucose Intolerance — An Integrated Index of Vascular Function?," *Circulation,* **106**, pp. 2085–2090.

[81] Blacher, J., Guerin, A. P., Pannier, B., Marchais, S. J., Safar, M. E., and London, G. M., 1999, "Impact of Aortic Stiffness on Survival in End-Stage Renal Disease," *Circulation,* **99**, pp. 2434–2439.

[82] Brands, P. J., Willigers, J. M., Ledoux, L. A. F., Reneman, R. S., and Hoeks, A. P. G., 1998, "A Noninvasive Method to Estimate Pulse Wave Velocity in Arteries Locally by Means of Ultrasound," *Ultrasound in Medicine and Biology,* **24**, pp. 1325–1335.

[83] Eriksson, A., Greiff, E., Loupas, T., Persson, M., and Pesque, P., 2002, "Arterial Pulse Wave Velocity with Tissue Doppler Imaging," *Ultrasound in Medicine and Biology,* **28**, pp. 571–580.

[84] Hartley, C. J., Taffet, G. E., Michael, L. H., Pham, T. T., and Entman, M. L., 1997, "Noninvasive Determination of Pulse-Wave Velocity in Mice," *American Journal of Physiology-Heart and Circulatory Physiology,* **42**, pp. H494–H500.

[85] Meinders, J. M., Kornet, L., Brands, P. J., and Hoeks, A. P. G., 2001, "Assessment of Local Pulse Wave Velocity in Arteries Using 2D Distension Waveforms," *Ultrasonic Imaging,* **23**, pp. 199–215.

[86] Rabben, S. I., Stergiopulos, N., Hellevik, L. R., Smiseth, O. A., Slordahl, S., Urheim, S., and Angelsen, B., 2002, "An Ultrasound-Based Method for Determining Pulse-Wave Velocity in the Carotid Artery," *European Heart Journal,* **23**, pp. 611–611.

[87] Zhang, X. M., Kinnick, R. R., Fatemi, M., and Greenleaf, J. F., 2005, "Noninvasive Method for Estimation of Complex Elastic Modulus of Arterial Vessels," *IEEE Transactions on Ultrasonics, Ferroelectrics and Frequency Control,* **52**, pp. 642–652.

[88] Xu, J. P., 2003, "Do We Need a Better Approach for Measuring Pulse-Wave Velocity?," *Ultrasound in Medicine and Biology,* **29**, pp. 1373–1373.

[89] Karamanoglu, M., 2003, "Errors in Estimating Propagation Distances in Pulse Wave Velocity," *Hypertension,* **41**, pp. E8–E8.

[90] Nichols, W. W., and O'Rourke, M. F., 2005, *McDonald's Blood Flow in Arteries*, 5th ed., Hodder Arnold, New York.

[91] Fujikura, K., Luo,J., Gamarnik, V., Pernot, M., Fukumoto, R., Tilson M. D., III, and Konofagou, E. E., 2007, "A Novel, Non-Invasive Technique for Pulse-Wave Imaging and Characterization of Clinically Significant Vascular Mechanical Properties in Vivo," *Ultrasonic Imaging,* **29**, pp. 137–154.

[92] Fujikura, K., Luo, J., Pernot, M., Fukumoto, R., Tilson, D., III, and Konofagou, E. E., 2006, "Pulse Wave Imaging in Murine Abdominal Aortas: A Feasibility Study," *Proceedings IEEE Ultrasonics Symposium,* pp. 868–871.

[93] Luo, J., Fujikura, K., Tyrie, L. S., Tilson, M. D., III, and Konofagou, E. E., 2007, "Pulse Wave Imaging of Abdominal Aortic Aneurysms: Comparison Between Control and Angiotensin II-Treated Mice in Vivo," *Proceedings IEEE Ultrasonics Symposium*, pp. 985–988.

[94] Pernot, M., Fujikura, K., Fung-Kee-Fung, S. D., and Konofagou, E. E., 2007, "ECG-Gated, Mechanical and Electromechanical Wave Imaging of Cardiovascular Tissues in Vivo," *Ultrasound in Medicine and Biology,* **33**, pp. 1075–1085.

[95] Cherin, E., Williams, R., Needles, A., Liu, G. W., White, C., Brown, A. S., Zhou, Y. Q., and Foster, F. S., 2006, "Ultrahigh Frame Rate Retrospective Ultrasound Microimaging and Blood Flow Visualization in Mice in Vivo," *Ultrasound in Medicine and Biology,* **32**, pp. 683–691.

[96] Liu, J. H., Jeng, G. S., Wu, T. K., and Li, P. C., 2006, "ECG Triggering and Gating for Ultrasonic Small Animal Imaging," *IEEE Transactions on Ultrasonics, Ferroelectrics and Frequency Control,* P. C., pp. 1590–1596.

[97] Pernot, M., and Konofagou, E. E., 2005, "Electromechanical Imaging of the Myocardium at Normal and Pathological States," *Proceedings IEEE Ultrasonics Symposium,* pp. 1091–1094.

[98] Williams, R., Needles, A., Cherin, E., Zhou, Y. Q., Henkelman, R. M., Adamson, S. L., and Foster, F. S., 2007, "Noninvasive Ultrasonic Measurement of Regional and Local Pulse-Wave Velocity in Mice," *Ultrasound in Medicine and Biology,* **33**, pp. 1368–1375.

[99] Luo, J., Fujikura, K., Homma, S., and Konofagou, E. E., 2007, "Myocardial Elastography at Both High Temporal and Spatial Resolution for the Detection of Infarcts," *Ultrasound in Medicine and Biology,* **33**, pp. 1206–1223.

[100] Mazhari, R., and McCulloch, A. D., 2000, "Integrative Models for Understanding the Structural Basis of Regional Mechanical Dysfunction in Ischemic Myocardium," *Annals of Biomedical Engineering,* **28**, pp. 979–990.

[101] Mazhari, R., Omens, J. H., Covell, J. W., and McCulloch, A. D., 2000, "Structural Basis of Regional Dysfunction in Acutely Ischemic Myocardium," *Cardiovascular Reseacrh,* **47**, pp. 284–293.

[102] Walker, W. F., and Trahey, G. E., 1995, "A Fundamental Limit on Delay Estimation Using Partially Correlated Speckle Signals," *IEEE Transactions on Ultrasonics, Ferroelectrics and Frequency Control,* **42**, pp. 301–308.

[103] Kallel, F., and Ophir, J., 1997, "Three-Dimensional Tissue Motion and Its Effect on Image Noise in Elastography," *IEEE Transactions on Ultrasonics, Ferroelectrics and Frequency Control,* **44**, pp. 1286–1296.

[104] Lai, W. M., Rubin, D., and Krempl, E., 1996, *Introduction to Continuum Mechanics*, 3rd ed., Pergamon Press, New York, NY.

[105] Kallel, F., and Ophir, J., 1997, "A Least-Squares Strain Estimator for Elastography," *Ultrasonic Imaging,* **19**, pp. 195–208.

[106] Stuber, M., Fischer, S. E., Scheidegger, M. B., and Boesiger, P., 1999, "Toward High-Resolution Myocardial Tagging," *Magnetic Resonance in Medicine,* **41**, pp. 639–643.

[107] Lee, W.-N., Ingrassia, C. M., Fung-Kee-Fung, S. D., Costa, K. D., Holmes, J. W., and Konofagou, E. E., 2007, "Theoretical Quality Assessment of Myocardial Elastography with in Vivo Validation," *IEEE Transactions on Ultrasonics, Ferroelectrics and Frequency Control,* **54**, pp. 2233–2245.

[108] Qian, Z., Metaxas, D. N., and Axel, L., 2006, "Boosting and Nonparametric Based Tracking of Tagged MRI Cardiac Boundaries," *Proceedings of MICCAI, LNCS,* **4190**, pp. 636–644.

[109] Wang, S., Lee, W.-N., Luo, J., and Konofagou, E. E., 2007, "An ECG-Gated Clinical Ultrasound System for High Frame-Rate Cardiovascular Imaging," *Proceedings IEEE Ultrasonics Symposium*, pp. 880–883.

[110] Lee, W.-N., Qian, Z., Tosti, C. L., Brown, T. R., Metaxas, D. N., and Konofagou E. E., "Preliminary Validation of Angle-Independent Myocardial elastography Using MR tagging in a Clinical Setting," Ultrasound in Medicine and Biology, 2008 (in press).

Chapter 7

HARMONIC MOTION IMAGING IN HIFU MONITORING

E. Konofagou and C. Maleke
Department of Biomedical Engineering, Columbia University, New York, NY, USA

ABSTRACT

Breast cancer is the most frequently diagnosed cancer (excluding skin cancer) and the no. 2 cancer killer in U.S. women. Due to recent advances in medical imaging, efficient screening and early detection of breast cancer have resulted to lower morbidity from the disease. Because of the successful detection of breast cancer at an early stage, treatment techniques have also improved. The premise of ablation techniques is that, if a tumor and its normal-tissue margin can be destroyed in situ, instead of being removed, the impact on the disease should be equivalent. In addition, if the mortality associated with operative intervention can be avoided, then the outcome using localized treatments may be more advantageous. Ablation techniques are, therefore, slowly emerging as less invasive, but equally effective, in the treatment of early-stage breast cancer, with high-intensity focused ultrasound (HIFU) being the only truly non-invasive, extracorporeal technique. In this chapter, the radiation force-based method of harmonic motion imaging (HMI) is described and proposed for monitoring of HIFU, an 1) entirely non-invasive (non-contact), 2) simple to implement, 3) real-time, 4) precise (estimating displacements of 1 μm to 10 μm), 5) fully integratable, and 6) low-cost technique for localized detection and in situ thermal treatment planning and monitoring of early-stage breast cancer. Most importantly, HMI's higher impact may lie in the fact that it can easily be applied in deeper-seated tumors with fewer alternative treatment options, such as in abdominal, e.g., liver and pancreatic cancers, preliminary feasibility of which will also be determined in vivo.

7.1 BREAST CANCER DETECTION

Since the dawn of medical practice, physicians have used palpation to detect the presence of abnormalities that could indicate pathologies, mainly because the mechanical properties of diseased tissue are typically different from those of the normal tissue surrounding them. For example, breast cancers have long been known to be harder than benign nodules on palpation [1]. Palpation, however, is limited to the detection of tumors that are close to the skin surface. In addition, other properties have been associated with diseased tissue, such as water content, acoustic tissue scattering and tissue density, giving rise to the field of medical imaging that allows imaging diagnosis well beyond the limits of palpation.

7.1.1 Clinical Detection and Diagnosis of Breast Cancer

In the clinic, mammography is the most reliable imaging technique with the highest sensitivity for tumor detection, since these tumors absorb X-ray radiation differently from normal tissue or through detection of microcalcifications, the latter being typically indicative of malignancies. In sonography, abnormalities, such as cancerous lesions, can be detected due to their hypoechoic appearance on the B-scan and/or associated acoustic "shadowing" or enhanced power Doppler responses. Magnetic resonance imaging (MRI) provides multiplanar imaging and excellent soft tissue contrast by distinguishing structures based on their water content. However, the aforementioned medical imaging techniques display tissue properties that are generally unrelated to the distinct hardness of pathological tissue, as detected by palpation. In order to take advantage of the distinct mechanical attributes of cancerous tissues compared to normal tissues, over the past 20 years, researchers have developed elasticity imaging techniques, mainly in the ultrasound and MRI imaging fields, with various forms of tissue perturbation for the successful detection of stiffer masses.

7.1.2 Elasticity Imaging Techniques for Breast Cancer Detection

Mechanical properties of tumor tissues are known to differ from the surrounding tissues as indicated by the use of palpation as a diagnostic tool. Infiltrating ductal carcinomas (accounting for ~75% of malignant breast tumors) have average Young's moduli of 558±180 kPa compared to 48±15 kPa and 20±8 kPa for normal glandular and fat tissue, respectively, in the breast (see Fig. 7-1) [2]. As a result, several imaging methods have been developed to estimate tissue stiffness, and thus detect cancer.

In the field of ultrasound, Krouskop et al [3] estimated vibration motion amplitudes and related them to estimates of elastic modulus by measuring the Doppler shifts of ultrasonic tone bursts. Parker et al [4] measured the tissue response to mechanical vibrations for "sonoelasticity imaging" (or, sonoelastography) and have reported results on the applications in in vivo human skeletal muscle [5] and ex vivo and in vivo prostate [6] among others Ophir et al [7] developed the method of elastography that applies a small external static compression (on the order of 1%) and uses cross-correlation techniques on radio frequency

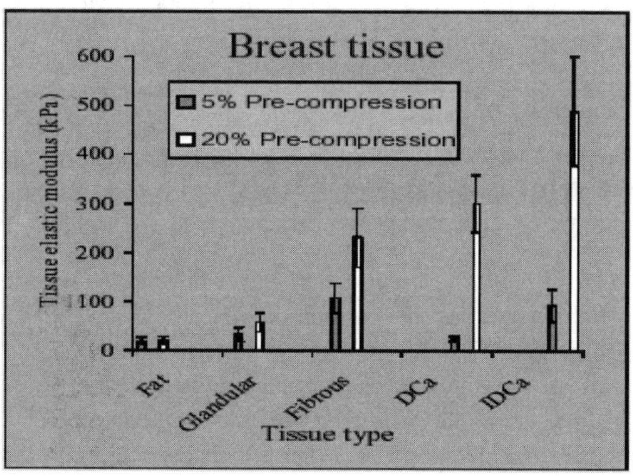

Figure 7-1. Elastic moduli (or stiffnesses) of normal and cancerous breast tissues; DCa: ductal carcinoma, IDCa: invasive ductal carcinoma [2] [Image courtesy of Thomas Krouskop, Baylor College of Medicine, Houston, Texas]

(RF) ultrasonic signals in order to estimate the displacement and strain in tissues resulting from the compression. This method has been proven to produce good quality strain images (or, elastograms) in muscle in vivo and in vivo breast cancer [8-11] , ex vivo normal prostate [11], ex vivo liver, and ex vivo kidney [12-14]. O'Donnell et al [15] demonstrated the use of phase shift techniques on baseband signals for reliable displacement and strain estimation during compression. In the field of MRI, quasi-static elastographic techniques have also been demonstrated feasible for detection of breast cancer ex vivo [16]. Several clinical trials have been initiated to bring ultrasound elastography to the clinic and the method is currently being tested worldwide regarding its potential to enhance breast cancer detection and diagnosis. Recent technical [17] and clinical [18, 19] studies have shown that ultrasound elastography can be successfully used in conjunction with standard B-mode ultrasonography for differentiation of benign from malignant tumors with 100% sensitivity and 99% specificity in 100 cases [20]. Despite the preliminary nature of these results, the premise of elasticity imaging lies in its potential of facilitating a significant reduction of unnecessary biopsies currently required in breast cancer.

The difficulties with the external deformation methods are due to the fact that they rely on knowledge of the boundary conditions for modulus reconstructions while the induced compression, or shear wave, may not sufficiently propagate to deeper-seated tissues, such as the liver. On the other hand, the internal methods can produce a concentrated force in a targeted region, at larger depths in the tissue, and can be used for probing and analyzing tissue properties within a millimeter-sized region, i.e., point by point, remotely within the focal region [21]. In 1990, Sugimoto et al [22] were the first to use a focused ultrasound transducer and produce an impulse radiation force that caused a localized static compression deep inside tissue specimens such as the human liver, kidney, and heart muscle. Their objective was to evaluate the localized tissue stiffness. The tissue displacement was measured as a function of time by using "pulse-echo-based" methods, and it exhibited an exponential decay over time after the applied force was removed. This measurement technique was the first to introduce and implement the idea of employing a focused ultrasound transducer to generate a force, namely the radiation force, deep inside the tissue. Several years later, various research groups proposed the use of impulse radiation force to induce brief mechanical excitations locally and image the resulting tissue response while RF data were collected during tissue relaxation (acoustic radiation force imaging [ARFI]) [23-25] or shear wave propagation, i.e., supersonic shear imaging [26, 27] and shear wave elasticity imaging (SWEI) [28]. In the field of MRI, magnetic resonance elastography (MRE) [29] utilizes shear excitation and shear wave generation in order to provide a direct estimate of the shear modulus through its direct dependence on the speed of the induced shear wave.

The first proposed technique for monitoring the response of tissues *during*, and at the site of, stimulus application at the site of the stimulus (as opposed to after the force is removed, such as in ARFI, or in regions away from the site of force application, such as SWEI and shear wave methods) was the ultrasound-stimulated acoustic emission (USAE) [30]. The USAE, or vibroacoustography, technique uses two confocal transducer elements of a spherically focused annular array driven at slightly different frequencies, f_1 and f_2 [30, 31]. The two beams interference at the focus produced a sinusoidal modulation of acoustic energy that resulted in an oscillatory radiation force. The resulting tissue displacement produced a localized acoustic source that emitted an acoustic signal recorded by a hydrophone during force application. Following the introduction of vibroacoustography, further studies have shown that an oscillatory acoustic radiation force can be generated in biological soft tissues at variable depths [32-35]. The remote oscillatory application of the mechanical stimulus has been of particular interest in the recent years [31] with applications on artery calcifications ex vivo [31], ex vivo porcine muscle [34, 35], ex vivo porcine liver [36, 37, 38], ex vivo breast cancer [38-40], and in vivo breast cancer [18]. More recently, HMI was introduced and shown that it can be used for both detection and ablation monitoring using a single-element, focused ultrasound transducer inducing an amplitude-modulated radiation force [36-38, 40, 41]

7.2 BREAST CANCER TREATMENT

There are several options for breast cancer treatment depending on the type, size, and metastatic state of the tumor, such as chemotherapy, radiation therapy, RF, and HIFU ablation.

7.2.1 Radiation Therapy and Chemotherapy

Radiation therapy is a clinically used breast cancer treatment method that applies high-energy X-rays to kill cancer cells. The radiation beam can be applied externally or internally (using radioactive seeds). Treatment is usually given 5 days a week over a period of about 6 weeks or 7 weeks. Each treatment lasts only a few minutes and is usually painless. Common side effects of radiation therapy are swelling and heaviness in the breast, skin burns in the treated area, and fatigue. Chemotherapy is another clinically used breast cancer treatment method that requires oral intake or intravenous injection of a drug. These drugs can kill cancer cells, but may also damage normal cells leading to several patient side-effects, such as fatigue, nausea and vomiting, loss of appetite, hair loss, mouth sores, changes in menstrual cycle, etc. Permanent side effects can include early menopause and infertility. The total treatment cycle usually lasts for 3 months to 6 months and is typically administered to patients with metastatic breast cancer.

7.2.2 Radio-Frequency (RF) Ablation

RF ablation is a rapidly emerging technology as a minimally invasive alternative to lumpectomy. For this treatment, a needlelike or starburst electrode is injected around the tumor tissue and the tip of the electrode is positioned at the center of the tumor. The tip causes frictional heat generated by intracellular ions moving in response to an alternating current. The electrode is connected to a function generator and the electrical current flows and raises the local temperature up to 95°C maintaining it for about 15 minutes. Since the first clinical feasibility study in 1999 [42], several clinical studies have followed indicating that the treated tissue shows that malignant cells can be completely destroyed by RF ablation [43-48].

7.2.3 High-Intensity Focused Ultrasound (HIFU) Ablation

The only truly noninvasive tumor ablation method is HIFU, or otherwise known as focused ultrasound surgery (FUS). Unlike RF ablation, HIFU does not require tissue incision or needle insertion, but instead induces heat extracorporeally. The general consensus from several studies used in vivo has been that extracorporeal HIFU is both safe and acceptable to patients [49]. The HIFU method utilizes acoustic waves that propagate through the tissue in the same fashion as diagnostic ultrasound. However, unlike diagnostic ultrasound, the acoustic waves are highly focused at a millimeter-sized region, i.e., the geometric focus of the ultrasound transducer. This focus can lie at several centimeters within the body. The focal spot diameter is typically 1 mm to 5 mm, and the highest acoustic energy can be deposited only at the focus for the localized ablative treatment [50]. There is no or very low heating along the beam path. High acoustic energy at the localized focus can cause temperature elevation that is sufficient to initiate coagulation necrosis in tissue (thermal lesions), while the surrounding tissues remain intact. The limitations of HIFU lie in efficiently monitoring temperature changes during treatment, and the current lack of ability to optimally stop the treatment upon lesion formation. The ability of HIFU to cause irreversible cell damage in tissues has been shown to treat early-stage breast cancer [51-55]. HIFU typically induces thermal and mechanical (or, cavitation) effects on cells in order to annihilate their function. Previous results have shown that HIFU induces coagulative necrosis of breast cancer cells [56] but that light microscopy may not be sufficient to demonstrate the associated changes [57]. Wu et al [52, 53] recently demonstrated this finding in 48 women with breast cancer (25 undergoing mastectomy only and 23 undergoing HIFU treatment followed by mastectomy), the molecular characteristics of the tumors, such as tumor expression of proliferating cell nuclear antigen (PCNA), were significantly altered and that a dramatic decrease of telomerase activity was observed in the HIFU-treated group.

7.2.4 Image Guidance of HIFU or Focused Ultrasound Surgery

In ultrasound-guided methods, the HIFU lesion can only be detected when cavitation or boiling occurs due to the high bubble concentration that causes high echo amplitudes [49]. However, bubble occurrence is usually unpredictable, and therefore unreliable for HIFU treatment monitoring. Ultrasound elastography has been proposed as an alternative method for monitoring the ablation of tumors due to its sensitivity to temperature-induced stiffness changes, as has been shown in bovine liver in vitro [58], but its required contact with, and external compression of, the breast may complicate the treatment planning and guidance in vivo, as Souchon et al [59] have shown in prostate applications. In the MRI-guided methods, T_1-weighted images are consecutively acquired to monitor the temperature rise during HIFU by measuring the spectral shifts occurring as a result of temperature change. This technique is known as MR-guided FUS, or MRgFUS [51, 54, 60]. In one of its initial clinical applications, nine patients (11 fibroadenomas, i.e., benign tumors) were treated. The treatment time was equal to 2 hours for 2-cm and 45 minutes for 1-cm lesion sizes with an average temperature rise of up to 45.9°C. Common side effects, such as skin burn or swelling following the treatment, were not observed 6 months after treatment [60] and there was no sign of tumor re-growth in the same treated region 3 years to 4 years following treatment [60]. Gianfelice et al [61] applied the MRI monitoring method and demonstrated that 79% of 24 patients had negative biopsy results after one to two HIFU sessions. More recently, Khiat et al [55] used a dynamic contrast-enhanced MRI (DCE-MRI) method, which could monitor the MR signal intensity and enhancement of residual tumors in 25 patients (26 invasive ductal carcinomas) after HIFU and tumor resection. After correlation with histopathology, it was concluded that in most of the patients, the tumors were eliminated by more than 90%.

7.2.5 Harmonic Motion Imaging (HMI) for High-Intensity Focused Ultrasound

Despite successful monitoring using the MRI-based imaging techniques in the clinic, the current lack of low-cost and reliable monitoring methods may result in the confinement of this promising treatment technique to large research centers worldwide with the exception of certain hospitals in the United States [49]. In other words, HIFU is in itself a low-cost treatment technique that requires an extremely costly monitoring device: an MRI system. As mentioned in Section 7.1.2, our research group has recently shown that the same therapeutic beam used in HIFU can be used to probe the tissue before, during, and after ablation. In other words, the same HIFU device can be used to detect [39], monitor [33, 36, 37, 62], and ablate [40] the tumor through the use of the radiation force that results during treatment. The main advantage is that the technique is stiffness-sensitive and in perfect alignment with the treatment reference frame. HMI is a radiation force-based technique that induces oscillatory displacements in the focal zone of the HIFU transducer for the detection of localized stiffness changes. HMI is thus an elasticity imaging technique, but, unlike elastography, it does not require any direct contact, external vibration, or compression of the tissues and can follow the stiffness-related changes where they are most essential, i.e., at the focus. The combination of detection and treatment into one device has been shown successful in monitoring lesion formation in real time in porcine liver ex vivo and detection of tumors ex vivo and in vivo. Due to its use of standard pulse-echo ultrasound methods, B-mode ultrasound images together with the HMI images can be used for both detection and monitoring. The integrated HMI technique may thus constitute the low-cost, real-time, fully integratable, and non-invasive image guidance technique for successful tumor ablation. In the following section, the significance of the development of such a technique is underlined. Note that tumor detection alone is *not* the sole purpose of the study or the technique. Instead, our studies aim at exhausting the potential of the technique for both detection and treatment monitoring in conjunction with more standard diagnostic and therapeutic ultrasound methods in order to more efficiently translate the latter in the clinic as well as expand its range of applications.

7.3 CLINICAL SIGNIFICANCE

Breast cancer is the most commonly diagnosed cancer in women worldwide with approximately 1 million new cases annually [63]. Due to recent advances in medical imaging, efficient screening and early detection of breast cancer have resulted to lower morbidity from the disease [52, 53]. Because of the successful detection of breast cancer at an early stage, treatment techniques have also improved. Mastectomy used to be considered as the gold standard for long-term survival rates of patients with breast cancer. However, toward the end of the last decade, it had been established that lumpectomy and breast conservation were as efficient in ensuring long-term survival rates in patients with early-stage disease [53]. The 1990s also witnessed an explosion of emergent technologies for a minimally invasive alternative to open surgery for localized malignancy [49]. Typically, open surgery has been linked to significant morbidity and mortality, occasionally causing immune response suppression and potential increase in the risk of perioperative metastatic tumor dissemination [49]. The idea behind the integrated HMI technique for HIFU is that, if a tumor, benign or malignant, and its normal-tissue margin can be destroyed in situ, instead of being surgically removed, the impact on the disease should be equivalent. In addition, if the mortality associated with operative intervention can be avoided, then the outcome using localized treatments is more advantageous. A recent example of such a successful minimally invasive technique is interstitial laser ablation for isolated colorectal metastases in the liver [64].

Ablation techniques are also slowly emerging as less invasive but equally effective in the treatment of early-stage breast cancer, with HIFU being the only truly non invasive, extracorporeal technique [53, 60, 61]. Despite its rapid development, HIFU has been shown successful in the treatment of breast cancer over the past 5 years and, therefore, the long-term efficacy of these techniques still remains to be established. However, the overall tendency of breast cancer treatment for reduced morbidity and rapid post-treatment recovery of the patients together with the HIFU devices becoming more and more readily available, the role of HIFU in oncology is expected to be strengthened, especially given the aforementioned preliminary but positive findings. Reliable functional and anatomical information during HIFU treatment can currently only be provided using MRI-based techniques. Despite its associated reliable monitoring, MRI can be extremely costly and restrictive in the patients treated (due to, e.g., pregnancy, claustrophobia, pacemaker, or

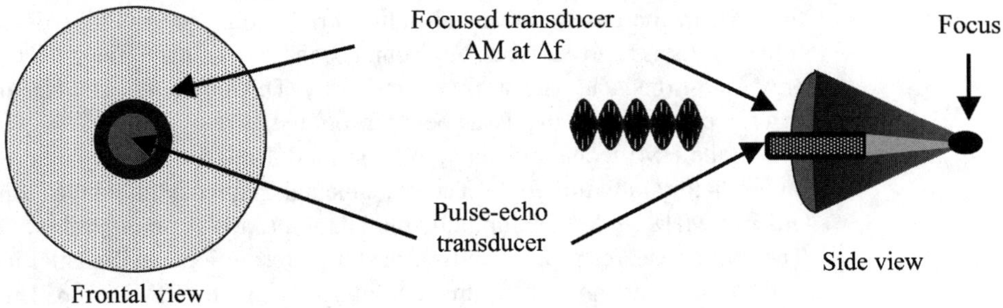

Figure 7-2. The focused transducer can generate both ablation and induce a localized vibration at the focus, as shown. The imaging (pulse-echo) transducer is then used to acquire the RF frames during the localized vibration of the tissue. The focus is on the order of 1 mm^3–3 mm^3

obesity issues), thus further limiting the application of HIFU, an inherently inexpensive technique.

The use of an all-ultrasound-based technique would utilize the same beam for both HIFU and localized vibration of tissues before, during, and after HIFU treatment to monitor the latter. The localized vibration will be achieved using amplitude-modulation on the therapeutic beam, which will not affect the treatment duration or efficacy but only induce a vibration at the focus of the transducer in order to detect the mechanical property changes of the treated tissue. The mechanical property of the treated tissue will change before, during, and after ablation [62, 65], and thus the vibration amplitude will change during treatment. The vibration amplitude can be monitored in real-time using cross-correlation techniques on the acquired pulse-echo (RF) signals using an ultrasound imaging transducer, confocal with the HIFU transducer. Additional advantages of the technique will be its capability of utilizing B-mode ultrasound for mapping the anatomical information of the breast and the same frame of reference between the therapy and imaging transducers given the fact that the transducers can be confocal and concentric (see Fig. 7-2). Due to the fact that an all-ultrasound system will be used, the real-time capability will be achievable, thus allowing real-time feedback for intraoperative improvement of the treatment plan. In summary, the HMI method combines the advantages of diagnostic ultrasound for high spatial and temporal resolution with the advantages of therapeutic ultrasound for non-invasive, low-cost treatment, and the advantages of elasticity imaging for sensitive, stiffness-based measurements during ablation. The combination results in a technique that is real-time, low-cost, and non-invasive, thus expanding the potential of the HIFU technique for treatment of early-stage breast cancer and simultaneously setting the foundations for future treatment of early-stage tumors in other organs such as liver and pancreas. An additional advantage of the proposed technique is the lack of prolonged registration and treatment planning procedures when separate monitoring and treatment systems are used. This may result in a more efficient treatment with the associated comfort improvement for the patient.

7.4 TUMOR DETECTION AND TREATMENT MONITORING

7.4.1 Harmonic Motion Imaging Technique

By using and focusing ultrasound waves in a specific region (see Fig. 7-2), it can induce and detect displacements of a few microns at variable depths within the tissue. More specifically, the HMI method uses ultrasound-induced radiation force to probe tissue properties. This force results as follows. As an ultrasound beam propagates through the tissue, part of its energy is absorbed and part of it scattered away. This momentum change of the beam results in a force that acts on the tissue being insonified.

In the HMI technique, an amplitude-modulated ultrasound beam is used to generate an oscillatory radiation force. The fundamental frequency of the ultrasound beam is on the order of MHz while the amplitude modulation (AM) is on the order of Hz (see Fig. 7-2). The former ensures probing of tissues at high resolution and the latter induces oscillations in the tissue that can be measured and imaged using the HMI setup. This oscillation is the largest at the focus of the transducer used, namely, the *focused* transducer (see Fig. 7-2). The diameter of the focus of the transducer in the MHz range is on the order of 1 mm to 3 mm. An object at the focal zone experiences an average energy density $<E>$ that fluctuates at the oscillation frequency equal to the AM frequency. An imaging, i.e., pulse-echo, imaging transducer (see Fig. 7-2) is then used to acquire the echo RF signals from the tissue being probed. The focused spectrum is then filtered from the imaging spectrum through standard bandpass filtering techniques, since the two ultrasound frequencies used are sufficiently distinct (e.g, 4.5 MHz and 7.5 MHz, respectively). Cross-correlation techniques are then used to estimate the resulting displacement during application of the oscillation [34, 37]. Finally, this can be repeated over a two-dimensional plane or a three-dimensional volume of the entire specimen or breast. Both temporal (see Fig. 7-3) measurements as well as two-dimensional (see Fig. 7-4) and three-dimensional (see Fig. 7-7) images of the localized mechanical responses, and thus localized mechanical properties, can then be generated. The HMI technique thus uses an all-ultrasound-based system for localized and remote vibration of the breast tissue as well as estimation of its mechanical properties *during* application of the radiation force. To our knowledge, this is the only technique that can provide highly lo-

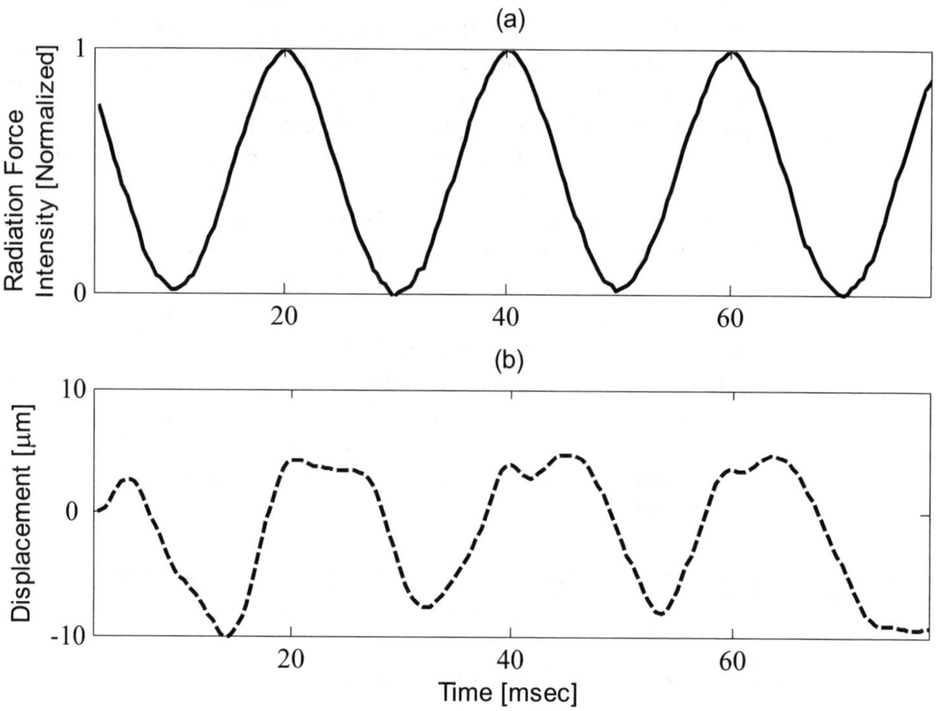

Figure 7-3. (a) Induced radiation force (normalized), (b) resulting displacement in a gelatin phantom

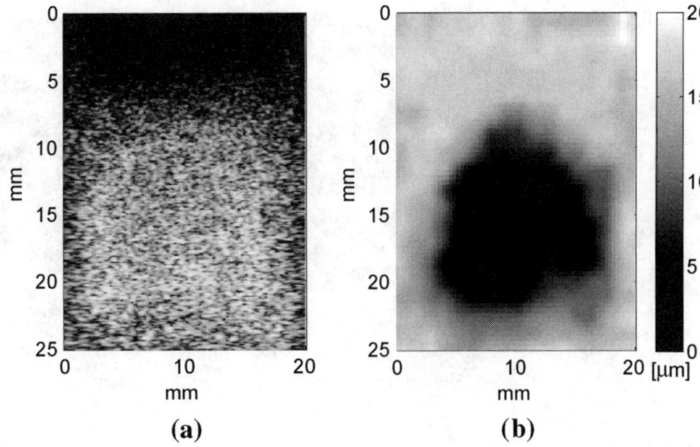

Figure 7-4. HMI imaging of a tissue-mimicking gel phantom with a cylindrical inclusion (40 kPa), so twice harder than the background (20 kPa). (a) B-mode and (b) two-dimensional HMI images. Note that deep-seated "tumor" can be detected using HMI even when they are not easily detectable on the B-mode

calized, mechanical measurements mechanical response *during* the application of the force, thus closely approximating the method of palpation but further allowing its application at larger depths. This has been verified using theoretical models [35], phantoms [35, 36] (see Fig. 7-4), and ex vivo liver and muscle tissues, before and after HIFU treatment [36, 37, 41]. In this study, we aim at developing and tailoring this technique for breast cancer detection and characterization.

7.4.2 Theoretical Framework for HMI Performance Assessment With Validation

The performance of HMI in estimating displacements in tissues was determined and validated in experimental phantoms [36]. In the phantom experiment, an oscillatory force was generated by a single-element focused ultrasound (FUS) transducer. Since the focus was highly localized and was harmonically varying, the motion characteristics can be directly related to the regional tissue modulus is similar to more established mechanical testing methods. This resulting motion was imaged simultaneously using a diagnostic (pulse-echo) transducer and a 1D cross-correlation technique on the acquired RF signals to estimate the resulting axial displacement. The FEM and experimental results of the phantom study with the same stiffness variations were compared and used to describe the tissue behavior during the application of the stimulus. These findings can provide the sensitivity and limitations of the HMI technique and could therefore be used to improve the design criteria in different HMI applications. Results showed that using simulated and experimental phantoms of similar moduli led to excellent agreement between the simulation model and the experiments (see Fig. 7-5).

7.4.3 HMI Technique in Breast Cancer Detection

Post-surgical breast cancer specimens were imaged using the HMI technique and the setup shown in Fig. 7-6. The diameter of a focused transducer (Riverside Research Institute, New York, NY, USA) and the radius of curvature are 84 mm and 90 mm, respectively. The oscillatory acoustic radiation force was generated by the 4.5-MHz focused transducer and

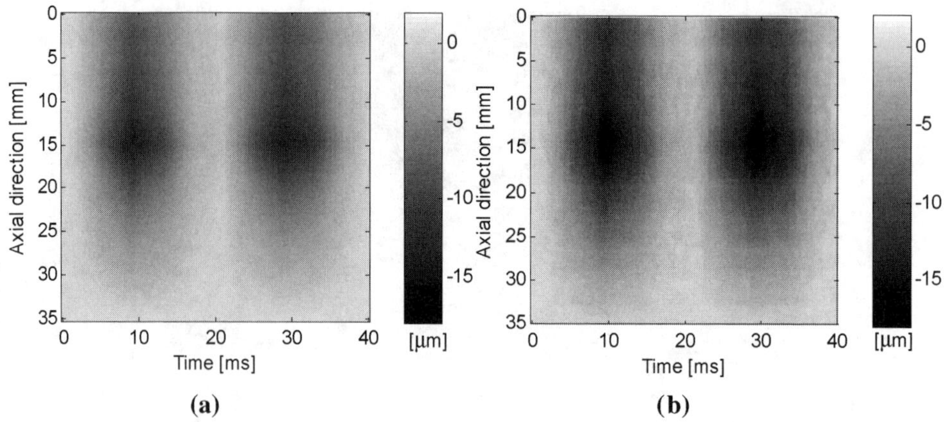

Figure 7-5. (a) Theoretical and (b) experimental M-mode HMI images over two full cycles at 25 Hz in phantoms at 20 kPa

a 50-Hz amplitude-modulated RF signal. A pulse-echo transducer with a center frequency of 7.5 MHz, a diameter of 12.5 mm, and a focal length of 60 mm (Panametrics, Waltham, MA, USA), was placed through the center of the focused transducer, so that the beams of the two transducers were properly aligned. Consecutive, filtered RF signals were acquired at a Pulse Repetition Frequency (PRF) of 6.5 kHz (Panametrics 5051PR, Panametrics). A degassed water chamber is attached to the focused/pulse-echo transducer assembly to ensure coupling without requiring water circulation, as implemented by others. There is a 2-cm distance between the chamber and the tissue specimen (immersed in degassed phos-

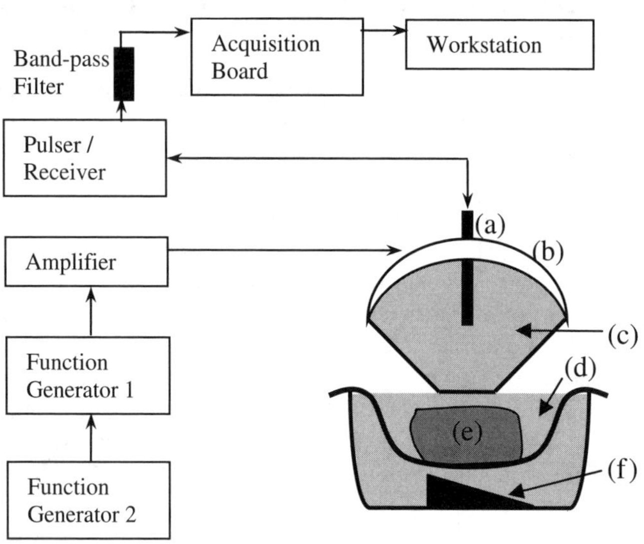

Figure 7-6. Block diagram of the experimental setup. (a) Pulse-echo transducer, (b) therapy transducer, and (c) degassed water. The specimen (or phantom gels) (e) rests on a membrane and (d) is submerged in degassed PBS (or, DI-water for the gels). In order to reduce possible reflections, i.e., specular reflection, an (f) absorber is placed on the bottom of the glass beaker

phate buffered saline [PBS]), rendering the method fully non-contact and thus without disadvantages such as trapped bubbles or tissue contact during raster scanning (see Fig. 7-7). A band-pass analog filter (Reactel, Inc., Gaithersburg, MD, USA) with cut-off frequencies of f_{c1} = 5.84 MHz and f_{c2} = 8.66 MHz was used to filter out the spectrum of the focused higher power beam. An acquisition board (CS14200, Gage Applied Technologies, Lachine, Canada) was used to capture filtered RF data with sampling frequency of 80 MHz at 14-bit digitization rate.

For the raster scanning process, the transducer was moved along a two-dimensional grid using a computer-controlled positioner (Velmex Inc., Bloomfield, NY, USA) at a step size of 1 mm. The time shift occurring between two successive RF images was calculated using a speckle-tracking technique, i.e., one-dimensional cross correlation was performed along the ultrasound beam axis with a window size of 1.3 mm and 85% overlap [35]. This method is simple to implement, computationally efficient, and provides an accurate estimation of small displacements (on the order of 10 μm).

Results from two breast specimens are used here as examples. Both masses were palpable. The first mass was a fibroadenoma (i.e., benign) that could be detected both on the mammogram and the sonogram. Its approximate dimension on the sonogram was 1.8 mm. The patient underwent lumpectomy and the specimen removed was of an approximate 3 mm to 5 mm in size. It was received for measurement 20 to 30 minutes following excision and before being sent to pathology. The entire HMI procedure (from setting up the sample to completing the scan) lasted approximately 30 minutes. The three-dimensional HMI image (obtained from combining multiple image planes from different depths in one image) is shown in Fig. 7-7a, which indicates the high uniformity of the benign tumor (darker, middle region) surrounded by normal tissue (lighter region). The second specimen was an atypical carcinoma (i.e., malignant) that could not be detected on the mammogram or ultrasound. Both the radiologist and the surgeon could palpate the mass and guessed an approximate size of 3 mm to 5 mm. The patient underwent lumpectomy and the specimen was again imaged following the same procedure as for the benign case. The three-dimensional HMI image in the malignant specimen is shown in Fig. 7-7b. This image spans over a larger volume as the specimen provided was much larger than in the benign case. The HMI result shows multiple regions of lower displacement (in black relative to the glandular (in gray) and fat (in light gray) tissue, indicating

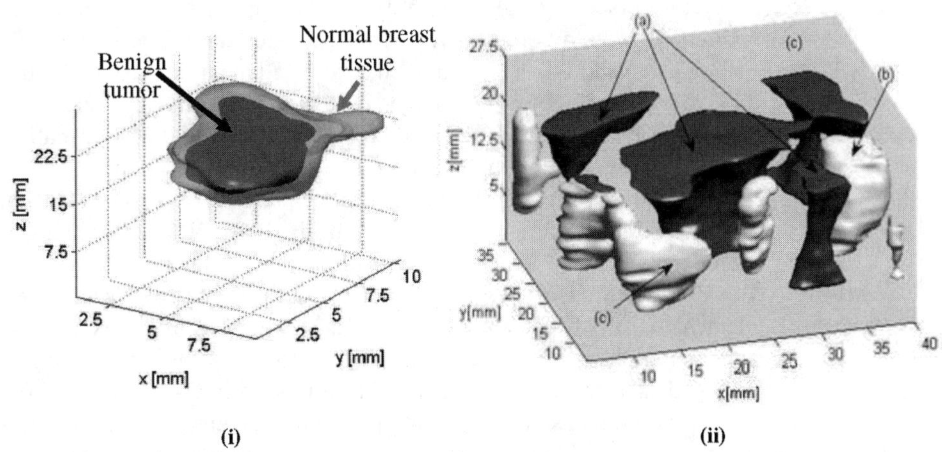

Figure 7-7. Ex vivo three-dimensional HMI imaging: (i) benign and (ii) malignant with (a) carcinoma (black) and (b) fat (light gray) are surrounded by (c) glandular (gray) tissues

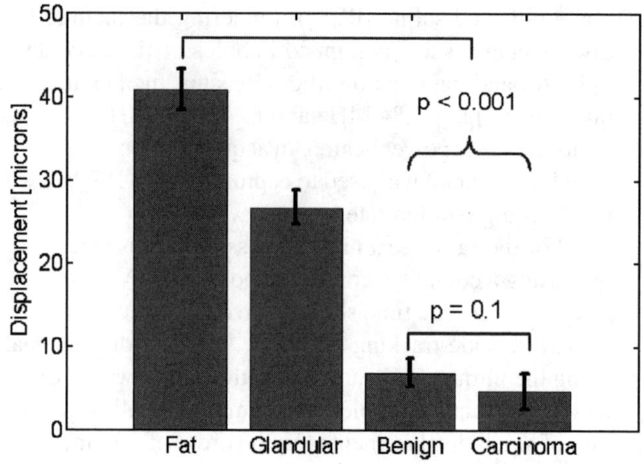

Figure 7-8. Preliminary ex vivo breast specimen results using the HMI technique

several regions of lower stiffness. All these regions combined span over a large volume of approximately 3 × 3 × 5 cm^3. This result was consistent with the palpation finding of a large palpable mass. Most importantly, the mass is visible on the HMI image (see Fig. 7-7ii) but not on the mammogram or ultrasound image obtained prior to surgery (images not shown). Finally, Fig. 7-8 summarizes the findings in these two cases. The fat and normal (glandular) breast tissue could be differentiated from the benign and malignant tissue at high significance ($p < 0.001$). The benign tumor experienced higher displacement (i.e., had lower hardness) than the malignant tumor but at lower significance ($p = 0.1$). These results are very preliminary but clearly show the potential of this novel technique to detect and map the extent of abnormal lesions in the breast. The tumors could not be ablated in this case so as to leave them intact before they were sent to pathology.

7.4.4 HMI Technique in the Detection of HIFU Ablation

It has been established that, during and after HIFU ablation, the stiffness can change up to a tenfold (see Fig. 7-9; [58, 65]). In order to verify the monitoring capability of the technique, ex vivo porcine liver was used, since the breast specimens had to be sent to pathology and could not be modified or affected in any way. A 60 × 60 × 30 mm^3 piece of ex vivo porcine liver was submerged in PBS solution and degassed for 30 minutes. The 20 × 50 mm^2 area was raster-scanned before and after the lesion formation. The intensity of the HIFU beam used in the raster scan process was 237 W/cm^2 with an AM frequency of 50 Hz. Raster scanning of the entire specimen in three dimensions led to three-dimensional HMI mapping of three lesions shown in Fig. 7-10a. Raster scanning lasted 30 minutes, which may not be efficient for real-time 2D mapping of the lesion size during ablation. However, methods for real-time monitoring at the HIFU focus have already been implemented. The displacement amplitudes at the focus of the ex vivo porcine liver or the ablated regions, were approximately equal to 8 μm while the normal (i.e., non-ablated) liver had an average displacement of 15 μm (see Fig. 7-10a). Fig. 7-10b shows the example of a two-dimensional slice out of the three-dimensional volume mapped in Fig. 7-10a identifying three HIFU lesions side-by-side at varying sizes. Both the locations and sizes of these lesions were verified using pathology (see Fig. 7-10b). These results indicate that the displacement amplitude decreases after lesion formation due to the associated higher tissue stiffness.

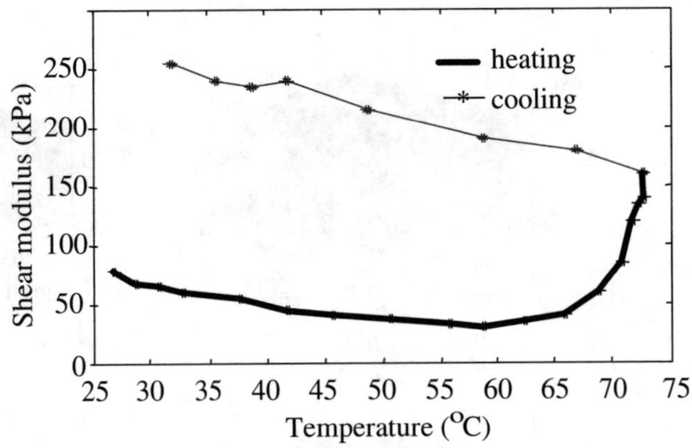

Figure 7-9. Stiffness variation as a result of temperature change during HIFU treatment as measured in porcine muscle ex vivo. (Adapted based on Wu et al [65])

7.4.5 Real-Time Monitoring of HIFU Using HMI

Monitoring of lesion formation was recently performed at approximately 1 frame/sec for M-mode HMI imaging using cross-correlation techniques built on Matlab and an M-mode display during HIFU ablation [41]. It is expected that once these techniques are built in a more efficient software platform, or built in hardware, they can be rendered in full real-time, i.e., 30 frames/sec. When used for monitoring of thermal ablation focused ultrasound (FUS), the integrated system was recently named HMIFU for 'Harmonic Motion Imaging for Focused Ultrasound' [40]. For the real-time monitoring application of the technique, the experiment was completed in ex vivo porcine liver with a sonication time varying between 10 and 47 seconds. Figure 7-11 shows the monitoring curve and image of the HMI tissue displacement amplitude at the focal depth versus temperature and sonication time. The focus is located at 16.5 mm and has displacement amplitude of 30 μm, which then decreases to about 15 μm (see Fig. 7-11a). The regions above and below the focus have approximately similar displacements (see Fig. 7-11b). This indicates that the highest acoustic radiation force is produced and maintained only at the focus and that after 20 seconds of sonication time, the properties of the liver tissue irreversibly change due to tissue coagulation (see Fig. 7-11b). Furthermore, it should be noted that the effect on the RF signals induced by the increase in the speed of sound with temperature also introduced a monotonically increasing slope in the HMI displacement in time, which could be easily removed due to the harmonic nature of the technique [41]. The HMI technique is, therefore, capable of accurately monitoring the stiffness-related heating process, possibly detecting the time of coagulation (see Fig. 7-11b) as well as separating the speed-of-sound effect from stiffness-related changes. It has also been established that the HMI technique alone does not induce a temperature rise higher than 1°C when used for before and after heating. The duration of sonication was equal to 47 seconds, which is a standard duration in HIFU, with the highest temperature reaching 62°C. The coagulation onset in this case was detected to be at 53°C, beyond which the slope of displacement variation was reversed indicating non-linear stiffening of the tissue (see Fig. 7-9). Note that the M-mode image in Fig. 7-11c clearly indicates that the displacement sharply decreases before the HIFU treatment is stopped, indicating that coagulation successfully occurred. The HMIFU cross-section of a liver vessel is also correctly indicated (near the top in both images in Fig. 7-11c). Table 7-1 shows the sensitivity of HMI in detecting small temperature variations and its specificity in detecting lesion formation. Note that during lesion formation, the HMI heating rate was switched from

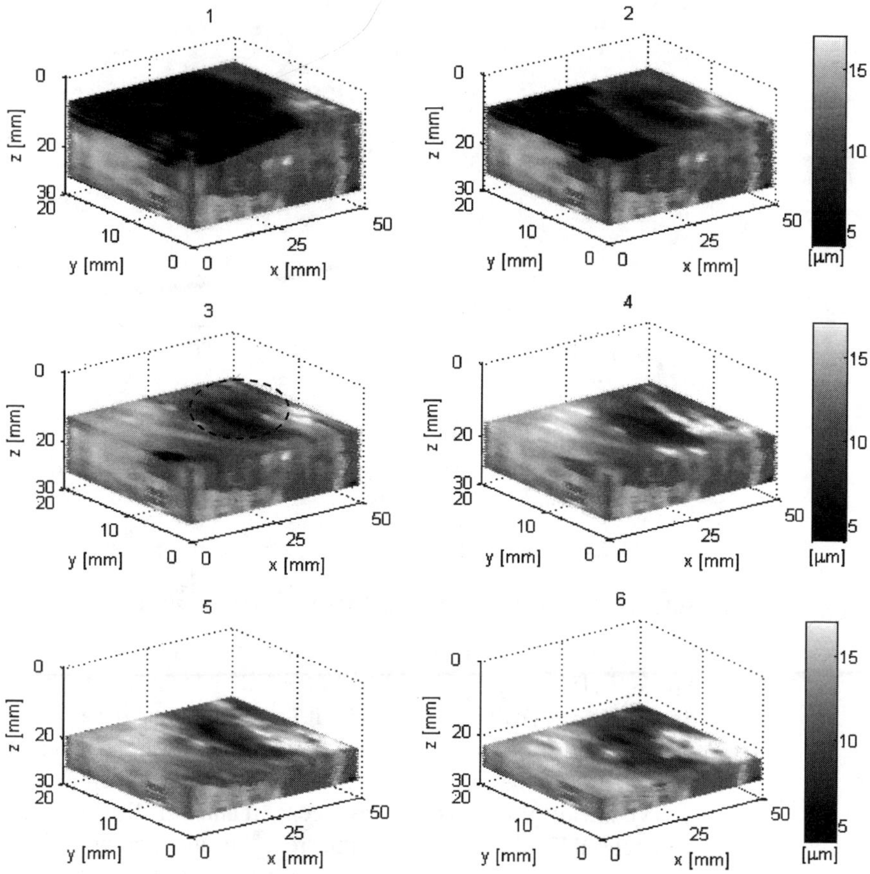

(a)

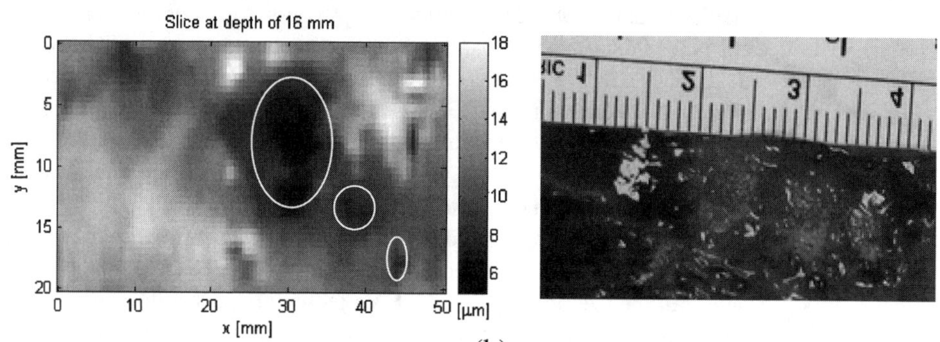

(b)

Figure 7-10. (a) Three-dimensional HMI imaging (cutting in progressively at 5 mm at each step) of the entire liver (in white, highest displacement), HIFU lesions (in black, lower displacement) and saline (null displacement), the latter used as the coupling medium; (b) HMI image through the dotted region in (a) showing three lesions side by side with the photograph after liver dissection (lower panel) of the liver showing the three lesions

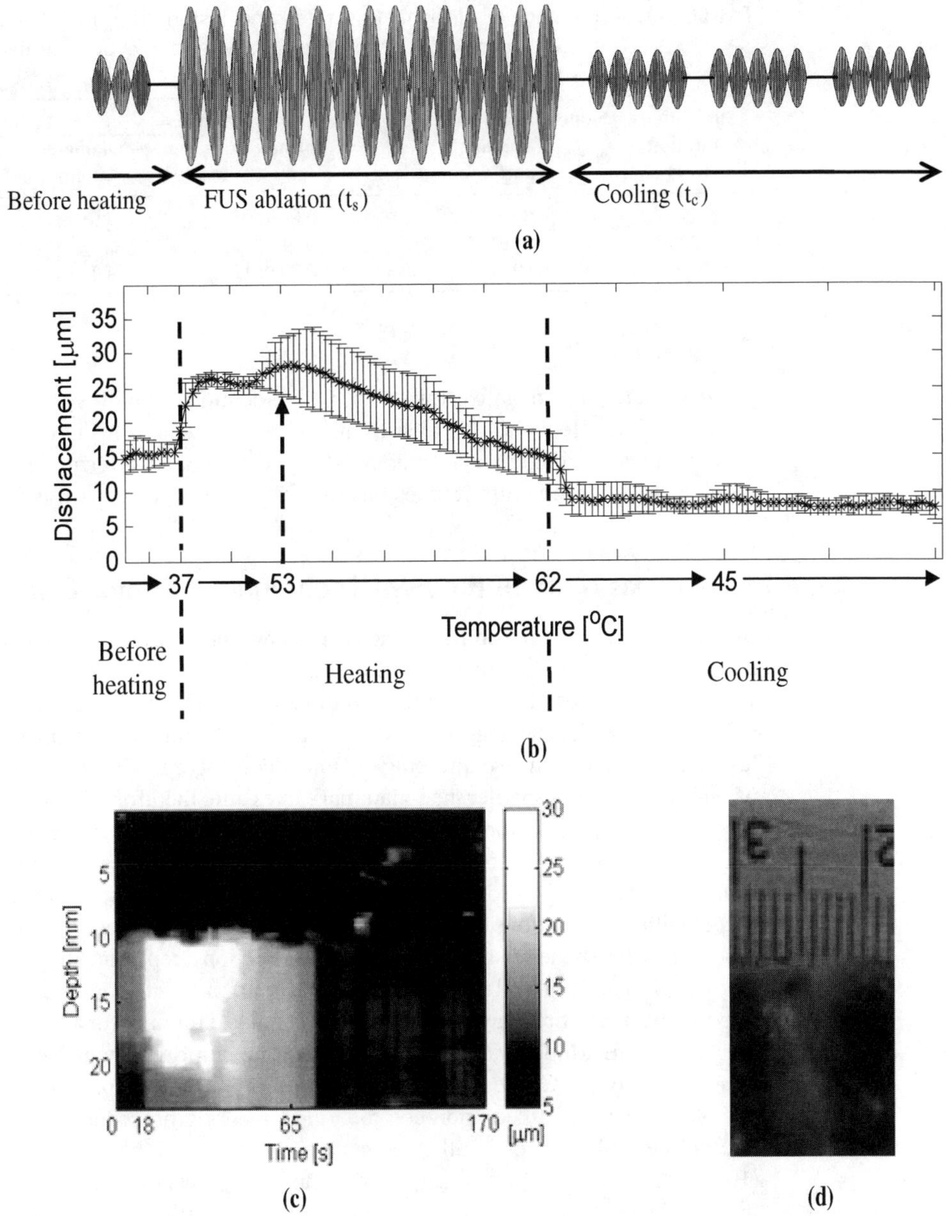

Figure 7-11. M-mode HMI, real-time monitoring: (a) a sequence of the HMIFU technique before, during, and after HIFU treatment; (b) HMI displacement variation with temperature before, during, and after HIFU ablation; (c) example of an M-mode HMI displacement image obtained in real-time during ablation (as in a) and (b, heating started at t = 18 sec and ended at t = 65 sec). The result of the ablation is shown in the photograph of the liver lesion (denoted by the dashed contour) in (d). A liver vessel running through the lesion was used as a registration reference between the images in (c) and (d). The curve in (b) was averaged over five different liver specimens and three locations in each liver, i.e., 15 locations total. The errorbars shown in (b) are equal to 1 standard deviation. Note that in (b) at 53°C, coagulation occurs and the HMI displacement changes from a positive to a negative rate indicating lesion formation. This was found consistent in all 21 lesions (see Table 7-1)

TABLE 7-1. Summary of HMI heating rates and lesion sizes for 10-second, 20-second, and 30-second HIFU exposures in seven ($n = 7$) porcine liver specimens, i.e., 21 sonication locations total

Sonication Duration (sec)	Slope in the Beginning of Heating (μm/°C)	Slope During Lesion Formation (μm/°C)	Lesion Size	
			Diameter (mm)	Length (mm)
10	0.87±0.08	–0.88±0.08	2.0±0.12	6.4±1.14
20	0.85±0.12	–0.71±0.16	3.4±0.13	13.1±1.15
30	0.80±0.14	–0.79±0.19	4.7±0.18	16.4±0.39

positive (before) to negative (after) in all 21 locations at high statistical significance ($p < 0.001$). This effect was due to stiffening of the tissue as a result of coagulation. Lesion formation was verified in a preliminary study that involved interrupting the treatment by a 1-second difference in different lesions to verify whether a lesion was formed.

7.4.6 In Vivo Feasibility of the HMI Technique in Tumor Detection

In order to determine feasibility of the HMI technique in vivo for tumor detection, planning and ablation, the authors performed preliminary results in four mice (4 weeks old), one wildtype mouse (used as a control), and three mice that were genetically engineered to develop tumors of the pancreas. In the transgenic model, animals carrying the conditional alleles were mated with mice expressing *cre* under control of the *Pdx1* promoter. Tumors of the pancreas are usually smaller than mammary ones and, therefore, feasibility on the pancreas was deemed sufficient for demonstrating feasibility in mammary tumors. Out of the four mice, one was the control and the other three had variant degrees of tumor size. HMI images using the same setup as for the aforementioned experimental studies was used (see Fig. 7-6), but with modifications to accommodate the live mice, such as comfortable bedding (soft absorber) beneath the scanned mice to avoid reflections, replacing the water chamber with a suspended water bath, and monitoring the mice heart rate with ECG and blood pressure. More importantly, the entire imaging duration was equal to 3 minutes for a 10×10 mm^2 region (see Fig. 7-12b). B-Mode images using the high-resolution system Vevo 770 (Visualsonics, Inc, Toronto, Canada) were used to locate the tumor prior to HMI imaging and pathology was performed upon completion of imaging procedures. The tumors were visible on the HMI images in all three cancerous mice and the results in the largest tumor ($10 \times 5 \times 3$ mm^3) (see Fig. 7-12ii) were compared against the control (see Fig. 7-12i). Note the excellent agreement between B-mode, HMI, and pathology images in Fig. 7-12.

7.5 CONCLUSION

The HMI technique was shown to offer the ability of real-time monitoring of temperatures and variation in tissue mechanical properties during the application of the acoustic radiation force in a fully integrated system (Fig. 7-13). Through real-time monitoring of the HMI response, the relative tissue stiffness changes during thermal treatment can be reliably indicated, unveiling thus the onset of coagulation. The M-mode HMI images may be used as a reliable tool for online visualization of lesion formation and size during thermal ablation. The ultimate goal of this technique is for the combined tumor detection and localization as well as real-time monitoring of tumor ablation based on the associated localized tissue stiffness changes in vivo.

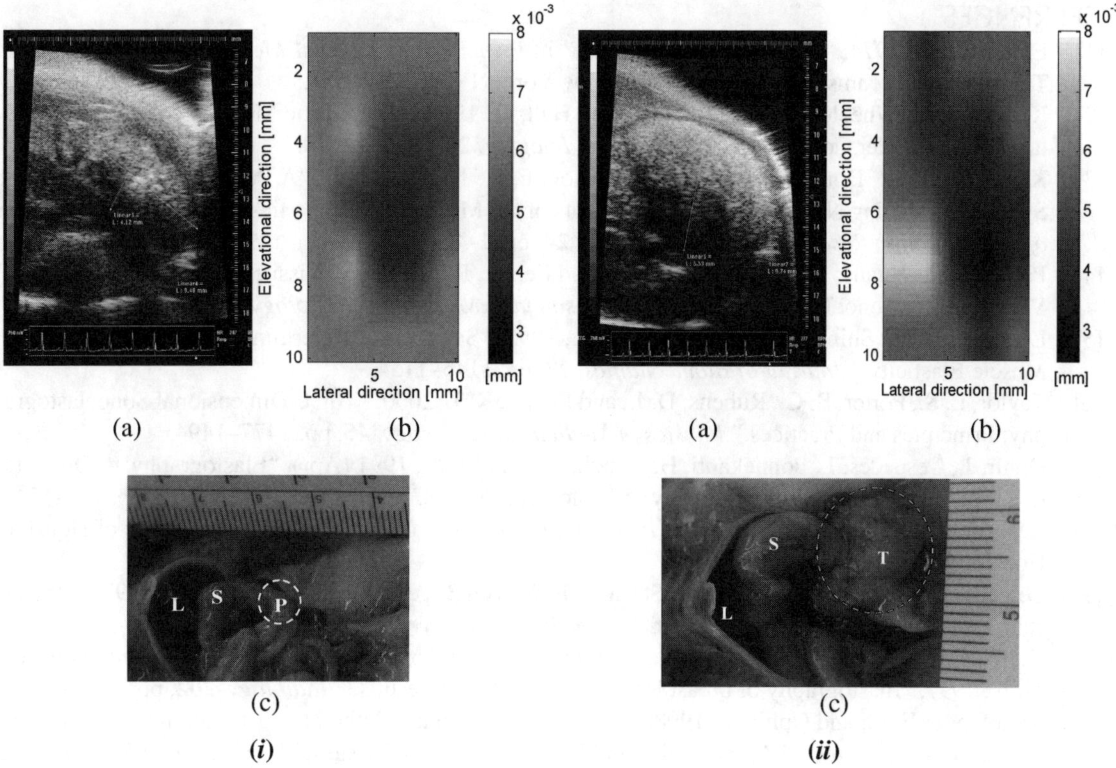

Figure 7-12. In vivo feasibility in (i) normal and (ii) tumorous murine pancreas with (a) B-mode image (with measurements and ECGs), (b) HMI image in an orthogonal plane to the B-mode, and (c) pathology of the specimen. [L: Liver, S: stomach, P: pancreas, T: tumor (in pancreas)]

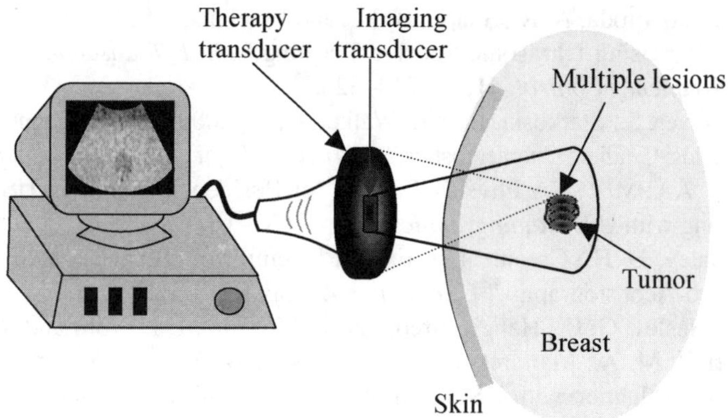

Figure 7-13. The HMIFU setup in in vivo applications in mice showing the field of view of the Ultrasonix (Sonix RP) array and an example of several lesions formed in the tumor seated under and close to the skin of the mouse

REFERENCES

[1] Emerson, K., 1974, "Diseases of the Breast," in *Principles of Internal Medicine*, M. Wintrobe, G. Thorn, and R. Adams, eds. McGraw-Hill, New York, NY, pp. 582–587.

[2] Krouskop, T., Wheeler, T. M., Kallel, F., and Hall, T., 1998, "The Elastic Moduli of Breast and Prostate Tissues Under Compression," *Ultrasonic Imaging,* **20**, pp. 151–159.

[3] Krouskop, T. A., Dougherty, D. R., and Vinson, F. S., 1987 (Spring), "A Pulsed Doppler Ultrasonic System for Making Noninvasive Measurements of the Mechanical Properties of Soft Tissue," *Journal of Rehabilitation Research and Development,* **24**, pp. 1–8.

[4] Parker, K. J., Huang, S. R., Musulin, R. A., and Lerner, R. M., 1990, "Tissue Response to Mechanical Vibrations for Sonoelasticity Imaging," *Ultrasound in Medicine and Biology,* **16**, pp. 241–246.

[5] Levinson, S. F., Shinagawa, M., and Sato, T., 1995, "Sonoelastic Determination of Human Skeletal Muscle Elasticity," *Journal of Biomechanics,* **28**, pp. 1145–1154.

[6] Taylor, L. S., Porter, B. C., Rubens, D. J., and Parker, K. J., 2000, "Three-Dimensional Sonoelastography: Principles and Practices," *Physics in Medicine and Biology,* **45**, pp. 1477–1494.

[7] Ophir, J., Cespedes, I., Ponnekanti, H., Yazdi, Y., and Li, X., 1991 (Apr), "Elastography: A Quantitative Method for Imaging the Elasticity of Biological Tissues," *Ultrasonic Imaging,* **13**, pp. 111–134.

[8] Céspedes, E. I., 1993, *Elastography: Imaging of Biological Tissue Elasticity*, University of Houston, Houston, TX.

[9] Céspedes, E. I., de Korte, C. L., van der Steen, A. F. W., von Birgelen, C., and Lancée, C. T., 1997, "Intravascular Elastography: Principle and Potentials," *Seminars in Interventional Cardiology,* **2**, pp. 55–62.

[10] Garra, B. S., Céspedes, E. I., Ophir, J., Spratt, R. S., Zuurbier, R. A., Magnant, C. M., and Pennanen, M. F., 1997, "Elastography of breast lesions: initial clinical results," *Radiology,* **202**, pp. 79–86.

[11] Konofagou, E. E., and Ophir, J., 1998, "A New Elastographic Method for Estimation and Imaging of Lateral Strains, Corrected Axial Strains and Poison's Ratios in Tissues," *Ultrasound in Medicine and Biology,* **24**, pp. 1183–1199.

[12] Kallel, F. and Ophir, J., 1998 (Oct), "Limits on the Contrast of Strain Concentrations in Elastography," *Ultrasound in Medicine and Biology,* **24**, pp. 1215–1219.

[13] Kallel, F., Ophir, J., Magee, K., and Krouskop, T., 1998, "Elastographic Imaging of Low Contrast Elastic Modulus Distributions in Tissue," *Ultrasound in Medicine and Biology.,* **24**, pp. 409–425.

[14] Ophir, J., Alam, S. K., Garra, B. S., Kallel, F., Konofagou, E. E., Krouskop, T., Merritt, C. R. B., Righetti, R., Souchon, R., Srinivasan, S., and Varghese, T., 2002, "Elastography: Imaging the Elastic Properties of Soft Tissues with Ultrasound," *Journal of Medical Ultrasound,* **29**, pp. 155–171.

[15] O'Donnell, M., Skovoroda, A. R., Shapo, B. M., and Emelianov, S. Y., 1994, "Internal Displacement and Strain Imaging Using Ultrasonic Speckle Tracking," *IEEE Transactions on Ultrasonics, Ferroelectrics, and Frequency Control,* **41**, pp. 314–325.

[16] Plewes, D. B., Silver, S., Starkoski, B., and Walker, C. L., 2000 (Apr), "Magnetic Resonance Imaging of Ultrasound Fields: Gradient Characteristics," *Journal of Magnetic Resonance Imaging,* **11**, pp. 452–457.

[17] Hall, T., 2003, "AAPM/RSNA Physics Tutorial for Residents: Topics in US: Beyond The Basics: Elasticity Imaging with US," *Radiographics,* **6**, pp. 1657–1671.

[18] Alizad, A., Whaley, D. H., Greenleaf, J. F., and Fatemi, M., 2006 (Dec), "Critical Issues in Breast Imaging by Vibro-Acoustography," *Ultrasonics,* **44**, pp. E217–E220.

[19] Regner, D. M., Hesley, G. K., Hangiandreou, N. J., Morton, M. J., Nordland, M. R., Meixner, D. D., Hall, T. J., Farrell, M. A., Mandrekar, J. N., Harmsen, W. S., and Charboneau, J. W., 2006 (Feb), "Breast Lesions: Evaluation with US Strain Imaging — Clinical Experience of Multiple Observers," *Radiology,* **238**, pp. 425–437.

[20] Barr, R., 2006, "Initial Results of Breast Real-time Elasticity Imaging to Characterize Lesions," in *RSNA Meeting*, Chicago, IL.

[21] Greenleaf, J. F., Fatemi, M., and Insana, M., 2003, "Selected Methods for Imaging Elastic Properties of Biological Tissues," *Annual Review of Biomedical Engineering,* **5**, pp. 57–78.

[22] Sugimoto, T. U., Itoh, S., K, 1990, "Tissue Hardness Measurement Using the Radiation Force of Focused Ultrasound," *Ultrasonics Symposium, 1990. Proceedings, IEEE 1990,* vol. 3, pp. 1377–1380, 4–7 Dec 1990.

[23] Nightingale, K. R., Kornguth, P. J., and Trahey, G. E., 1999 (Jan), "The Use of Acoustic Streaming in Breast Lesion Diagnosis: A Clinical Study," *Ultrasound in Medicine and Biology,* **25**, pp. 75–87.

[24] Nightingale, K. R., Palmeri, M. L., Nightingale, R. W., and Trahey, G. E., 2001 (Jul), "On the Feasibility of Remote Palpation Using Acoustic Radiation Force," *Journal of the Acoustical Society of America,* **110**, pp. 625–634.

[25] Lizzi, F. L., Muratore, R., Deng, C. X., Ketterling, J. A., Alam, S. K., Mikaelian, S., and Kalisz, A., 2003 (Nov), "Radiation-Force Technique to Monitor Lesions During Ultrasonic Therapy," *Ultrasound in Medicine and Biology,* **29**, pp. 1593–1605.

[26] Bercoff, J., Tanter, M., and Fink, M., 2004 (Apr), "Supersonic Shear Imaging: A New Technique for Soft Tissue Elasticity Mapping," *IEEE Transactions on Ultrasonics, Ferroelectrics and Frequency Control,* **51**, pp. 396–409.

[27] Bercoff, J., Pernot, M., Tanter, M., and Fink, M., 2004 (Apr), "Monitoring Thermally-Induced Lesions with Supersonic Shear Imaging," *Ultrasonic Imaging,* **26**, pp. 71–84.

[28] Sarvazyan A. P., Rudenko O. V., Swanson S. D., Fowlkes J. B., and Emelianov S. Y., 1998, "Shear wave elasticity imaging: a new ultrasonic technology of medical diagnostics," *Ultrasound in Medicine and Biology*, **24**, pp. 1419–1435.

[29] Sinkus, R., Lorenzen, J., Schrader, D., Lorenzen, M., Dargatz, M., and Holz, D., 2000 (Jun), "High-Resolution Tensor MR Elastography for Breast Tumour Detection," *Physics in Medicine and Biology,* **45**, pp. 1649–1664.

[30] Fatemi, M., and Greenleaf, J. F., 1998 (Apr), "Ultrasound-Stimulated Vibro-Acoustic Spectrography," *Science,* **280**, pp. 82–85.

[31] Fatemi, M., and Greenleaf, J. F., 2000 (Jun), "Probing the Dynamics of Tissue at Low Frequencies with the Radiation Force of Ultrasound," *Physics in Medicine and Biology,* **45**, pp. 1449–1464.

[32] Michishita, K., Hasegawa, H., and Kanai, H., 2003 (Jul), "Ultrasonic Measurement of Minute Displacement of Object Cyclically Actuated by Acoustic Radiation Force," *Japanese Journal of Applied Physics Part 1-Regular Papers Short Notes & Review Papers,* **42**, pp. 4608–4612.

[33] Konofagou, E., Thierman, J., and Hynynen, K., 2001 (Nov), "A Focused Ultrasound Method for Simultaneous Diagnostic and Therapeutic Applications — A Simulation Study," *Physics in Medicine and Biology,* **46**, pp. 2967–2984.

[34] Konofagou, E. E., and Hynynen, K., 2002, "Localized Harmonic Motion Imaging: Theory, Simulations and Experiments," in *IEEE Symposium Ultrasonics*, Munich, Germany, pp. 1895–1898.

[35] Konofagou, E. E., and Hynynen, K., 2003 (Oct), "Localized Harmonic Motion Imaging: Theory, Simulations And Experiments," *Ultrasound in Medicine and Biology,* **29**, pp. 1405–1413.

[36] Maleke, C., Pernot, M., and Konofagou, E. E., 2005, "A Single-Element Focused Transducer Method for Harmonic Motion Imaging," in *IEEE Symposium Ultrasonics*, Rotterdam, Netherlands, pp. 17–20.

[37] Maleke, C., Pernot, M., and Konofagou, E. E., 2006 (Jul), "A Single-element focused ultrasound transducer method for harmonic motion imaging," *Ultrasonic Imaging,* **28**, pp. 144–158.

[38] Maleke, C., Luo, J., Pelegri, M., and Konofagou, E., 2007, "Mapping of Regional Tissue Mechanical Changes Using Harmonic Motion Imaging," in *ASME International Mechanical Engineering Congress and Exposition*, Seattle, WA.

[39] Maleke, C., and Konofagou, E., 2007, "Amplitude Modulated Harmonic Motion Imaging (AM-HMI) Technique for Breast Cancer Early Detection and Treatment," in *American Institute of Ultrasound in Medicine Conference* (Abstract).

[40] Maleke C., and Konofagou E.E., 2008, "Harmonic Motion Imaging for Focused Ultrasound (HMIFU): A Fully Integrated Technique for Sonication and Monitoring of Thermal Ablation in Tissues," *Physics in Medicine and Biology*, **53**(6), pp. 1773–1793.

[41] Maleke, C., and Konofagou, E. E., 2006, "An All-Ultrasound-Based System for Real-Time Monitoring and Sonication of Temperature Change and Ablation," in *IEEE Symposium Ultrasonics*, Vancouver, Canada, pp. 204–207.

[42] Jeffrey, S. S., Birdwell, R. L., Ikeda, D. M., Daniel, B. L., Nowels, K. W., Dirbas, F. M., and Griffey, S. M., 1999 (Oct), "Radiofrequency Ablation of Breast Cancer — First Report of an Emerging Technology," *Archives of Surgery,* **134**, pp. 1064–1068.

[43] Izzo, F., Thomas, R., Delrio, P., Rinaldo, M., Vallone, P., DeChiara, A., Botti, G., D'Aiuto, G., Cortino, P., and Curley, S. A., 2001 (Oct), "Radiofrequency Ablation in Patients with Primary Breast Carcinoma — A Pilot Study in 26 Patients," *Cancer,* **92**, pp. 2036–2044.

[44] Elliott, R. L., Rice, P. B., Suits, J. A., Ostrowe, A. J., and Head, J. F., 2002 (Jan), "Radiofrequency Ablation of a Stereotactically Localized Nonpalpable Breast Carcinoma," *American Surgeon,* **68**, pp. 1–5.

[45] Burak, W. E., Agnese, D. M., Povoski, S. P., Yanssens, T. L., Bloom, K. J., Wakely, P. E., and Spigos, D. G., 2003 (Oct), "Radiofrequency Ablation of Invasive Breast Carcinoma Followed by Delayed Surgical Excision," *Cancer,* **98**, pp. 1369–1376.

[46] Hayashi, A. H., Silver, S. F., van der Westhuizen, N. G., Donald, J. C., Parker, C., Fraser, S., Ross, A. C., and Olivotto, I. A., 2003 (May), "Treatment of Invasive Breast Carcinoma with Ultrasound-Guided Radiofrequency Ablation," *American Journal of Surgery,* **185**, pp. 429–435.

[47] Fornage, B. D., Sneige, N., Ross, M. I., Mirza, A. N., Kuerer, H. M., Edeiken, B. S., Ames, F. C., Newmanj, L. A., Bariera, G. V., and Singletary, S. E., 2004 (Apr), "Small (≤2-cm) Breast Cancer Treated with US-Guided Radiofrequency Ablation: Feasibility Study," *Radiology,* **231**, pp. 215–224.

[48] Noguchi, M., Earashi, M., Fujii, H., Yokoyama, K., Harada, K. I., and Tsuneyama, K., 2006 (Feb), "Radiofrequency Ablation of Small Breast Cancer Followed by Surgical Resection," *Journal of Surgical Oncology,* **93**, pp. 120–128.

[49] Kennedy, J. E., 2005 (Apr), "High-Intensity Focused Ultrasound in the Treatment of Solid Tumours," *Nature Reviews Cancer,* **5**, pp. 321–327.

[50] Ter Haar, G., 1995, "Ultrasound Focal Beam Surgery," *Ultrasound in Medicine and Biology,* **21**, pp. 1089–1100.

[51] Huber, P. E., Jenne, J. W., Rastert, R., Simiantonakis, I., Sinn, H. P., Strittmatter, H. J., von Fournier, D., Wannenmacher, M. F., and Debus, J., 2001 (Dec), "A New Noninvasive Approach in Breast Cancer Therapy Using Magnetic Resonance Imaging-Guided Focused Ultrasound Surgery," *Cancer Research,* **61**, pp. 8441–8447.

[52] Wu, F., Wang, Z. B., Cao, Y. D., Chen, W. Z., Zou, J. Z., Bai, J., Zhu, H., Li, K. Q., Jin, C. B., Xie, F. L., Su, H. B., and Gao, G. W., 2003 (Oct), "Changes in Biologic Characteristics of Breast Cancer Treated with High-Intensity Focused Ultrasound," *Ultrasound in Medicine and Biology,* **29**, pp. 1487–1492.

[53] Wu, F., Wang, Z., Cao, Y., Chen, W., Bai, J., Zou, J., and Zhu, H., 2003, "A Randomized Clinical Trial of High-Intensity Focused Ultrasound Ablation for the Treatment of Patients with Localized Breast Cancer," *British Journal of cancer*, **12**, pp. 2227–2233.

[54] Furusawa, H., Namba, K., Thomsen, S., Akiyama, F., Bendet, A., Tanaka, C., Yasuda, Y., and Nakahara, H., 2006 (Jul), "Magnetic Resonance-Guided Focused Ultrasound Surgery of Breast Cancer: Reliability and Effectiveness," *Journal of the American College of Surgeons,* **203**, pp. 54–63.

[55] Khiat, A., Gianfelice, D., Amara, M., and Boulanger, Y., 2006 (Apr), "Influence of Post-Treatment Delay on the Evaluation of the Response to Focused Ultrasound Surgery of Breast Cancer by Dynamic Contrast Enhanced MRI," *British Journal of Radiology,* **79**, pp. 308–314.

[56] Wu, F., Chen, W. Z., Bai, J., Zou, J. Z., Wang, Z. L., Zhu, H., and Wang, Z. B., 2001 (Aug), "Pathological Changes in Human Malignant Carcinoma Treated with High-Intensity Focused Ultrasound," *Ultrasound in Medicine and Biology,* **27**, pp. 1099–1106.

[57] Ter Haar, G., 2001 (Dec), "Acoustic Surgery," *Physics Today,* **54**, pp. 29–34.

[58] Righetti, R., Kallel, F., Stafford, R. J., Price, R. E., Krouskop, T. A., Hazle, J. D., and Ophir, J., 1999 (Sep), "Elastographic Characterization of HIFU-Induced Lesions in Canine Livers," *Ultrasound in Medicine and Biology,* **25**, pp. 1099–1113.

[59] Souchon, R., Rouviere, O., Gelet, A., Detti, V., Srinivasan, S., Ophir, J., and Chapelon, J. Y., 2003 (Jul), "Visualisation of HIFU Lesions Using Elastography of the Human Prostate in Vivo: Preliminary Results," *Ultrasound in Medicine and Biology,* **29**, pp. 1007–1015.

[60] Hynynen, K., Pomeroy, O., Smith, D. N., Huber, P. E., McDannold, N. J., Kettenbach, J., Baum, J., Singer, S., and Jolesz, F. A., 2001 (Apr), "MR Imaging-Guided Focused Ultrasound Surgery of Fibroadenomas in the Breast: A Feasibility Study," *Radiology,* **219**, pp. 176–185.

[61] Gianfelice, D., Khiat, A., Amara, M., Belblidia, A., and Boulanger, Y., 2003 (Nov), "MR Imaging-Guided Focused Ultrasound Surgery of Breast Cancer: Correlation of Dynamic Contrast-Enhanced MRI with Histopathologic Findings," *Breast Cancer Research and Treatment,* **82**, pp. 93–101.

[62] Konofagou, E. E., Thierman, J., Karjalainen, T., and Hynynen, K., 2002 (Mar), "The Temperature Dependence of Ultrasound-Stimulated Acoustic Emission," *Ultrasound in Medicine and Biology,* **28**, pp. 331–338.

[63] Parkin, D. M., Pisani, P., and Ferlay, J., 1993 (Jun), "Estimates of the Worldwide Incidence of 18 Major Cancers in 1985," *International Journal of Cancer,* **54**, pp. 594–606.

[64] Vogl, T. J., Mack, M. G., Balzer, J. O., Engelmann, K., Straub, R., Eichler, K., Woitaschek, D., and Zangos, S., 2003 (Nov), "Liver Metastases: Neoadjuvant Downsizing with Transarterial Chemoembolization Before Laser-Induced Thermotherapy," *Radiology,* **229**, pp. 457–464.

[65] Wu, T., Felmlee, J. P., Greenleaf, J. F., Riederer, S. J., and Ehman, R. L., 2001, "Assessment of thermal tissue ablation with MR elastography," *Magnetic Resonance in Medicine,* **45**, pp. 80–87.

[66] T. J. B. Hall, M.; Insana, M.F.; Krouskop, T.A., 1997 (November), "Phantom Materials for Elastography," *IEEE Transactions on Ultrasonics, Ferroelectrics, and Frequency Control,* **44**, pp. 1355–1365.

[67] Jensen, J. and Svendsen, N., 1992, "Calculation of Pressure Fields from Arbitrarily Shaped, Apodized, and Excited Ultrasound Transducers," *IEEE Transactions on Ultrasonics, Ferroelectrics, and Frequency Control,* **39**, pp. 262–267.

[68] Bamber, J., and Dickinson, R., 1980, "Ultrasonic B-Scanning: A Computer Simulation," *Physics in Medicine and Biology*, **25**, pp. 463–479.

[69] Seggie, D., Leeman, S., and Burge, R., 1983, "Realistic Simulation of B-Scan Images," in *Proceedings of the IEEE Ultrasonics Symposium*, Atlanta, GA, pp. 714–717.

Section 2:

MAGNETIC RESONANCE METHODS

Chapter 8

CHARACTERIZATION OF SKELETAL MUSCLE ELASTICITY USING MAGNETIC RESONANCE ELASTOGRAPHY

Qingshan Chen[1], Armando Manduca[2], and Kai-Nan An[1]
[1]Biomechanics Laboratory, Department of Orthopedic, Mayo Clinic, Rochester, MN 55905, USA
[2]MRI Research Laboratory, Department of Radiology, Mayo Clinic, Rochester, MN 55905, USA

ABSTRACT

Magnetic resonance elastography (MRE) is a non-invasive phase-contrast MR technique that directly visualizes and quantitatively measure propagating strain waves in tissue-like materials subject to harmonic mechanical excitation. A phase-contrast MRI technique is used to spatially map the shear wave displacement field, and local quantitative values of shear modulus are then calculated through various inversion algorithms, and elastogram generated. This information can be clinically useful to understand the effects of pathologies on the mechanical properties of skeletal muscle and to quantify the effects of treatment. In this chapter, the principles of MRE image acquisition and stiffness inversion algorithm are reviewed, specific instrumention to apply MRE on skeletal muscles are introduced, and unique challenges to MRE when applied on skeletal muscles are discussed.

8.1 CLINICAL BACKGROUND

Information on elasticity (i.e., resistance to deformation) of skeletal muscle has clinical applications in several fields including orthopedics, sports medicine, physical medicine and rehabilitation, endocrinology, and rheumatology. Often, pathological skeletal muscle tissues exhibit a noticeable difference in elastic and/or other constitutive tissue properties compared to normal healthy tissue. Technically, Young's modulus is the quantity delineating the elasticity of biological tissues. Since in vitro measurement of Young's modulus of biological tissue is often impossible in clinical settings, in vivo image based methods that assesses Young's modulus of biological tissues have been developed, known as elastography.

Conventionally, skeletal muscles were often clinically assessed with functional examinations (e.g., the manual palpatory examination), validated clinical scales (e.g., modified Ashworth scale), force measurements using handheld and isokinetic dynamometers, and surface and fine wire electromyography (EMG) [1-5]. While these measurements provided information that clinicians can use to track the changes in muscle function in their patients, no quantities on mechanical properties of the skeletal muscles were given, and high subjectivity and poor repeatability of the above examinations were often reported [1].

Magnetic resonance elastography (MRE) is a non-invasive phase-contrast MR technique that directly visualizes and quantitatively measure propagating strain waves in tissue-like materials subject to harmonic mechanical excitation [6, 7]. A phase-contrast MRI technique is used to spatially map the shear wave displacement fields. From this data, local quantitative values of shear modulus can be calculated and images (elastograms) that depict tissue elasticity can be generated, allowing for quantitative characterization of tissue elasticity.

Compared to the conventional examinations, MRE stiffness imaging on skeletal muscle offers the following advantages [6]:

- Direct visualization and quantitative measurement of tissue vibratory displacements;
- High sensitivity to small motions in the order of microns;
- A filed of view unencumbered by acoustic window requirements;
- The ability to obtain full three-dimensional displacement information throughout a three-dimensional volume.

8.2 PRINCIPLE OF MRE IMAGING

MRE uses propagating mechanical waves rather than static stress as a probe. The major advantage of doing so is that it does not require estimation of the regional static stress distribution [6]. High-frequency longitudinal acoustic waves (ultrasound) are not directly suitable for use as a probe because their propagation is governed by the bulk modulus, which varies little in soft tissue [6]. In MRE, a phase-contrast MRI technique is used to spatially map and measure displacement patterns corresponding to harmonic shear waves with amplitudes of microns or less [7]. A conventional MRI system is used with an additional motion-sensitizing gradient $G_r(t)$ imposed along a specific direction $\vec{r}$, switched in polarity at some adjustable frequency. Trigger pulses synchronize an oscillator amplifier unit that drives an electromechanical actuator coupled to the surface of the object to be imaged, inducing shear waves in the object at the same frequency as the motion-sensitizing gradient. Any cyclic motion of the spins in the presence of these motion-sensitizing gradients causes a measurable accumulative phase shift in the received MR signal [7]. Accumulation of the phase shift is based on the well-known Larmor equation:

$$\omega = \frac{\mathrm{d}\phi}{\mathrm{d}t} = \gamma B \tag{8-1}$$

where ω is the spin frequency, ϕ the phase of the spins, γ the electron gyromagnetic ratio, and B the magnetic filed strength. Therefore, the phase of the spins can be expressed as

$$\phi(t) = \int \omega(t)\,\mathrm{d}t = \int \gamma B(t)\,\mathrm{d}t \tag{8-2}$$

When the magnetic field gradient $G_r(t)$ is superimposed on the main magnetic field B_0 along direction $\vec{r}$, then the phase of spins moving along the direction $\vec{r}$ is given by

$$\phi(t) = \gamma \int [B_0 + G_r(t) \cdot r(t)]\,\mathrm{d}t \tag{8-3}$$

The first term in the above equation is a constant, since B_0 is non time-varying and does not reflect information about the motion. It is the second term that is of interest in motion detection. Consider a sinusoidal mechanical wave at a specific frequency ω propagating within a biological material. The position vector, $\vec{r}(t)$, describing the instantaneous spin position thus is given by

$$\vec{r}(t) = \vec{r}_0 + \vec{\xi}_0 e^{j(\vec{k}\cdot\vec{r} - \omega t + \theta)} \tag{8-4}$$

where $\vec{r}_0$ is the mean position of the spin, $\vec{\xi}_0$ is the peak displacement of the spin from its mean position, $\vec{k}$ is the wave number, θ is the initial phase offset of the wave. Because the motion is sinusoidal, it is possible to estimate the displacement with a single realization of an appropriate gradient waveform. Consider a rectangular gradient function that is switched in polarity at the same frequency as the position vector. The magnitude of this gradient could be described by

$$|G_r(t)| = \begin{cases} +|G|; t \in [nT, (2n+1)\,T/2) \\ -|G|; t \in ((2n+1)\,T/2, (n+1)T) \end{cases} \tag{8-5}$$

where $n = 0, 1, 2, \ldots, N - 1$, $T = \dfrac{2\pi}{\omega}$, and N is the number of gradient cycles. The time duration $\tau = NT$ is chosen such that the integral of the dot product of the position vector and the gradient vector over τ is zero. The observed phase shift in the measured signal is then given by [7]

$$\phi(\vec{r},\theta) = \gamma \int_0^{\tau = NT < TE} G_r(t)\cdot \xi_0 e^{j(\vec{k}\cdot\vec{r} - \omega t + \theta)} dt$$

$$= \frac{2\gamma NT(\vec{G}_r \cdot \vec{\xi}_0)}{\pi} \sin(\vec{k}\cdot\vec{r} + \theta) \tag{8-6}$$

Alternatively, if sinusoids are chosen as the basis set, a gradient waveform can be considered, given by

$$|G_r(t)| = G_0 \cos\omega t; t \in [0, NT] \tag{8-7}$$

By following similar analysis for squared gradients, the measured phase shift obtained is [7]

$$\phi(\vec{r},\theta) = \frac{\gamma NT(\vec{G}_r \cdot \vec{\xi}_0)}{2} \cos(\vec{k}\cdot\vec{r} + \theta) \tag{8-8}$$

Equations (8-6) and (8-8) related a measurable phase shift of the received signal to the displacement amplitude vector $\vec{\xi}_0$ and the wave number $\vec{k}$, with a set of experimental controllable variables such as the motion-sensitizing magnetic gradient vector $\vec{G}_r$ and duration NT, which makes it possible to achieve extreme sensitivity to small-amplitude synchronous motion by accumulating phase shifts over multiple cycles of mechanical excitation and the motion-sensitizing gradient waveform [6, 7]. Thus, the resulting spin phase images reflect the displacement of spins due to strain wave propagation in the tissue and are, therefore, termed "wave images." The local wavelength of the propagating waves depends on the elasticity of the material at each location in the object, therefore allowing the construction of the tissue elastogram.

Generally, rectangular motion-sensitizing gradient is utilized, and two acquisitions are made for each repetition, in an interleaved fashion reversing the polarity of the motion-sensitizing gradients. This reduces systematic phase errors and doubles the sensitivity to small displacements. The direction of the motion-sensitizing gradients is collinear with the direction of vibratory motion [6]. Therefore, a single MRE acquisition captures only one component of the displacement data; i.e., it is sensitized to motion in a single direction.

However, the experiment can be repeated and the sensitization direction varied, to capture all three orthogonal components of displacement. Thus, full three-dimensional displacement information can be acquired [6].

Too much amplitude can cause phase wrap, i.e., large displacement may cause accumulated phase shifts outside the range of $\pm\pi$, which are ambiguous. To overcome this, standard phase unwrapping algorithms can be applied to MRE data with good success, since the wrapped regions are usually fairly well localized. Other common pre-filtering techniques include directional filtering [8], and first harmonic component extraction [6].

In terms of hardware realization, the motion-sensitizing gradient can be readily incorporated into gradient echo (GRE) pulse sequence usually for two-dimensional imaging (see Fig. 8-1), or more recently echo-planar imaging (EPI) pulse sequence allowing for fast imaging time that is desirable for three-dimensional imaging. Typical acquisition parameters for skeletal muscle tissues were: pulse repetition times (T_R) of 100 milliseconds, minimal echo delay (T_E) allowing for motion encoding at the given excitation frequency, slice thickness of 5 mm, flip angle of 60 deg., 256 frequency encoding views, 128 to 256 phase encoding views, and four to eight phase offsets [1, 9-12] Choice of excitation frequency for a particular application depends on a tradeoff: On one hand, increasing frequency would cause higher resolution, since the propagating frequency is small; on the other hand, increasing frequency would also lead to higher attenuation, as the tissue viscoelasticity increases at higher frequency. In practice, excitation frequency in the range of 90 Hz to 200 Hz has been generally used for skeletal muscle MRE imaging.

Choices of driver that create the mechanical excitation on skeletal muscles in vivo include: electromagnetic coil driver, piezoelectric driver, pneumatic driver, and focused ultrasound. An electromagnetic coil driver applies alternating currents to an annular coil to generate motion. It is easy to construct, costs less, and can be customized easily to account for special in vivo applications such as muscle, breast, or brain MRE [13, 14]. However, artifacts may be produced due to the magnetic interference of the driver coil. A piezoelectric driver has the advantage of arbitrary orientated with respect to the main magnet filed, but its fabrication is elaborate and time costly, and has a limited maximum displacement [6, 15]. A pneumatic driver is easy to construct, free of artifact, low-cost, has good frequency

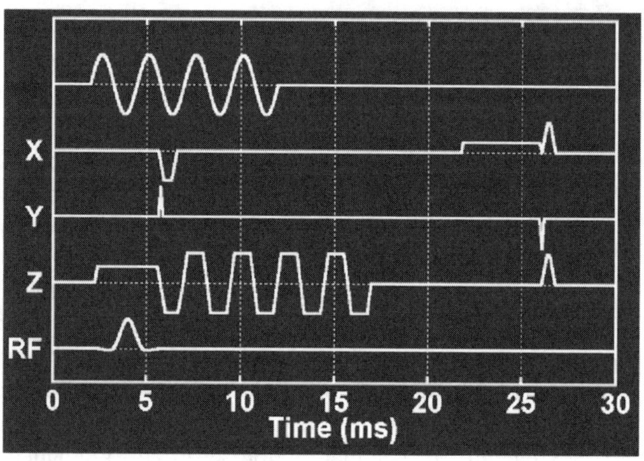

Figure 8-1. Schematic diagram of magnetic resonance elastography system. The square wave motion-sensitizing magnetic gradient is superimposed along the conventional MRI imaging gradients and RF pulses. The phase offset (θ) between the mechanical vibration and the motion-sensitizing gradient can be varied

response. The ultrasound beam is temporally modulated to create cyclic variation in acoustic radiation pressure at the focus of the ultrasound source, which can be localized deep within an object [16]. Regardless of the type of driver chosen, trigger pulses are always provided by the sequencing computer of the MR scanner, and are fed to a function generator to generate a coupled motion waveform that is amplified and applied to the driver.

8.3 INTRODUCTION TO ELASTICITY INVERSION ALGORITHMS

After the wave images are obtained through the MRE experiment, image of tissue elasticity distribution, often termed "stiffness image," is determined via a process generally termed "stiffness inversion," one of the key processes in MRE. While numerous approaches have been developed to invert MRE displacement data to reconstruct the underlying mechanical properties of the tissue, in this chapter three typical algorithms are introduced. Each represents a major category of inversion algorithms: Algebraic Inversion for Helmholtz equation based inversion algorithm, Phase Gradient approach based on wavelength evaluation, and Overlapping Subzone Technique for finite element based inversion algorithm.

8.3.1 Wave Motion in Elastic Solids

Governing equations for stress wave propagation in a solid medium is [17].

$$\sigma_{ij,j} + f_i = \rho \ddot{u}_i \qquad (8-9)$$

where σ_{ij} is the Cauchy stress tensor, f_i the body force vector, u_i the displacement vector, and ρ the material density. As the case for most biological tissues, the body force term is ignored in the following discussions.

Due to the small vibratory displacements (on the order of microns) utilized in MRE, linear material elasticity is often assumed. In the general linearity case, stress tensor σ_{ij} and strain tensor ε_{ij} are related by

$$\sigma_{ij} = C_{ijkl}\varepsilon_{kl} \qquad (8-10)$$

where C_{ijkl} in general is a rank 4 tensor with up to 36 independent quantities. If one assumes material isotropy, the relationship reduces to two independent quantities, namely, Lame constants, λ and μ (shear modulus)

$$\sigma_{ij} = 2\mu\varepsilon_{ij} + \lambda\delta_{ij}\varepsilon_{kk} \qquad (8-11)$$

or

$$C_{ijkl} = \lambda\delta_{ij}\delta_{kl} + \mu(\delta_{ik}\delta_{jl} + \delta_{il}\delta_{jk}) \qquad (8-12)$$

where δ_{ij} denotes the Kronecker delta. Under the assumption of infinitesimal deformation, strain tensor ε_{ij} is determined by

$$\varepsilon_{ij} = \frac{1}{2}(u_{i,j} + u_{j,i}) \qquad (8-13)$$

Substituting Eqs. (8-11) and (8-12) into Eq. (8-9) and assuming simple harmonic motion, the general equation for harmonic motion in an isotropic, linearly elastic medium is obtained.

$$[\lambda u_{j,j}]_{,i} + [\mu(u_{i,j} + u_{j,i})]_{,j} = -\rho\omega^2 u_i \qquad (8-14)$$

with ω being the frequency of the simple harmonic motion. Alternatively, Eq. (8-14) can be rewritten in the form of differential operator (∇)

$$\mu\nabla^2\vec{u}+(\lambda+\mu)\nabla(\nabla\cdot\vec{u})=-\rho\omega^2\vec{u} \qquad (8-15)$$

If one further assumes incompressibility (i.e., $u_{j,j}=\nabla\cdot\vec{u}=0$), which is often a valid assumption for most biological tissues, Eq. (8-15) then simplifies to the Helmholtz equation [6].

$$\mu\nabla^2\vec{u}=-\rho\omega^2\vec{u} \qquad (8-16)$$

In such condition it can be seen that the terms involving components in the different orthogonal are now decoupled, and each component satisfies the equation separately. Thus, measurements in only one sensitization direction suffice to determine the shear modulus μ [6]. Such a property is utilized in the development of the Algebraic Inversion algorithm discussed in below.

8.3.2 Algebraic Inversion of the Differential Equation (AIDE)

Consider a simplest version of AIDE algorithm that starts with rewriting Eq. (8-14) into [6, 18]

$$A\begin{bmatrix}\lambda(a)+\mu(a)\\ \mu(a)\end{bmatrix}=-\rho\omega^2\begin{bmatrix}u_1(a)\\ u_2(a)\\ u_3(a)\end{bmatrix} \qquad (8-17)$$

where

$$A=\begin{bmatrix}u_{i,i1}(a) & u_{1,ii}(a)\\ u_{i,i2}(a) & u_{2,ii}(a)\\ u_{i,i3}(a) & u_{3,ii}(a)\end{bmatrix} \qquad (8-18)$$

Using the concept of pseudo-inverse, Eq. (8-17) can then be inverted to obtain an estimate of the Lame constants for each position [18]

$$\begin{bmatrix}\lambda(a)+\mu(a)\\ \mu(a)\end{bmatrix}=-\rho\omega^2(A^{H}A)^{-1}A^{H}\begin{bmatrix}u_1(a)\\ u_2(a)\\ u_3(a)\end{bmatrix} \qquad (8-19)$$

where A^{H} denotes the conjugate transpose of matrix A. If, as in the case of Helmholtz equation, material incompressibility is further assumed, then $A_{i1}=0$, and Eq. (8-19) decouples into three separate expressions for shear modulus $\mu(a)$ [18]

$$\begin{bmatrix}u_{1,ii}(a)\,\mu(a)\\ u_{2,ii}(a)\,\mu(a)\\ u_{3,ii}(a)\,\mu(a)\end{bmatrix}=-\rho\omega^2\begin{bmatrix}u_1(a)\\ u_2(a)\\ u_3(a)\end{bmatrix} \qquad (8-20)$$

Therefore [18],

$$\mu(a)=-\rho\omega^2\frac{u_i(a)}{u_{i,11}(a)+u_{i,22}(a)+u_{i,33}(a)} \qquad (8-21)$$

where index i can be either 1, 2, or 3. Thus, in the case of Helmholtz equation, measurements in only one sensitization direction suffice to determine μ, the simplest version of AIDE algorithm.

8.3.3 Phase Gradient

Phase gradient algorithm is a stiffness inversion approach that is based on wavelength evaluation [6]. That is, after extracting the harmonic component at the driving frequency, there results both an amplitude and a phase (relative to an arbitrary zero point) that characterize the harmonic oscillation at each pixel in the MRE phase image. The gradient of the phase (i.e., local wave number) $\vec{k}$ of this harmonic shear wave is then the precise quantity sought: phase gradient is easily convertible to local wavelength, given the well-known relationship between the wave length and wave number in a given direction $\lambda = \frac{2\pi}{k}$. In principle, this phase gradient method has a very high resolution. However, in practice such a method can be very vulnerable to imaging noise, wave reflections, mode conversions [17], etc., making its application rather limited [6].

8.3.4 A Finite Element Based Inversion Algorithm: Overlapping Subzone Technique

Despite the success of the above inversion algorithms in "bench validation" using gel phantoms, they often show limits when applied to skeletal muscle MRE. Specifically, the simplification assumptions that the aforementioned inversion algorithms are based on, namely, infinite geometry, isotropic, and elastic, are often violated by skeletal muscles. First of all, skeletal muscle tissues often have complicated boundary conditions depending on the anatomical connections and constraints between the tissue of interest and the surrounding tissues, anatomical geometry of the tissue of interest, and the manner in which mechanical vibratory motion is introduced into the tissue of interest. Like any partial derivative equation (PDE), a particular solution to the wave equation depends on the given boundary conditions. Physically, boundary conditions usually affect the wave propagation in elastic solids through two complexities: reflection and mode conversion [17]. Both pose challenges for the above mentioned inversion algorithms. Secondly, skeletal muscles are well known to be anisotropic, with maximum active stress generated parallel to the fiber axis [19]. Finally, the viscoelasticity property of skeletal muscles often has important clinical relevance and it is therefore a pity to simply ignore this during stiffness inversion. An advanced stiffness inversion algorithm would be expected to estimate the viscoelastic property of the tissue as well.

An inversion algorithm based on forward finite element analysis coupled with various optimization algorithms has been proposed to solve the above problems. A typical example of such an inversion algorithm, Overlapping Subzone Technique, is introduced in this section. The typical strategy is to define a single objective to be minimized that is the sum of the squared differences between measured and calculated quantities in all subzones [20, 21]. That is

$$\min F(E) = \min\left[\sum_{k=1}^{M} \min F_k(E_k)\right] \qquad (8-22)$$

where

$$F_k(E_k) = \sum_{a=1}^{N_k}\left[\left(u^m_{a_1} - u^c_{a_1}\right)^2 + \left(u^m_{a_2} - u^c_{a_2}\right)^2 + \left(u^m_{a_3} - u^c_{a_3}\right)^2\right] \qquad (8-23)$$

where $u^m_{a_1}, u^m_{a_2}, u^m_{a_3}$, are the x, y, and z vector components of the measured displacement at location a in subzone Ω_k, while $u^c_{a_1}, u^c_{a_2}, u^c_{a_3}$, are the calculated vector components at the same location, for a total of N_k different locations. E_k is an M-dimensional vector of property parameters in subzone Ω_k [20, 21].

Determination of the subzone elastic properties requires differentiation of in Eq. (8-23) with respect to each of the property parameters contained in E_k, which produces the nonlinear system [20].

$$\frac{\partial F_k}{\partial E_{k_1}} = \sum_{a=1}^{N_k}(u^m_{a_1} - u^m_{a_1})\frac{\partial u^c_{a_1}}{\partial E_{k_1}} + \sum_{a=1}^{N_k}(u^m_{a_2} - u^m_{a_2})\frac{\partial u^c_{a_2}}{\partial E_{k_1}} + \sum_{a=1}^{N_k}(u^m_{a_3} - u^m_{a_3})\frac{\partial u^c_{a_3}}{\partial E_{k_1}} = 0$$

$$\frac{\partial F_k}{\partial E_{k_2}} = \sum_{a=1}^{N_k}(u^m_{a_1} - u^m_{a_1})\frac{\partial u^c_{a_1}}{\partial E_{k_2}} + \sum_{a=1}^{N_k}(u^m_{a_2} - u^m_{a_2})\frac{\partial u^c_{a_2}}{\partial E_{k_2}} + \sum_{a=1}^{N_k}(u^m_{a_3} - u^m_{a_3})\frac{\partial u^c_{a_3}}{\partial E_{k_2}} = 0$$

$$\dots \quad \dots$$

$$\frac{\partial F_k}{\partial E_{k_M}} = \sum_{a=1}^{N_k}(u^m_{a_1} - u^m_{a_1})\frac{\partial u^c_{a_1}}{\partial E_{k_M}} + \sum_{a=1}^{N_k}(u^m_{a_2} - u^m_{a_2})\frac{\partial u^c_{a_2}}{\partial E_{k_M}} + \sum_{a=1}^{N_k}(u^m_{a_3} - u^m_{a_3})\frac{\partial u^c_{a_3}}{\partial E_{k_M}} = 0 \quad (8-24)$$

Solution of this equation set by Newton's method leads to iterative improvements in the elastic property profile such that

$$E_k^{(n+1)} = E_2^{(n)} + \Delta E_k \quad (8-25)$$

where ΔE_k is the property update vector in subzone Ω_k, determined from the solution of the regularized matrix system [20]

$$[H_k^{(n)} + \alpha I]\,\{\Delta E_k\} = \{-f_k^{(n)}\} \quad (8-26)$$

with $H_k^{(n)}$ having the elements [20]:

$$h_{ij}^{(n)} = \frac{\partial fi}{\partial E_{kj}^{(n)}} = -\sum_{a=1}^{N_k}\left[\frac{\partial u^c_{a_1}}{\partial E_{ki}^{(n)}}\frac{\partial u^c_{a_1}}{\partial E_{kj}^{(n)}} + \frac{\partial u^c_{a_2}}{\partial E_{ki}^{(n)}}\frac{\partial u^c_{a_2}}{\partial E_{ki}^{(n)}} + \frac{\partial u^c_{a_3}}{\partial E_{ki}^{(n)}}\frac{\partial u^c_{a_3}}{\partial E_{ki}^{(n)}}\right] \quad (8-27)$$

and α is a scalar regularization parameter added to the diagonal of H to facilitate its inversion, as H is known to be poorly conditioned. Solution of Eq. (8-26) requires a vehicle for calculating the subzone displacement field and its derivatives with respect to each property parameters (i.e., $\vec{u}_j' \equiv \frac{\partial \vec{u}}{\partial E_{kj}}$) given the current estimates of the property parameters. Again, the displacement field can be obtained through solving Eq. (8-15) using Galerkin method finite element analysis. The required derivatives $\vec{u}_j'$ are determined by differentiating Eq. (8-15) directly with respect to each property parameter and solve the resulting partial differential equation on the same finite element discretization [20]:

$$\mu\nabla^2\vec{u}_j' + \nabla(\lambda+\mu)\nabla\cdot\vec{u}_j' = -\rho\omega^2\vec{u}_j' - \nabla\cdot\frac{\partial G}{\partial E_{kj}}\nabla\vec{u} - \nabla\frac{\partial}{\partial E_{kj}}(\lambda+\mu)\nabla\cdot\vec{u} \quad (8-28)$$

Now, because Eq. (8-28) is evaluated at the current property estimate, $\vec{u}$ can be computed through the finite element solution of Eq. (8-15), thus leaving $\vec{u}_j'$ the only unknown in Eq. (8-28).

In practice, the subzone inversion algorithm begins with an initial estimate of the elastic property, E^0, defined over the entire problem space Ω. From this estimate, a global displacement filed, u^c, is computed from Eq. (8-15) with the finite element method based on known displacement conditions from the MRE measurements, u^m, applied along the global boundary, Γ. The squared error between the resulting displacement solution and the measured MRE data is then calculated for each element. By using this error metric, a hierarchical list of element centroids is generated whose order is based on a decreasing squared-error contribution. A subzone domain, Ω_k, is then formed about an element centroid in the list by including all nearby elements whose centroids are within a user-defined range. Once the subzone is identified, a finite element method is used to calculate the displacement field on the subzone by using the latest property parameter estimates. The MRE displacement measurements on the subzone boundary Γ_k are used as the boundary conditions in the finite element analysis. The subzone property parameters are iteratively updated with the inversion process embodied in Eq. (8-26) until a local convergence criterion between the computed and measured displacement fields has been reached. At this point, the next element centroid in the error contribution list will be used to define another subzone, and the process of local convergence in the displacement field is repeated. The zoning process continues until every element in the global mesh has been included in at least one subzone. After that, another global displacement field calculation will be executed with the latest property parameters profile, and the zoning process repeats until a minimal number of element iterations is reached [20, 21] (see Fig. 8-2).

The above example is the simplest version of Overlapping Subzone Technique under the assumptions of linear isotropic elasticity. Nevertheless, one can see that the concerns on boundary conditions have already been eased, as the boundary conditions have been inputted into the forward finite element analysis to solve the displacement field. More complicated wave equations can be developed to model anisotropy and viscoelasticity, and solved by finite element analysis to gain the displacement filed $\vec{u}_j$ and its derivatives $\vec{u}_j'$. Therefore, in principle a fully developed finite element based inversion algorithm would be able to address all the three primary challenges mentioned above: boundary conditions, anisotropy, and viscoelasticity. Mechanical characterization of skeletal muscles in the aspects of material anisotropy and viscoelasticity will be discussed in the following section.

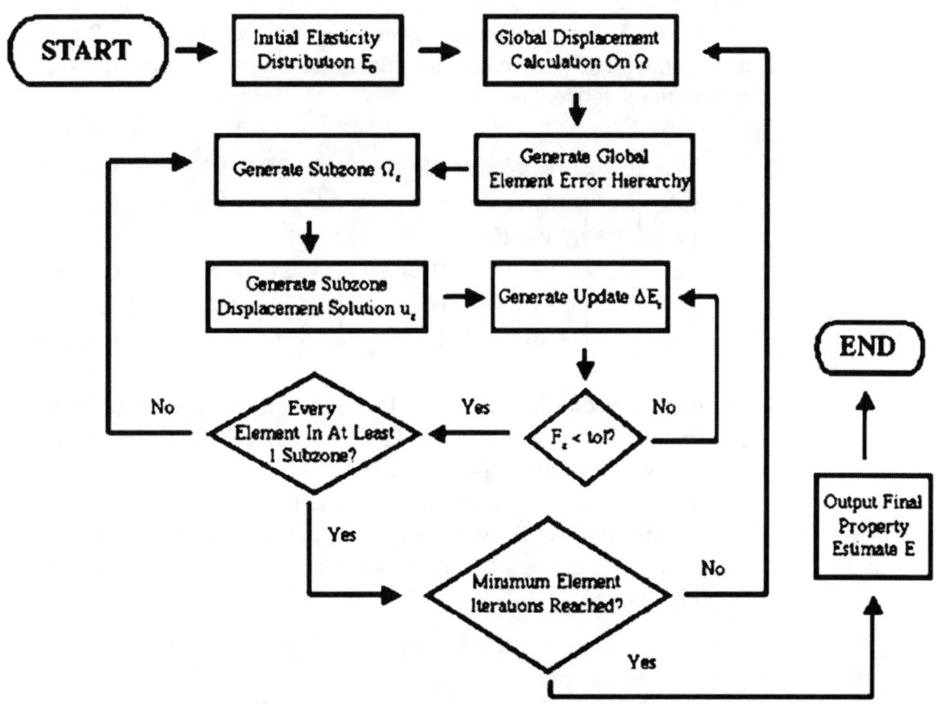

Figure 8-2. Flowchart of the subzone inversion algorithm. Courtesy of Van Houten et al. 1999

8.4 MECHANICAL CHARACTERIZATION OF SKELETAL MUSCLES

Skeletal muscles are often evaluated by MRE under two conditions: relaxed condition and contracted condition. Significance of each is discussed in below.

Resistance to stretching a muscle comes primarily from two sources, titin and the extracellular matrix (ECM). The relative contribution of each of these sources may vary, but each has a significant role. Hill's [22] conceptual model provides a schema for the understanding of the mechanical properties of muscle. The Hill model is composed of three elements: two which are arranged in series which, in turn, are in parallel with the third element. The contractile element is freely extendable when at rest, but capable of shortening when activated by an electrical stimuli. The contractile element is connected to an elastic serial element. The serial element accounts for the muscle elasticity during isometric (constant muscle length) force conditions. These two elements are then joined in parallel with another elastic element used to account for the elasticity of the muscle at relaxed condition. The parallel element accounts for the ECM surrounding the muscle fibers while the series elastic element accounts for the elasticity of the cross-bridges within the muscle.

ECM of the muscle occupies the intercellular space and is indirectly connected to the contractile proteins of the muscle cell through the subsarcolemmal cytoskeleton. In addition to providing mechanical support for the tissue and passive resistance to stretching, the ECM has a number of important roles. Two are of particular importance: the transmission of force from muscle fibers to the tendon and the transfer of mechanical signaling to the muscle cells. The structure, composition, and function of the ECM are influenced by many factors including the activity, age, and pathology associated with the muscle. Therefore, stiffness measurement of muscle under relaxed condition provides important information in studying these changes.

Even under the relaxed condition, skeletal muscle tissues show significant anisotropy, which is often simplified as transverse isotropy. Assuming a Cartesian coordinate system where axis 3 along the muscle fiber direction and axes 1 and 2 being the two orthogonal axes in the plane perpendicular to the muscle axis direction, transverse isotropic elasticity can be characterized as

$$\begin{bmatrix} \varepsilon_{11} \\ \varepsilon_{22} \\ \varepsilon_{33} \\ \varepsilon_{12} \\ \varepsilon_{13} \\ \varepsilon_{23} \end{bmatrix} = \begin{bmatrix} 1/E_{\mathrm{p}} & -v_{\mathrm{p}}/E_{\mathrm{p}} & -v_{\mathrm{tp}}/E_{\mathrm{t}} & 0 & 0 & 0 \\ -v_{\mathrm{p}}/E_{\mathrm{p}} & 1/E_{\mathrm{p}} & -v_{\mathrm{tp}}/E_{\mathrm{t}} & 0 & 0 & 0 \\ -v_{\mathrm{tp}}/E_{\mathrm{t}} & -v_{\mathrm{tp}}/E_{\mathrm{t}} & 1/E_{\mathrm{t}} & 0 & 0 & 0 \\ 0 & 0 & 0 & 2(1+v_{\mathrm{p}})/E_{\mathrm{p}} & 0 & 0 \\ 0 & 0 & 0 & 0 & 1/G_{\mathrm{t}} & 0 \\ 0 & 0 & 0 & 0 & 0 & 1/G_{\mathrm{t}} \end{bmatrix} \begin{bmatrix} \sigma_{11} \\ \sigma_{22} \\ \sigma_{33} \\ \sigma_{12} \\ \sigma_{13} \\ \sigma_{23} \end{bmatrix} \quad (8-29)$$

assuming isotropy in 1-2 plane. The five independent elasticity parameters are: v_{tp} the Poisson's ratio between axis 3 and axis 1 (or axis 2), v_p the Poisson's ratio in 1-2 plane, E_{p} the Young's modulus in 1-2 plane, E_{t} the Young's modulus along axis 3, and G_{t} the shear modulus along axis 3. Extracting of elasticity parameters of an anisotropic material generally requires three-dimensional MRE imaging for stiffness inversion.

Like many biological soft tissues, viscoelasticity of skeletal muscle is often modeled by Voigt model:

$$\sigma_{ij} = C_{ijkl}\varepsilon_{kl} + D_{ijkl}V_{kl} \quad (8-30)$$

where V_{kl} is the strain rate tensor ($V_{ij} = \dot{\varepsilon}_{ij}$), and D_{ijkl} is the viscosity coefficient tensor. The one-dimensional representation of the Voigt model is simply:

$$\sigma = E\varepsilon + \eta\dot{\varepsilon} \quad (8-31)$$

where E is the Young's modulus and η is the viscous coefficient. Dynamic mechanical analysis (DMA) is often utilized to characterize the elastic component (storage modulus E') and viscous component (loss modulus E'') and tan Delta ($\tan\delta = \frac{E''}{E'}$) of a linear viscoelastic material, respectively. For Voigt model, (8-19) simply

$$E' = E$$
$$E'' = \omega\eta \quad (8-32)$$

In the contracted condition, MRE study on human biceps has reported increased shear wavelength with both passive and active tension T [23, 24]. Such increase in wavelength can be achieved through two different mechanisms: (1) direct change in skeletal muscle material property through, for example, change in number of cross-link, or change in fiber pennation angle; (2) effect of tension on flexural wavelength. While the former is self-explanatory, the later can be illustrated by a simplified model of flexural wave propagation in one-dimensional elastic beam [17]:

$$\lambda = \frac{2\pi}{\left[-\frac{T}{2EI} + \left(\frac{T^2}{4E^2I^2} + \frac{\omega^2\rho A}{EI}\right)^{\frac{1}{2}}\right]^{\frac{1}{2}}} \quad (8-33)$$

with E being the Young's modulus, I the moment of inertia of the beam, A the cross-sectional area of the beam, and ρ the material density. The denominator is a monotonically decreasing function of T. Therefore, the higher the tension is, the longer the shear wavelength would result. In addition, flexural wave velocity in the beam increases with applied frequency, suggesting a dispersive system [17]. Although a simplified one-dimensional model, the above principle of flexural wave propagation may apply to in vivo skeletal muscles.

8.5 APPLICATIONS OF MRE TO SKELETAL MUSCLES

MRE has been successfully applied to skeletal muscles in the biceps brachii [11, 25], flexor digitorum profundus [26], thigh muscles [9, 10, 26], and upper trapezius [12]. While some studies developed methods for collecting MRE data in specific muscles and reported a database of muscle stiffness, other studies used MRE for specific tasks such as correlating MRE to EMG, and more clinically importantly, determination of changes in shear stiffness in pathological muscles.

Positioning and loading devices are often used to acquire MRE image data at three muscle conditions: relaxed, passively loaded, and actively contracted at 5% to 20% of maximal voluntary contracture (MVC). For example, to test the biceps under load, a pulley system consisting of a hand grip at one end and weights at the other end was used [26]. The muscles in the lower limb has been tested in the relaxed configuration as well as actively contracted state using three types of positioning and loading devices: 1) a positioning and loading device capable of adjusting the position of the ankle joint where weights connected to a cable and pulley system applied plantarflexion and dorsiflexion moments (see Fig. 8-3a), thus investigating an isotonic contraction [26]; 2) a foot plate with a strain gauge incorporated to measure isometric force with the ankle fixed in neutral [23, 27]; and 3) a positioning and loading device with an MR-compatible torque cell (see Fig. 8-3b) to measure isometric moments and allowing for passive joint positioning (see Fig. 8-3c) [9, 10].

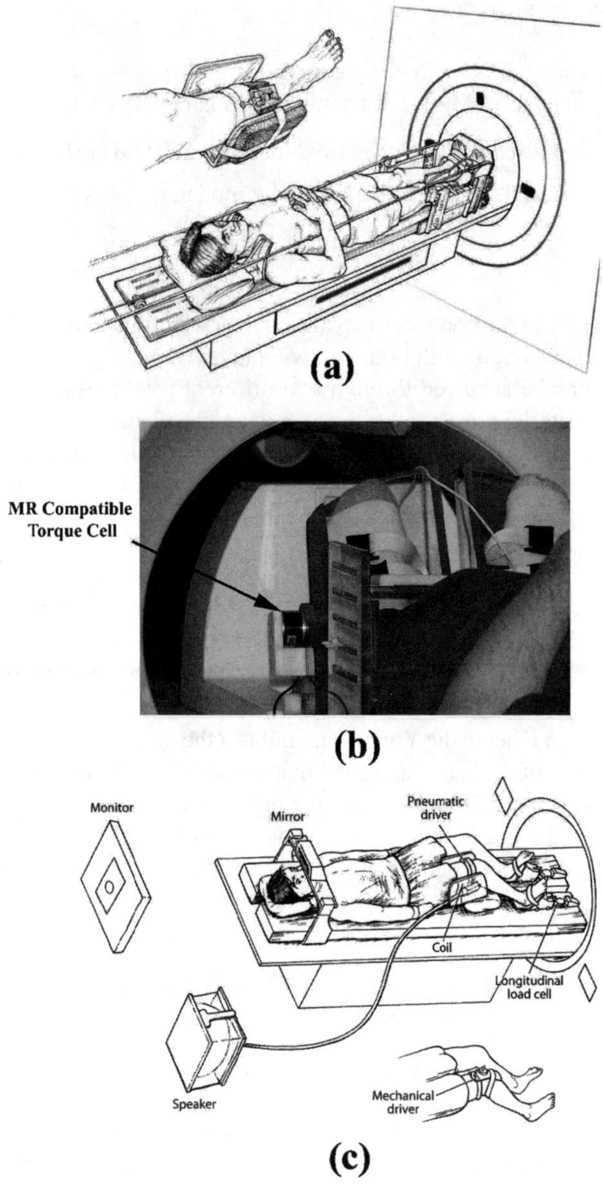

Figure 8-3. Examples of positioning and loading devices used to collect muscle MRE data. (a) Weights connected to a cable and pulley system applied plantarflexion and dorsiflexion moments to the ankle joint; (b) MR-compatible torque cell; (c) leg press and examples of mechanical and pneumatic drivers used to induce shear waves into the proximal leg muscles

8.5.1 Databases of Muscle Stiffness Using MRE

The repeatability of muscle MRE acquisition was assessed in the biceps brachii [26] and lateral gastrocnemius. Repeat data were collected in the relaxed biceps brachii of two volunteers over 7 days. The mean and standard deviations of the stiffness values of the repeat trials were calculated [26]. Repeat data were collected from nine volunteers in the relaxed lateral gastrocnemius. The coefficient of variation was 15.7%. The mean standard deviation was 2.32 kPa, which was comparable to the standard deviations reported in the biceps brachii. When the lateral gastrocnemius was contracted, the coefficient of variation increased to 19.4% and the standard deviation increased to 5.2 kPa.

An increase in shear stiffness was observed when the muscles were isometrically contracting [10, 11, 23, 27]. For example, when the volunteers extended their proximal legs at 10% of the MVC, there was a significant increase ($p < 0.05$) in the shear stiffness of the vastus lateralis. At 20% of the MVC, there was a significant increase in the shear stiffness of the vastus medials [10].

8.5.2 Correlation of MRE Data With a Functional Examination

It was hypothesized that the shear stiffness of distal leg muscles were correlated to muscle activity that is measurable by EMG [1, 23]. The shear stiffness measured with MRE was compared to the EMG activity of the muscles under the same loading conditions. EMG data were collected with surface electrodes for the medial gastrocnemius, lateral gastrocnemius, and tibialis anterior, while fine wire EMG data were collected from the soleus. The R^2 values ranged from 0.82 to 0.90 [23]. The highest correlation was found in the soleus, suggesting that the correlations in the other three muscles may improve if a fine wire EMG electrode was used [23].

8.5.3. Assessing Pathologic Muscle With MRE

Three typical pathologies affecting skeletal muscles are discussed in this section: hyperthyroid myopathy, myositis, and myofascial pain.

Graves' disease is a common cause of hyperthyroidism and leads to a variety of clinical symptoms including skeletal muscle weakness (i.e., hyperthyroid myopathy) that may be quite profound but is reversible following correction of hyperthyroidism [28, 29]. In hyperthyroidism, a decrease in the number of slow twitch muscle fibers and an increase in fast twitch muscle fibers were observed [30]. This change in skeletal muscle composition may result in change in muscle stiffness. The in vivo MRE examinations on stiffness of proximal lower limb muscle (vastus medialis) have shown to be capable of detecting stiffness differences between healthy and the pathologic muscles. Mechanical vibration was applied to the proximal lower limb via a pneumatic tube driver, and muscle contracture was monitored by the MR-compatible torque cell assembled to the foot plate. MRE scans on the patients with hyperthyroid myopathy showed a lower shear stiffness in relaxed condition (2.11±0.61 kPa) compared to the shear stiffness following treatment of hyperthyroidism (5.52±1.52 kPa). Pre-treatment muscle stiffness was also significantly lower than that of the age-matched healthy volunteers (4.56±0.40 kPa) [9, 10] (see Fig. 8-4).

Idiopathic inflammatory myopathy (IIM or myositis) is a heterogeneous group of disorders characterized by symmetric proximal muscle weakness and elevated serum levels of enzymes derived from skeletal muscle. Types of myositis include dermatomyositis, polymyositis, and inclusion body myositis. All have in common lymphocytic infiltration of muscles, edema within the muscle, altered muscle composition, and impaired muscle function. Associated with these changes is a decrease in muscle endurance and/or strength. Traditional assessments of disease activity include manual muscle strength testing, measurements of muscle enzymes, electromyography studies, muscle biopsy, and disease activity measures as defined in the literature [31]. While these methods are useful clinical tools, questions still arise about the affects of the disease and its treatment on the muscle.

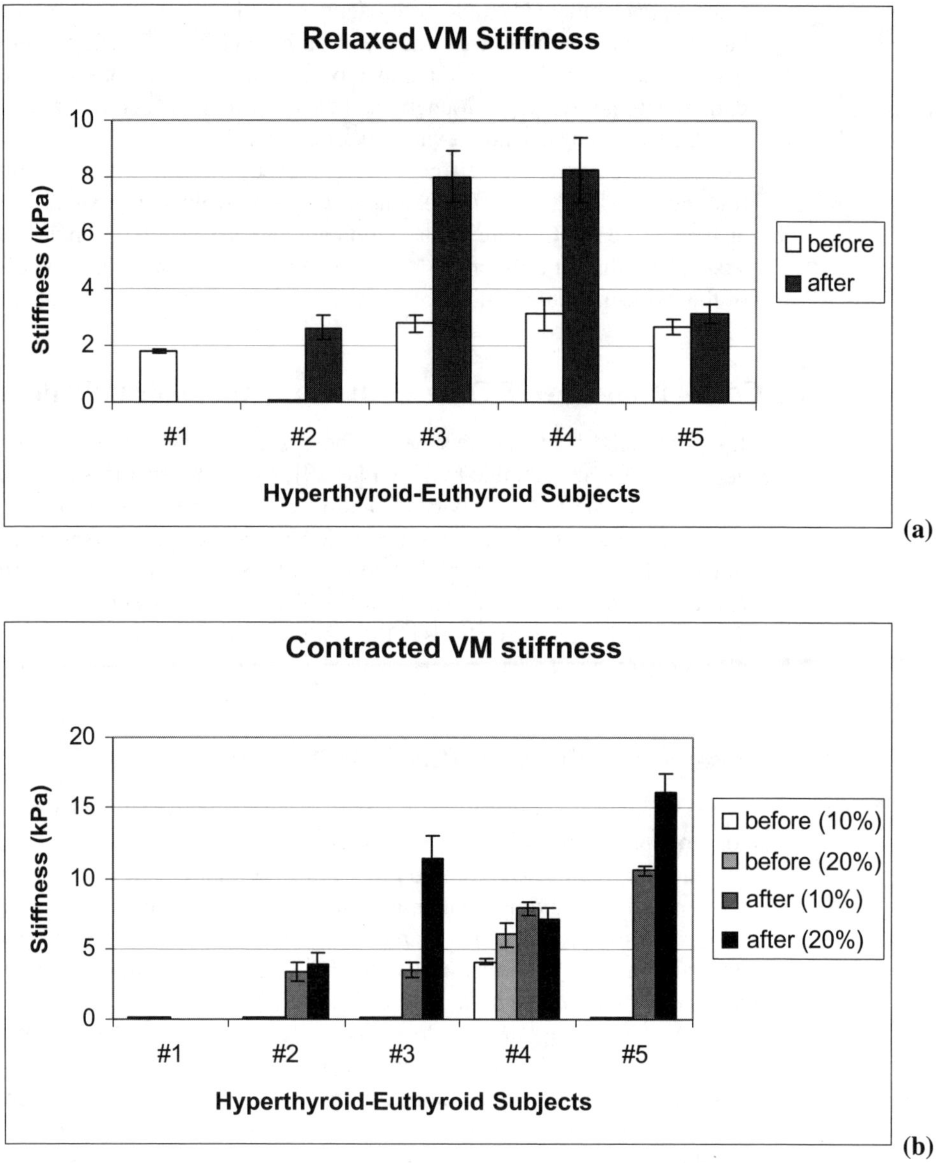

Figure 8-4. Stiffness of vastus medialis (VM) measured at relaxed and contracted state before and after treatment of hyperthyroidism

Specifically, what are the effects of myositis on the muscle fibers and the surrounding matrix, as well as what is the cause of the continued lack of endurance, in many, despite treatment? It is hypothesized that muscle extracellular matrix, reflected by stiffness, is important not only for the force transmission, thus the function of the muscle, but also the mechano-biology of the cell, thus the response to treatment. Therefore, evaluating the stiffness of muscle tissue would reveal information about the conditions of muscle extracellular matrix, which in turn may provide insight on the effects of myositis on muscle extracellular matrix. To test the above hypothesis, MRE is adopted to assess the shear stiffness of the muscles of the proximal leg. Muscle stiffness was examined following the same MRE scan protocol as hyperthyroid myopathy. Preliminary results on five patients with myositis have shown decreased shear stiffness values in vastus medialis in both the relaxed and active states when compared to healthy controls.

Myofascial pain syndrome (MPS) is a painful musculoskeletal condition, a common cause of musculoskeletal pain [32, 33]. The main findings about myofascial pain are localized taut bands of increased tone which contain even more circumscribed points of tenderness (i.e., trigger points) that when compressed produce stereotypical patterns of referred pain. The identification of taut bands and trigger points was not only important for diagnosis, but also potential treatment [34]. Despite their importance, there still lacks either a laboratory test or imaging technique capable of objectively confirming either their nature or location, while criticisms remain on the repeatability and subjectivity issues of conventional manual palpation examination. Because of its ability to differentiate tissue stiffness, MRE is proposed as an imaging method to objectively detect the location of myofascial taut band and quantify the stiffness of myofascial taut band which therefore facilitates the myofascial pain diagnosis. First, the myofascial pain patients underwent manual palpatory examination by an experienced physician, and location of myofascial taut band detected by palpation was marked. MRE scan was then performed on patients. Mechanical vibration (planar wave) was applied on the muscle-tendon junction at spin of scapula allowing a shear wave to propagate through upper trapezius, one the most frequent locations of myofascial taut band and trigger point. MRE phase images showed chevron-shaped wave front in the vicinity of myofascial taut band detected by manual palpation (see Fig. 8-5a). Stiffness images showed a statistically significant 50% to 100% ($p = 0.01$) increase of shear stiffness (8.4 kPa) in the taut band regions of the involved subjects relative to that of the controls (4.2 kPa) or in nearby uninvolved muscle (4.8 kPa) (see Fig. 8-5 b).

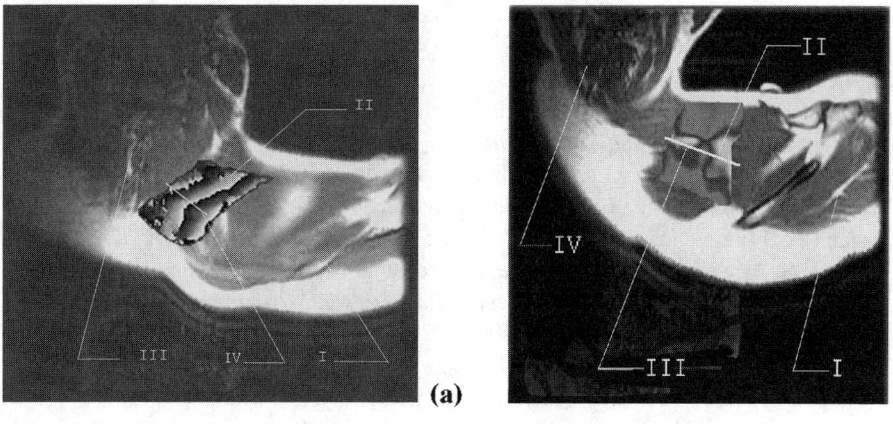

(a) (b)

Figure 8-5. (a) Typical MRE phase image of upper trapezius of a patient with myofascial pain superimposed to the MR image of the same subject, showing chevron-shaped wave fronts under the planar wave mechanical vibration. (I) Spine of scapular; (II) MRE phase image with chevron-shaped wave fronts observed in the region of taut band; (III) myofascial taut band identified by the palpation; (IV) cervical spine. (b) Typical MRE stiffness image of upper trapezius of a patient with myofascial pain, superimposed to the MR image of the same subject. (I) Spine of scapular; (II) MRE stiffness image showing taut band region; (III) location of the taut band identified through the palpation examination; (IV) cervical spine

REFERENCES

[1] Ringleb, S. I., Bensamoun, S. F., Chen, Q., Manduca, A., An, K. N., and Ehman, R. L., 2007, "Applications of Magnetic Resonance Elastography to Healthy and Pathologic Skeletal Muscle," *Journal of Magnetic Resonance Imaging*, **25**(2), pp. 301–309.

[2] Leonard, C. T., Deshner, W. P., Romo, J. W., Suoja, E. S., Fehrer, S. C., and Mikhailenok, E. L., 2003, "Myotonometer Intra- and Interrater Reliabilities," *Archives of Physical Medicine and Rehabilitation*, **84**(6), pp. 928–932.

[3] Leonard, C. T., Brown, J. S., Price, T. R., Queen, S. A., Mikhailenok, E. L., 2004, "Comparison of Surface Electromyography and Myotonometric Measurements During Voluntary Isometric Contractions," *Journal of Electromyography and Kinesiology*, **14**(6), pp. 709–714.

[4] Bizzini, M., and Mannion, A. F., 2003, "Reliability of a New, Hand-Held Device for Assessing Skeletal Muscle Stiffness," *Clinical Biomechanics*, **8**(5), pp. 459–461.

[5] Arokoski, J. P., Surakka, J., Ojala, T., Kolari, P., and Jurvelin, J. S., 2005, "Feasibility of the Use of a Novel Soft Tissue Stiffness Meter," *Physiological Measuerment*, **26**(3), pp. 215–228.

[6] Manduca, A., Oliphant, T. E., Dresner, M. A., et al, 2001, "Magnetic Resonance Elastography: Non-Invasive Mapping of Tissue Elasticity," *Medical Image Analysis*, **5**(4), pp. 237–254.

[7] Muthupillai, R., Lomas, D. J., Rossman, P. J., Greenleaf, J. F., Manduca, A., and Ehman, R. L., 1995, Magnetic Resonance Elastography by Direct Visualization of Propagating Acoustic Strain Waves, *Science*, **269**(5232), pp. 1854–1857.

[8] Manduca, A., Lake, D. S., Kruse, S. A., and Ehman, R. L., 2003, "Spatio-Temporal Directional Filtering for Improved Inversion of MR Elastography Images," *Medical Image Analysis*, **7**(4), pp. 465–473.

[9] Bensamoun, S. F., Ringleb, S. I., Chen, Q., and Ehman, R. L., An, K. N., Brennan, M., 2007, "Thigh Muscle Stiffness Assessed with Magnetic Resonance Elastography in Hyperthyroid Patients Before and After Medical Treatment," *Journal of Magnetic Resonance Imaging*, **26**(3), pp. 708–713.

[10] Bensamoun, S. F., Ringleb, S. I., Littrell, L., et al, 2006, "Determination of Thigh Muscle Stiffness Using Magnetic Resonance Elastography" *Journal of Magnetic Resonance Imaging*, **23**(2), pp. 242–247.

[11] Dresner, M. A., Rose, G.H., Rossman, P. J., Muthupillai, R., Manduca, A., and Ehman, R. L., 2001, "Magnetic Resonance Elastography of Skeletal Muscle," *Journal of Magnetic Resonance Imaging*, **13**(2), pp. 269–276.

[12] Chen, Q., Bensamoun, S. F., Basford, J. R., Thompson, J., and An, K. N., 2007 (In press), "Identification and Quantification of Myofascial Taut Bands with Magnetic Resonance Elastography," *Archives of Physical Medicine and Rehabilitation.*

[13] Braun, J., Buntkowsky, G., Bernarding, J., Tolxdorff, T., and Sack, I., 2001, "Simulation and Analysis of Magnetic Resonance Elastography Wave Images Using Coupled Harmonic Oscillators and Gaussian Local Frequency Estimation," *Magnetic Resonance Imaging*, **19**(5), pp. 703–713.

[14] Kruse, S. A., Smith, J. A., Lawrence, A. J., et al, 2000, "Tissue Characterization Using Magnetic Resonance Elastography: Preliminary Results," *Physics in Medicine & Biology*, **45**(6), pp. 1579–1590.

[15] Braun, J., Braun, K., and Sack, I., 2003, "Electromagnetic Actuator for Generating Variably Oriented Shear Waves in MR Elastography," *Magnetic Resonance in Medicine*, **50**(1), pp. 220–222.

[16] Wu, T., Felmlee, J. P., Greenleaf, J. F., Riederer, S. J., and Ehman, R. L., 2000, "MR Imaging of Shear Waves Generated by Focused Ultrasound," *Magnetic Resonance in Medicine*, **43**(1), pp. 111–115.

[17] Graff, K. F., 1991, *Wave Motion in Elastic Solids*, Dover Publications, Inc., New York, US.

[18] Oliphant, T. E., Manduca, A., Ehman, R. L., and Greenleaf, J. F., 2001, "Complex-Valued Stiffness Reconstruction for Magnetic Resonance Elastography by Algebraic Inversion of the Differential Equation," *Magnetic Resonance in Medicine*, **45**(2), pp. 299–310.

[19] Papazoglou, S., Braun, J., Hamhaber, U., and Sack, I., 2005, "Two-Dimensional Waveform Analysis in MR Elastography of Skeletal Muscles," *Physics in Medicine & Biology*, **50**(6), pp. 1313–1325.

[20] Van Houten, E. E., Paulsen, K. D., Miga, M. I., Kennedy, F. E., and Weaver, J. B., 1999, "An Overlapping Subzone Technique for MR-Based Elastic Property Reconstruction," *Magnetic Resonance in Medicine*, **42**, pp. 779–786.

[21] Van Houten, E. E., Miga, M. I., Kennedy, F. E., and Paulsen, K. D., 2000, "Elasticity Reconstruction from Experimental MR Displacement Data: Initial Experiences with an Overlapping Subzone Finite Element Inversion Process," *Medical Physics*, **27**(1), pp. 101–107.

[22] Hill, A. V., 1970, *First and Last Experiments in Muscle Mechanics,* Cambridge University Press, Cambridge, UK.

[23] Heers, G., Jenkyn, T., Alex Dresner M, et al, 2003, "Measurement of Muscle Activity with Magnetic Resonance Elastography," *Clinical Biomechanics*, **18**(6), pp. 537–542.

[24] Basford, J. R., Jenkyn, T. R., An, K. N., Ehman, R. L., Heers, G., and Kaufman, K. R., 2002, "Evaluation of Healthy and Diseased Muscle with Magnetic Resonance Elastography," *Archives of Physical Medicine and Rehabilitation*, **83**(11), pp. 1530–1536.

[25] Sack, I., Bernarding, J., and Braun, J., 2002, "Analysis of Wave Patterns in MR Elastography of Skeletal Muscle Using Coupled Harmonic Oscillator Simulations," *Magnetic Resonance Imaging*, **20**(1), pp. 95–104.

[26] Uffmann, K., Maderwald, S., Ajaj, W., et al, 2004, "In Vivo Elasticity Measurements of Extremity Skeletal Muscle with MR elastography," *NMR in Biomedicine*, **17**(4), pp. 181–190.

[27] Jenkyn, T. R., Ehman, R. L., and An, K. N., 2003, "Noninvasive Muscle Tension Measurement Using the Novel Technique of Magnetic Resonance Elastography (MRE)," *Journal of Biomechanics*, **36**(12), pp. 1917–1921.

[28] Celsing, F., Westing, S. H., Adamson, U., et al, 1990, "Muscle Strength in Hyperthyroid Patients Before and After Medical Treatment," *Clinical Physiology*, **10**, pp. 545–550.

[29] Zurcher, R. M., Horber, F. F., Grunig, B. E., et al, 1989, "Effect of Thyroid Dysfunction on Thigh Muscle Efficiency," *Journal of Clinical Endocrinology and Metabolism*, **69**, pp. 1082–1086.

[30] Brennan, M. D., Coenen-Schimke, J. M., Bigelow, M. L., et al, 2006, "Changes in Skeletal Muscle Protein Metabolism and Myosin Heavy Chain Isoform Messenger Ribonucleic Acid Abundance After Treatment of Hyperthyroidism," *Journal of Clinical Endocrinology and Metabolism*, **91**, pp. 4650–4656.

[31] Oddis, C. V., Rider, L. G., Reed, A. M., et al, 2005, "International Consensus Guidelines for Trials of Therapies in the Idiopathic Inflammatory Myopathies," *Arthritis and Rheumatology*, **52**(9), pp. 2607–2615.

[32] Alvarez, D. J., and Rockwell, P. G., 1002, "Trigger Points: Diagnosis and Management," *American Family Physician*, **65**(4), pp. 653–660.

[33] Gerwin, R. D., 2001, "Classification, Epidemiology, and Natural History of Myofascial Pain Syndrome," *Current Pain & Headache Reports*, **5**(5), pp. 412–420.

[34] Graboski, C. L., Gray, D. S., Burnham, R. S., 2005, "Botulinum Toxin A Versus Bupivacaine Trigger Point Injections for the Treatment of Myofascial Pain Syndrome: A Randomised Double Blind Crossover Study," *Pain*, **118**(1-2), pp. 170–175.

Part **2**:

CHARACTERIZATION

Section 1:

VESSEL CHARACTERIZATION

Chapter 9

ACOUSTIC RESPONSE OF THE HUMAN ARTERIES

Essa El-Aklouk[1], Ahmed Al-Jumaily[1], and Andrew Lowe[2]
[1]Institute of Biomedical Technologies (IBTec), Auckland University of Technology (AUT), Auckland, New Zealand
[2]Pulsecor Ltd., Auckland, New Zealand

ABSTRACT

In hypertension and aging, central elastic arteries become stiffer, and hence, the central pulse pressure is augmented because of the increase in the pulse wave velocity and the early return of reflected waves to the heart from the periphery. Valuable information on arterial properties such as stiffness can be obtained from both central (aortic) and peripheral (radial) pressure waveforms. A feasibility study for the non-invasive estimation of arterial stiffness using pressure waves detected by a pneumatic cuff wrapped around the upper arm is presented. The propagation and reflection of arterial pressure waves (generated by the heart) in the central elastic arteries are simulated using a simplified water hammer acoustic model. Furthermore, a lumped parameter model is used to describe the transmission of the pressure waves from the brachial artery to the cuff external wall. By combining the two models, it is possible to simulate the pressure contours in the brachial artery and illustrate how these pressures transmit to the cuff's external wall. The effects of aortic stiffness, aortic geometrical variations, heart rate, and cuff pressure were investigated by simulating the model using different parameters and observing the pressure and strain augmentation and the timing of the return of the reflected wave. The model results obtained in this work are in agreement with published experimental results.

9.1 BASIC CONCEPTS

Cardiovascular disease is a major problem worldwide. It is the number one cause of death and disability in the United States and in the majority of the developed world. The total cost of cardiovascular disease in the United States was estimated at $368 billion in 2004 [1]. Recent statistics show that 42% of deaths in New Zealand are caused by cardiovascular diseases [2].

Arteriosclerosis (hardening of the arteries) is the major underlying cause of all deaths resulting from cardiovascular disease [3]. One common form of arteriosclerosis is known as atherosclerosis, which refers to the deposition of atheromas (fats and other substances often known as plaque) inside the arterial wall. This can lead to complete calcification of the artery wall, which renders it extremely stiff [3].

Plaque deposition causes arterial stenosis (narrowing) that leads to blood flow obstruction. Deposits can also rupture, which may lead to the formation of blood clots that quickly obstruct or completely stop blood flow in the affected artery [3, 4]. Complete obstruction of blood flow will lead to the necrosis of the tissue supplied by the artery. This can happen to any organ or limb and cause sudden death in the case of coronary artery thrombosis, which leads to a heart attack [3, 4], or stroke in the case of arteries supplying the brain.

Excessive weakening of the artery wall caused by atherosclerosis can lead to aneurysms. An aneurysm is a localized, blood-filled dilation of a blood vessel. These aneurysms usually affect the arteries of the brain and the aorta. Rupture of an aneurysm can happen at any time and may lead to sudden death [3, 5].

Another common category of cardiovascular disease is that relating to the function of the heart itself. One determinant of heart function is known as afterload, which is effectively the force (caused by blood pressure in the ascending aorta) against which the heart pumps. The afterload is, in large part, affected by the stiffness of the arterial system by the following mechanism: The rhythmic action of the heart introduces pressure waves into the arterial tree. These waves are partially reflected when they experience a change in the characteristic impedance such as at the arterial bifurcations [3, 6]. Interference between the forward-traveling waves and the reflected waves determines the shape of the pressure wave. The timing of these interference patterns is related to the pulse wave velocity (PWV). High PWV (which occurs in stiff arteries) leads to the reflected wave augmenting the pressure at the aortic valve during systole, rather than diastole, thereby increasing the afterload. A chronically increased afterload can lead to conditions such as left ventricular hypertrophy and heart failure [3].

9.1.1 Non-Invasive Stiffness Detection Methods

As is evident from the discussion above, arteriosclerosis is the underlying reason for a large number of cardiovascular events. The main problem lies in the fact that arteriosclerosis usually goes unnoticed. For 65% of men and 47% of women diagnosed in the United States, the first sign of atherosclerosis is a heart attack or cardiac death [7]. Arteriosclerosis begins early in life and progresses with age [8]. The poor rate of diagnosis is caused by the small number of symptoms associated with the onset of the disease and the lack of regular monitoring.

Deaths from cardiovascular disease and specifically arteriosclerosis can be greatly reduced with proper monitoring and treatment [9].

There has been an increasing emphasis on the need to accurately diagnose arterial stiffness in medical literature [9]. Unfortunately, conventional invasive methods to diagnose arterial stiffness such as catheterization are often expensive, difficult, and involve risks to the patient of a magnitude that cannot be justified by the potential benefits of the additional information.

For this reason, demand has developed for accurate and non-invasive methods to measure arterial stiffness. This has resulted in a number of non-invasive methods to assess arterial stiffness becoming available. These methods employ various principles and generally make use of hydraulic and/or elastic theories.

Non-invasive stiffness detection methods can be divided into three main methods: local methods that make use of the pressure-area analysis of a specific region of the central or peripheral arteries, regional methods that make use of the PWV in a segment of the arterial tree, and waveform analysis methods that use of the shape of the pressure pulse at central and peripheral locations.

Local methods are based on the principle that the distension of a region in the arteries caused by a given pressure is dependent on the stiffness/compliance at this specific location as shown in Fig. 9-1. The compliance of the artery can be described as [10]

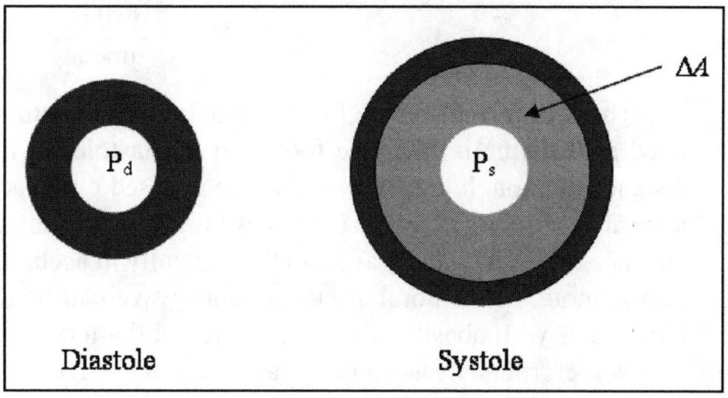

Figure 9-1. A schematic diagram showing the change in area of the artery with blood pressure

$$\text{compliance} = \frac{\Delta A}{A_d \cdot P_p}, \qquad (9-1)$$

where

ΔA = change in the cross-sectional area between the diastolic and systolic pressures
A_d = diastolic cross-sectional area
P_p = pulse pressure

These methods do not require the development of models of the circulation and can be used non-invasively and directly to estimate the arterial stiffness at various sites along the arterial tree. The local arterial stiffness of most of the peripheral arteries can be measured directly by simultaneously measuring artery distension and blood pressure [10]. Ultrasonic devices are commonly used to measure the diameter of the artery, whereas applanation tonometry is used to measure the local blood pressure. Recently, researchers were also able to measure the stiffness of central arteries such as the aorta using magnetic resonance imaging [9].

The main advantage of this technique is that it does not require the development of circulatory models and that the local stiffness is directly measured. However, this method requires high levels of expertise and is expensive and time-consuming to perform. Furthermore, the technique is normally limited by the accuracy of the arterial diameter measurement because a video or image analysis system is usually used. Also, the local blood pressure in the central arteries cannot be measured directly in a non-invasive manner, and thus, the pulse pressure is usually estimated from the brachial or radial blood pressure or by using a transfer function that predicts the central pressure waveform from applanation tonometry measurements at the radial artery [6]. This is introduces further limitations in accuracy [11].

Regional methods for arterial stiffness estimation are based on the principle that PWV is directly dependent on the stiffness of the artery through which the wave is propagating. PWV-based methods are simple, non-invasive, cheap, and reasonably accurate in determining arterial stiffness. The regional stiffness of the aorta is usually of major interest for a number of reasons. First, the aorta is the biggest contributor to the arterial buffering function [12, 13]; second, aortic PWV has been shown to be an independent predictor of cardiovascular disease in a number of populations because it has the largest effect on the left ventricular function [14, 15]. The aortic PWV is usually determined by measuring pressure, artery distension, or Doppler waveforms at the carotid and femoral arteries. The distance traveled by the waves is usually estimated by measuring the body surface distance between the two measuring sites. The PWV can be estimated as

$$\text{PWV} = \frac{\text{distance}}{\text{time lag}} \quad (9-2)$$

The foot-to-foot method is commonly used to estimate the transit time, which is defined as the time it takes the foot (end of diastole) of the wave to travel over a defined distance. Although the PWV technique is based on an accepted propagative model of the arterial system, there are inaccuracies introduced by the difficulty in estimating the actual distance traveled by the wave and the difficulty in accurately locating the wave's foot [16]. Furthermore, the femoral artery pressure wave can be difficult to accurately measure in individuals with obesity, diabetes, and femoral artery disease [16].

Waveform analysis methods are based on the fact that arterial pressure waveforms are a composite of forward-traveling waves produced by the contraction of the left ventricle and backward-traveling reflected waves from peripheral sites of impedance mismatch [17, 18]. The timing of the arrival of the reflected wave to the ascending aorta depends on both the reflection site and the PWV. In elastic arteries, PWV is relatively low, and hence, the reflected wave arrives at the ascending aorta during diastole [17, 18]. As the stiffness of the central arteries increases, the PWV increases and the reflected waves arrive at the ascending aorta earlier, where it augments the systolic pressure [17, 18]. This phenomenon is used as an indicator of arterial stiffness where the augmentation index (AI) is used to quantify its magnitude as shown in Fig. 9-2. In the central arteries, the AI is defined as [17-19]

$$\text{AI} = \frac{P_s - P_i}{P_s - P_d}, \quad (9-3)$$

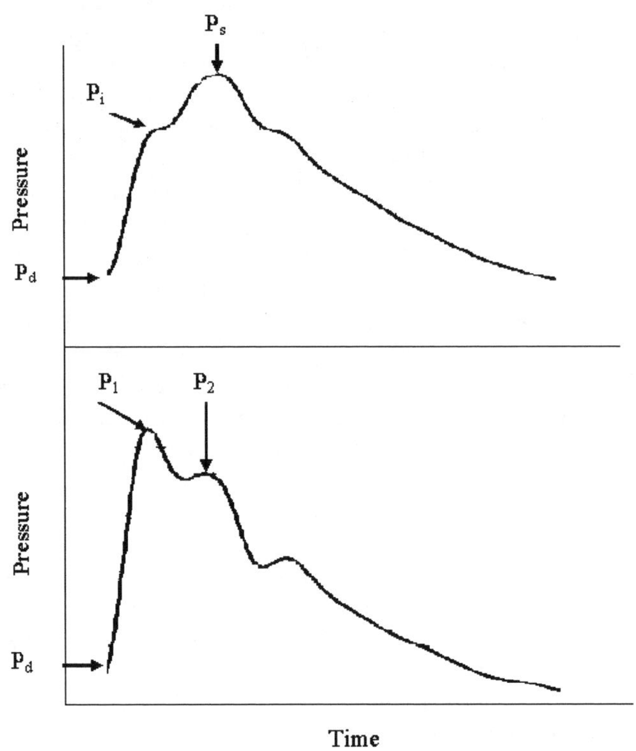

Figure 9-2. (Top) Central pressure waveform showing wave reflection from the lower body. (Bottom) Radial artery waveform showing wave reflection from the lower body

where

P_i = pressure at the inflection point corresponding to the arrival of the reflected wave
P_s = systolic pressure
P_d = diastolic pressure

The ascending aorta pressure waveform is an indicator of the real load that the left ventricle has to pump against, making it a primary target for analysis. Direct non-invasive measurement of ascending aortic pressure is not currently possible. Hence, ascending aortic pressure is generally either estimated using the common carotid pressure waveform directly or from the radial artery pressure in conjunction with a transfer function [20, 21]. Other sites such as the finger [22] have also been used. The transfer functions used to estimate aortic waveforms can be either individualized or generalized [20, 21]. The accuracy of these transfer functions has been the topic of much debate and, especially for high frequencies, is disputed [21]. Although common carotid artery tonometry does not require the use of a transfer function, it requires a high degree of technical expertise and cannot be applied in obese patients and patients with an advanced degree of atherosclerosis.

Recently, pressure wave augmentation in the radial artery and finger has been identified as an indicator of arterial stiffness [22-24]. As the stiffness of the elastic arteries increases, PWV increases and the reflected waves from the lower body arrive at the radial artery and the finger faster, which increases pressure augmentation [22, 24]. A different method is used to calculate the radial artery and finger AI. This is because of the different characteristic waveforms as shown in Figs. 9-2a and b. The radial AI is defined as [22]

$$AI_r = \frac{P_2}{P_1} \qquad (9-4)$$

This index has been shown to be related to arterial stiffness and is methodologically more easily obtained [22].

In summary, there are limitations relating to all three schemes for non-invasive measurement of arterial stiffness. Local methods are unable to estimate properties of the central arteries of interest. Regional PWV methods rely on estimates of pulse-path length and accurate detection of pulse onset and can be difficult to apply in practice. Waveform analysis methods are reliant on uncertain transformations of peripheral waveforms, which can be difficult to obtain or neglect any transformation on the basis that the central and peripheral augmentations are statistically correlated.

The current work was prompted by another waveform analysis technology (Pulsecor, Auckland, New Zealand) that intends to make use of a brachial artery pulse measured using an inflatable cuff to non-invasively detect arterial stiffness. Hence, there is a need to develop a physiologically based mathematical model of the arterial system and the interactions between the pneumatic cuff with the upper arm soft tissue and the brachial artery. The central aim of this model is to be able to replicate the brachial artery pressure pulse and describe the transmission of this pressure through the pneumatic cuff. The developed model should be able to predict the effect of arterial stiffness on the shape of the obtained waveforms and also simulate the effect of a number of other artery and cuff parameters on the waveforms.

9.1.2 Cardiovascular Models

Numerous attempts have been made to mathematically model the human arterial tree including lumped [25-27], one-dimensional [25, 28-30] and multidimensional models [31]. Lumped and one-dimensional models are frequently used to simulate pressure wave propagations in the arteries, and they provide a good understanding of wave behavior at reasonable computational cost. Multi-dimensional models are able to simulate the complete system and are mainly used to give a better insight into the complex flow patterns that are noticed in the blood vessels [31]. However, they result in complex equations that require relatively high computational power and are unnecessary to study pressure wave propagation in the system.

Lumped models are often used to describe blood pressure and blood flow in the arterial system [25-27]. The lumped models are based on an electrical circuit analogy where voltage represents blood pressure, current represents flow, resistance represents arterial resistance, capacitors represent the compliance of the vessels, and inductors represent the inertia of the blood [25-27]. The Windkessel model was originally based on the analogy of a fire-hose system. This analogy resulted in the development of the electrical version of the Windkessel model. The Windkessel model is widely used to model blood flow and blood pressure waveforms in the arterial system. It has the advantage of being easy to solve because it results in simple ordinary differential equations (ODEs).

However, the Windkessel model has some limitations. First, it is not clear how to estimate the model parameters such as the resistances and other elements from measured blood pressure and blood flow [27]. Furthermore, the model assumes that the system consists of separate conduits and cushioning functions, which is not the case because the major arteries have both features. The model also assumes that the PWV is infinite throughout the system. PWV is not only finite; it has been measured to vary within parts of the arterial system [3]. For these reasons, the Windkessel model is only appropriate in modeling the arterioles [25]. One-dimensional distributed propagative models are more suited to describe pressure and flow in the major arteries [28, 30]. These models describe the arterial system as a network of simple distensible tubes where the tubes' distributed elastic properties allow for the finite speed propagation of pressure waves [30].

Wave reflections arise from changes in the impedance in the arterial tree. There has been a lot of debate on the origin of wave reflections in the cardiovascular system. Branching points in the major arteries and the high resistance arterioles are considered to be the major sites of reflection in the cardiovascular system [3, 32-35]. Substantial evidence suggests that the iliac bifurcation (or aortic bifurcation) is the main source of wave reflection [33-35]. Even in other studies where the high resistance arterioles are considered to be the major reflection site, the effective reflection site is measured to be in the region of the iliac bifurcation [3]. Hence, in this work, the iliac bifurcation is assumed to be the site of wave reflection from the lower body.

9.1.3 Cuff-Soft Tissue Artery Models

During the past couple of decades, there have been a number of models developed to describe the mechanics and dynamics of blood pressure, pneumatic cuff, arm soft tissue, and brachial artery hemodynamics. All of these models are mainly one dimensional or lumped. These models were developed to study the oscillometric technique of blood pressure measurement. Mauck et al [36] developed a simple one-dimensional model of the artery-soft tissue-cuff system. A more complete one-dimensional model has been developed by Forster and Turney [37], where the collapse of the brachial artery caused by the occlusive cuff is described in more detail.

Drzewiecki et al [38] further improved the accuracy of the models by using an improved model for the pneumatic cuff and the brachial artery collapse. Finally, Ursino and Cristalli [39-41] developed separate lumped models for the pneumatic cuff, arm soft tissue, and brachial artery. These models were based on previous work in the field with some improvements in the physical and mathematical description [38-40]. The cuff-soft tissue-brachial artery model developed in this work is based on these models.

9.2 THEORETICAL FORMULATION

This section is divided into two sub-sections. The first deals with the acoustic model, which describes pressure wave propagation in the large systemic arteries. A number of assumptions were made to produce a practical model for wave propagation in the arteries so that it could be solved analytically. The final equations relate the pressure wave propagating velocity (c) to arterial geometry and material properties. This section also discusses the wave reflection behavior as these propagating waves reach a bifurcation or an area of large impedance mismatch. The resulting equations relate the reflection coefficient (R) and the transmission coefficient (T) to the impedance mismatch resulting from artery bifurcation or an abrupt change in geometry or material properties.

The second section deals with lumped parameter mathematical models developed for the brachial artery, upper arm soft tissue, and blood pressure cuff. These three models are then linked together to relate the intravascular brachial artery pressure to the circumferential strain on the outer cuff wall.

9.2.1 Acoustic Model

In this section, an acoustic model that describes the pressure wave propagation in the large systemic arteries is formulated.

Wave Propagation. The large central arteries are principally branching, compliant tubes filled with fluid. In this section, the general equations governing pressure wave propagation and reflection are developed. Wave propagation in a compliant tube conveying fluid is governed by partial differential equations. A number of assumptions can be made and justified to describe wave propagation by a set of ODEs.

It is assumed that fluid contained in the arteries is homogenous and is non-viscous. Taylor's work [42] has shown that the effects of non-homogeneity and viscosity of blood are only significant in arteries with very small diameters such as arterioles. In addition, it assumed that the arteries are cylindrical elastic tubes. Fung [43] showed that the effect of nonlinear viscoelasticity on wave propagation is minimal.

It is also assumed that the length of the tube is long compared with its diameter. Chang and Atabek [44] showed that this assumption is justified. The above assumptions infer that the flow can be treated as one dimensional.

From inspection of the free body diagram shown in Fig. 9-3, the basic equations of conservation of mass and momentum can be derived by considering mass conservation in the segment of an artery of length dx as shown in Fig. 9-3a, which leads to Eq. (9-5). Balancing the forces acting in the axial direction on a fluid element of length dx and cross-sectional area A as shown in Fig. 9-3b leads to Eq. (9-6)

$$\frac{\partial A}{\partial t}+\frac{\partial}{\partial x}(Au)=0 \qquad (9-5)$$

$$\frac{\partial u}{\partial t}+u\frac{\partial u}{\partial x}+\frac{1}{\rho}\frac{\partial p}{\partial x}=0, \qquad (9-6)$$

where

A = area of the artery
u = velocity of the fluid
ρ = density of the fluid

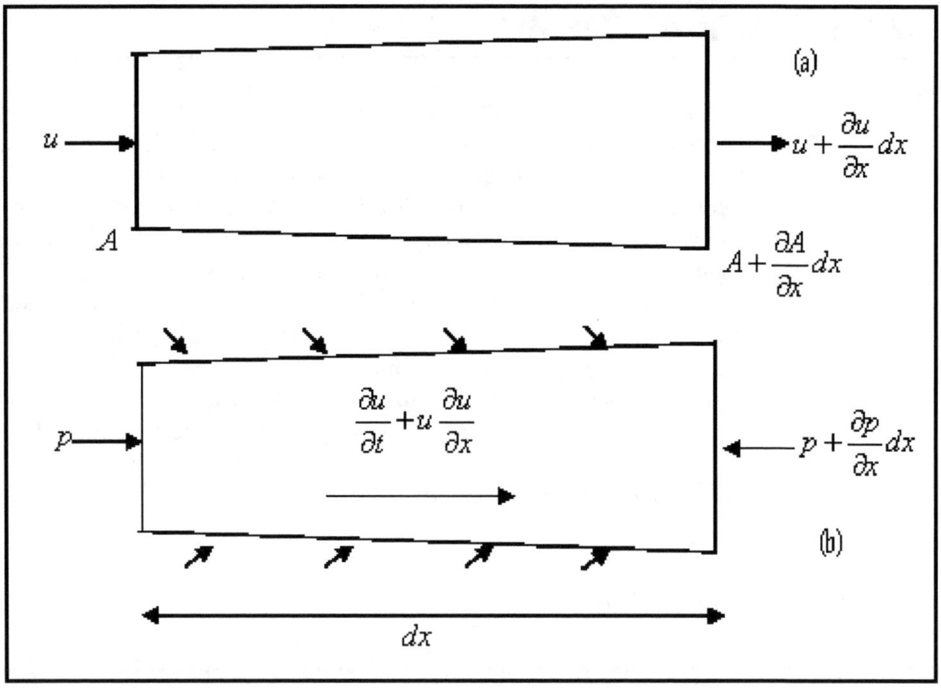

Figure 9-3. Arterial segment free body diagram showing mass and momentum conservation

It is assumed that the tube is thin-walled because the thickness-to-diameter ratio is usually less than 1:10 in the arteries. The bulk modulus of blood is around 100 times the elastic modulus of the artery wall, which leads to the assumption of incompressibility. Any small change in transmural pressure dp is balanced by a change in the artery's circumferential tension $\frac{Eh\mathrm{d}r}{r}$, where E is Young's modulus of elasticity, h is artery wall thickness, and r is the artery radius. A force balance gives

$$\frac{Eh\mathrm{d}r}{r} = r\mathrm{d}p \qquad (9-7)$$

Equations (9-5) to (9-7) can be used to derive an equation governing steady-state flow through an elastic artery, and with further simplifications, the theoretical wave speed within the artery can be determined.

Equation (9-5) is linearized by substituting $A = \pi r^2$ and assuming that the wave amplitude is much smaller than the wave length. Neglecting second-order terms reduces the equation to

$$\frac{\partial u}{\partial x} + \frac{2}{\mathrm{r}}\frac{\partial r}{\partial t} = 0 \qquad (9-8)$$

Equation (9-6) can be linearized by assuming small disturbances and noting that the blood velocity is low compared with the wave velocity in the arteries. The $u\frac{\mathrm{du}}{\mathrm{dx}}$ term is, therefore, not significant, and Equation (9-6) can be written as

$$\frac{\partial u}{\partial t} + \frac{1}{\rho}\frac{\partial p}{\partial x} = 0 \qquad (9-9)$$

Combining Eqs. (9-7) and (9-9) results in

$$\frac{\partial u}{\partial x}+\frac{2r}{Eh}\frac{\partial p}{\mathrm{d}t}=0 \qquad (9-10)$$

Differentiating Eq. (9-8) with respect to x and Eq. (9-10) with respect to t, then neglecting the second-order terms and substituting gives

$$\frac{\partial^2 p}{\partial x^2}-\frac{1}{c^2}\frac{\partial^2 p}{\partial t^2}=0 \qquad (9-11)$$

Equation (9-11) is the wave equation where c is the wave speed,

$$c=\sqrt{\frac{Eh}{2\rho r}} \qquad (9-12)$$

The solution of Eq. (9-11) can be written as

$$P(x,t)=P_0 f(x-ct)+P_0' f(x+ct), \qquad (9-13)$$

where

$P_0 f(x-ct)$ = incident pressure wave traveling in the positive x direction
$P_0' f(x+ct)$ = pressure wave traveling in the negative x direction

The above formulation demonstrates that pressure wave speed c through the artery depends on the arterial elasticity, thickness, radius, and the density of the blood. This will be used in Section 9.3 to develop the complete wave propagation model in the central arterial system of interest.

Wave Reflection. Pressure waves are partially reflected when they experience a sudden change in the medium of transmission such as a bifurcation or a sudden change in the artery's geometry or material properties. Central arteries such as the aorta bifurcate into a number of smaller arteries. Also, in some diseases, such as aneurysms, discrete sections of the arteries and arterial tree degenerate and cause a large change in material properties and geometry. When a pressure wave arrives at such a change in transmission medium, part of that wave is transmitted, whereas the remainder is reflected.

Consider a tube that branches into two daughter tubes T_1 and T_2, as shown in Fig. 9-4. At the junction, it is assumed that pressure p is a single valued function, which can be written as

$$p_I+p_R=p_{T_1}=p_{T_2}, \qquad (9-14)$$

where

p_I = incident pressure wave
p_R = reflected pressure wave in the parent tube
p_{T_1} = pressure transmitted into the first daughter tube T_1
p_{T_2} = pressure transmitted into the second daughter tube T_2

Also, it can be assumed that the flow is continuous:

$$Q_I-Q_R=Q_{T_1}+Q_{T_2}, \qquad (9-15)$$

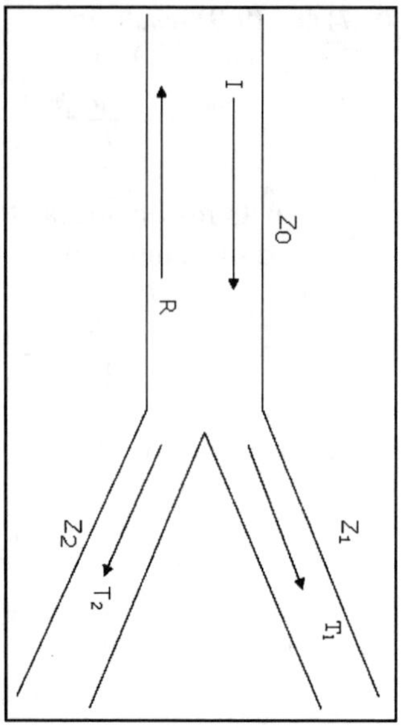

Figure 9-4. A schematic figure showing conditions at a bifurcation

where

Q_I = incident flow wave
Q_R = reflected flow wave
Q_{T_1} = flow into the first daughter tube T_1
Q_{T_2} = flow wave into the second daughter tube T_2

Equations (9-14) to (9-15) describe the pressure and flow conditions at the junction. Now a relationship between the pressure and flow is required. The characteristic impedance of the artery is defined by

$$Z_i = \frac{\rho_i c_i}{Ai}, \qquad (9-16)$$

where

Z_i = ratio of oscillatory pressure to oscillatory flow when the wave travels in the positive x direction

$$Z_i = \frac{p_i}{Q_i} \qquad (9-17)$$

Using Eq. (9-17), Eq. (9-15) can be written as

$$\frac{p_I - p_R}{Z_0} = \frac{p_{T_1}}{Z_1} + \frac{p_{T_2}}{Z_2} \qquad (9-18)$$

Using Eqs. (9-14) and (9-18), the reflection coefficient R can be expressed as

$$R = \frac{p_R}{p_I} = \frac{Z_0^{-1} - (Z_1^{-1} + Z_2^{-1})}{Z_0^{-1} + (Z_1^{-1} + Z_2^{-1})}, \qquad (9-19)$$

and the transmission coefficient T can be written as

$$T = \frac{p_{T_1}}{p_1} = \frac{p_{T_2}}{p_1} = \frac{2Z_0^{-1}}{Z_0^{-1} + (Z_1^{-1} + Z_2^{-1})} \qquad (9-20)$$

Using Eqs. (9-19) and (9-20), the amplitude of the reflected and transmitted pressure waves at a junction can be calculated from knowledge of the impedance of the parent and daughter arteries, which is determined by the geometry and material properties of the respective arteries.

9.2.2 Cuff-Soft Tissue-Brachial Artery Model

This section deals with lumped parameter mathematical models for the upper arm soft tissue and blood pressure cuff.

Cuff-Soft Tissue Model. In this section, a mathematical model for an upper arm compressed by a pneumatic cuff is developed. A literature shows a variety of models that have been proposed to study this structure [36-41]. However, all of these models were intended to study the oscillometric method of blood pressure measurement, whereas the main purpose of this work is to study the interactions among the pneumatic cuff, upper arm soft tissue, and brachial artery hemodynamics. The model developed in this work is based on recent work proposed by Ursino and Cristalli [39-41]. Separate models will be developed for the pneumatic cuff, soft tissue, and the brachial artery. Finally, these models will be linked together as a complete representation of the interactions between the three structures to estimate the brachial artery pressure from the circumferential strain on the cuff's outer wall. The upper arm is modeled as a cylinder of elastic homogenous tissue with a pneumatic cuff wrapped around it. The bone is modeled as a rigid cylinder embedded in the center of the upper arm. The brachial artery is assumed to be embedded within the arm soft tissue as shown in Fig. 9-5.

In this work, the upper arm soft tissue is modeled as three adjacent segments as shown in Fig. 9-5a, where the middle segment L_2 is subjected to an external pressure applied by the pneumatic cuff. The pneumatic cuff is modeled as a cylinder made of inner and outer compliant sheets containing air as shown in Fig. 9-6. Current commercial cuffs have a relatively stiff external wall compared with their internal walls [39, 41]. Also, the compliance of the walls is inversely proportional to the cuff's internal pressure [39, 41].

Using the above assumptions, the following relation between the cuff's internal pressure and the total volume enclosed by the external wall can be written as

$$\frac{dV_e}{dt} = C_e \frac{dp_c}{dt} = \frac{1}{\beta_e (p_c + p_{e0})} \frac{dp_c}{dt}, \qquad (9-21)$$

where

V_e	=	volume enclosed by the cuff external wall
C_e	=	compliance of the cuff's external wall
p_c	=	cuff internal pressure
β_e and p_{e0}	=	constants that govern the cuff's external wall mechanics

The relation among the cuff's internal pressure, pressure on the outside surface of the arm, and the total volume enclosed by the cuff's internal wall can be expressed as

$$\frac{dV_i}{dt} = C_i \left(\frac{dp_b}{dt} - \frac{dp_c}{dt} \right) = \frac{1}{\beta_i (p_c - p_b + p_{i0})} \left(\frac{dp_b}{dt} - \frac{dp_c}{dt} \right) \qquad (9-22)$$

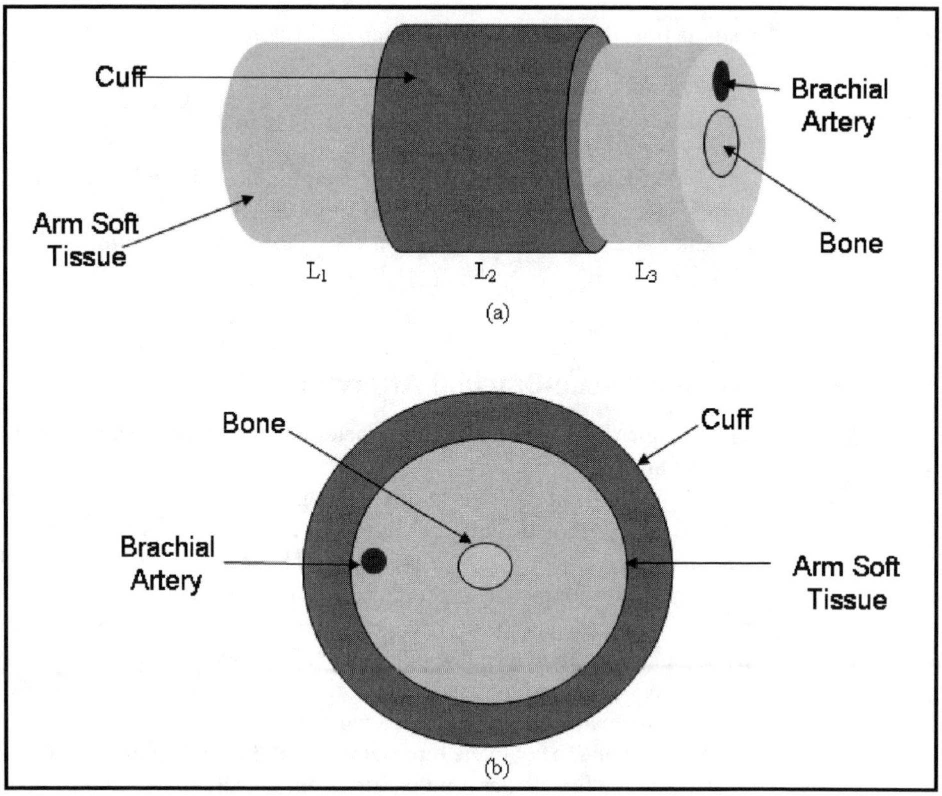

Figure 9-5. A schematic of the upper arm with a pneumatic cuff wrapped around it

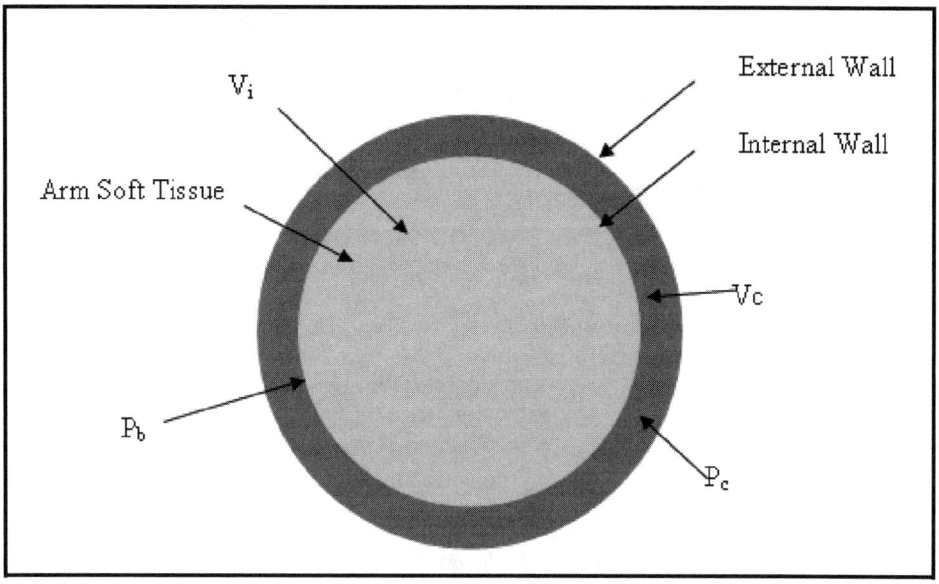

Figure 9-6. A transverse section of the upper arm with a pneumatic cuff wrapped around it showing pressures and the volumes

where

V_i = volume enclosed by the cuff's internal wall
C_i = compliance of the cuff's internal wall
p_b = pressure on the arm's external surface
β_i and p_{i0} = constants that govern the cuff's internal wall mechanics

If the cuff's internal wall thickness is assumed negligible, one can write

$$V_e = V_i + V_c, \qquad (9-23)$$

where

V_c = volume of air inside the cuff

Differentiating Eq. (9-23) and using Eqs. (9-21) and (9-22) results in

$$\frac{1}{\beta_e(p_c + p_{e0})}\frac{dp_c}{dt} = \frac{1}{\beta_i(p_c - p_b + p_{i0})}\left(\frac{dp_b}{dt} - \frac{dp_c}{dt}\right) \qquad (9-24)$$

Equation (9-24) expresses the relationship between the cuff's internal pressure and the pressure on the outside surface of the arm.

As mentioned earlier, the arm's middle segment L_2 is subjected to an external pressure applied by the pneumatic cuff, whereas segments L_1 and L_3 are not subjected to any pressure. Hence, it can be assumed that the only interaction among the three segments is through the longitudinal stress σ_z. The assumption that the stress distribution is symmetric around the axis leads to the following equations describing stresses and displacements in the upper arm tissue. For each segment, the stresses can be written as

$$\sigma_r = \lambda\varepsilon_z + 2A(G + \lambda) - 2G\frac{B}{r^2} \qquad (9-25)$$

$$\sigma_\theta = \lambda\varepsilon_z + 2A(G + \lambda) - 2G\frac{B}{r^2} \qquad (9-26)$$

$$\sigma_z = 2A\lambda + (\lambda + 2G)\varepsilon_z, \qquad (9-27)$$

where

σ_r, σ_θ, and σ_z = radial, tangential, and longitudinal stresses, respectively
u_r and u_z = radial and longitudinal displacements, respectively
ε_z = strain in the longitudinal direction
A and B = parameters that depend on the boundary conditions and will be derived hereafter by imposing two boundary conditions

The displacements can be written as

$$u_r = Ar + \frac{B}{r} \qquad (9-28)$$

$$u_z = \varepsilon_z z \qquad (9-29)$$

G and λ are the shear and bulk moduli, respectively:

$$G = \frac{E_s}{2(1 + \upsilon_s)} \qquad (9-30)$$

$$\lambda = \frac{2G\upsilon_s}{1-2\upsilon_s}, \tag{9 – 31}$$

where

E_s = soft tissue elasticity
υ_s = soft tissue's poisons ratio

The first boundary condition imposed in this model considers the pressure on the outside surface of the arm as equal to the radial stress:

$$\sigma_r = -P_b \tag{9 – 32}$$

The second boundary condition imposed in this model is that the volume changes due to blood pressure fluctuations in the upper arm are concentrated around the center of the arm near the bone, which is considered to be undeformable. This is done to obtain a simple model that preserves the symmetry of the arm. Hence,

$$u_r(r_i) = \frac{\mathrm{d}V_b}{2\pi r_i} \tag{9 – 33}$$

where

u_r = radial tissue displacement
r_i = radius of the bone
$\mathrm{d}V_b$ = change in brachial artery volume per unit length

Expressions for A and B can now be found by using Eqs. (9-25), (9-27), (9-28), (9-32), and (9-33):

$$A = \frac{-p_b}{N} - \frac{\lambda\sigma_z}{(\lambda+2G)N} + \frac{G}{\pi r_e^2 N}\mathrm{d}V_b \tag{9 – 34}$$

$$B = \frac{r_i^2}{N}p_b + \frac{\lambda r_i^2}{(\lambda+2G)N}\sigma_z + \frac{Nr_e^2 - 2Gr_i^2}{2\pi r_e^2 N} \tag{9 – 35}$$

$$N = 2G\left(1+\frac{r_i^2}{r_e^2}\right) + 4\lambda\frac{G}{\lambda+2G} \tag{9 – 36}$$

Substituting Eq. (9-27) into Eqs. (9-34) and (9-35) then substituting Eqs. (9-34) to (9-36) into Eq. (9-28) results in an expression describing the changes in the arm's radius $\frac{\mathrm{d}r_e}{\mathrm{d}t}$:

$$\frac{\mathrm{d}r_e}{\mathrm{d}t} = \frac{1}{2\pi r_e^2 \mathrm{D}}\left(2G\left(\frac{r_e^2 - r_i^2}{r_e}\right) + r_e D\right)\frac{\mathrm{d}V_b}{\mathrm{d}t} - \frac{1}{\mathrm{D}}\left(\frac{r_e^2 - r_i^2}{r_e}\right)\frac{\mathrm{d}p_b}{\mathrm{d}t}$$

$$-\frac{\lambda}{D(\lambda+2G)}\left(\frac{r_e^2 - r_i^2}{r_e}\right)\frac{\mathrm{d}\sigma_z}{\mathrm{d}t} \tag{9 – 37}$$

Furthermore, an expression describing changes in the segment's length $\frac{\mathrm{d}l}{\mathrm{d}t}$ can be established for each segment by substituting Eq. (9-27) into Eqs. (9-34) and (9-35) then substituting Eqs. (9-34) to (9-36) into Eq. (9-29) and assuming small soft tissue displacements:

$$\frac{dl}{dt} = \left(\frac{2\lambda l}{(\lambda+2G)D}\right)\frac{dp_b}{dt} + \left(\frac{l}{\lambda+2G} + \frac{2l\lambda^2}{D(\lambda+2G)^2}\right)\frac{d\sigma_z}{dt}$$

$$-\left(\frac{2\lambda Gl}{(\lambda+2G)\pi r_e^2 D}\right)\frac{dV_b}{dt} \qquad (9-38)$$

The extravascular pressure is defined as the external pressure acting on the brachial artery. The brachial artery lies longitudinally in the arm, and it can be assumed that it does not receive any significant longitudinal stresses. The extravascular pressure can be estimated to be equal to the average of radial and tangential stresses in each segment:

$$p_e = \frac{\sigma_r + \sigma_\theta}{2} \qquad (9-39)$$

From Eq. (9-39), an expression describing changes in the extravascular pressure $\frac{dp_e}{dt}$ in each segment can be established by substituting Eq. (9-27) into Eqs. (9-34) and (9-35) then substituting Eqs. (9-34) to (9-36) into Eq. (9-39). This results in

$$\frac{dp_e}{dt} = \left(\frac{2G(3\lambda+2G)}{D(\lambda+2G)}\right)\frac{dp_b}{dt} + \left(\frac{2G\lambda(3\lambda+2G)}{D(\lambda+2G)^2} - \frac{\lambda}{\lambda+2G}\right)\frac{d\sigma_z}{dt}$$

$$-\left(\frac{2G^2(3\lambda+2G)}{\pi r_e^2 D(\lambda+2G)}\right)\frac{dV_b}{dt} \qquad (9-40)$$

In Eqs. (9-37), (9-38), and (9-40), D is equal to

$$D = \frac{4G\lambda}{\lambda+2G} + 2G \quad 1 + \left(\frac{r_i^2}{r_e^2}\right) \qquad (9-41)$$

Differential Eqs. (9-37), (9-38), and 9-40) establish expressions for $\frac{dr_e}{dt}$, $\frac{dl}{dt}$, and $\frac{dp_e}{dt}$ in terms of $\frac{dV_b}{dt}$, $\frac{dp_b}{dt}$, and $\frac{d\sigma_z}{dt}$. Considering the three tissue segments, Eqs. (9-37), (9-38), and (9-40) constitute 9 differential equations and 18 variables. Hence, nine additional constrains need to be imposed for the system to be solvable. These constrains are considered below.

The only external pressure on the upper arm is the pressure applied by the cuff on segment L_2 and is equal to p_b:

$$p_{b2} = p_b \qquad (9-42)$$

$$p_{b1} = p_{b3} = 0 \qquad (9-43)$$

From Newton's second law, we can assume that longitudinal stress on each of the three segments is the same:

$$\sigma_z = \sigma_{z1} = \sigma_{z2} = \sigma_{z3} \qquad (9-44)$$

It can be assumed that there is no significant change in the blood volume in segments L_1 and L_3 because the brachial artery does not collapse in these segments:

$$\frac{dV_{b1}}{dt} = \frac{dV_{b3}}{dt} = 0 \qquad (9-45)$$

In addition, it can be assumed that the total length of the upper arm remains constant:

$$\frac{dl_1}{dt}+\frac{dl_2}{dt}+\frac{dl_3}{dt}=0 \tag{9-46}$$

Now, an expression describing changes in the upper arm longitudinal stress can be found by substituting Eq. (9-38) into Eq. (9-46) and making use of Eqs. (9-42) to (9-45):

$$\frac{d\sigma_z}{dt}=\frac{\left(\frac{2\lambda l}{(\lambda+2G)D}\right)_2\frac{dp_b}{dt}-\left(\frac{2\lambda Gl}{(\lambda+2G)\pi r_e^2 D}\right)_2\frac{dV_b}{dt}}{\sum_{i=1}^{3}\left(\frac{l}{\lambda+2G}+\frac{2l\lambda^2}{D(\lambda+2G)^2}\right)i}, \tag{9-47}$$

where

i = arm segment number

By assuming that the cuff is always in contact with the outside surface of the arm and that the cuff's longitudinal displacement is small, a relationship between the arm's radius r_{e2} and the volume enclosed by the cuff's internal surface V_i can be expressed as

$$V_i=\pi r_{e2}^2 l_2 \tag{9-48}$$

Differentiating Eq. (9-48) and using Eq. (9-22) results in

$$2\pi r_{e2}\frac{dr_{e2}}{dt}l_2=\frac{1}{\beta_i(p_c-p_b+p_{i0})}\left(\frac{dp_b}{dt}-\frac{dp_c}{dt}\right) \tag{9-49}$$

Considering the three tissue segment L_1, L_2, and L_3, Eqs. (9-24), (9-37), (9-38), (9-40), (9-47), and (9-49) form a system of 12 differential equations in 12 unknowns $l_{1,2,3}$, $r_{e1,2,3}$, $p_{e1,2,3}$, p_c, p_b, and σ_z. This system can be integrated numerically using MATLAB. However, to avoid recurrence, Eqs. (9-24) and 9-49) can re-written. Substituting Eq. (9-47) into Eq. (9-37) and substituting the result into Eq. (9-49) results in

$$\frac{dp_b}{dt}=\left(\frac{1}{J\beta_i(p_c-p_b+p_{i0})}\right)\frac{dp_c}{dt}-\frac{2\pi r_{e2}l_2}{J}\left(\frac{1}{2\pi r_e^2 D}\left(2G\left(\frac{r_e^2-r_i^2}{r_e}\right)+r_eD\right)_2\right)\frac{dV_b}{dt}$$
$$-\frac{2\pi r_{e2}l_2}{J}\left(\frac{\left(\frac{\lambda}{D(\lambda+2G)}\right)_2\left(\frac{2\lambda Gl}{(\lambda+2G)\pi r_e^2 D}\right)_2\left(\frac{r_e^2-r_i^2}{r_e}\right)_2}{\sum_{i=1}^{3}\left(\frac{l}{\lambda+2G}+\frac{2l\lambda^2}{D(\lambda+2G)^2}\right)i}\right) \tag{9-50}$$

where

$$J=\frac{1}{\beta_i(p_c-p_b+p_{i0})}+2\pi r_{e2}l_2\left(\frac{1}{D}\left(\frac{r_e^2-r_i^2}{r_e}\right)\right)_2$$
$$-2\pi r_{e2}l_2\left(\frac{\frac{\lambda}{D(\lambda+2G)}\left(\frac{r_e^2-r_i^2}{r_e}\right)\left(\frac{2\lambda l}{(\lambda+2G)D}\right)}{\sum_{i=1}^{3}\left(\frac{l}{\lambda+2G}+\frac{2l\lambda^2}{D(\lambda+2G)^2}\right)i}\right) \tag{9-51}$$

Substituting Eq. (9-50) into Eq. (9-24) results in

$$\frac{\mathrm{d}p_c}{\mathrm{d}t} = \frac{2\pi r_{e2} l_2}{JK\beta_i(p_c - p_b + p_{i0})}\left(\frac{1}{2\pi r_e{}^2 D}\left(2G\left(\frac{r_e^2 - r_i^2}{r_e}\right) + r_e D\right)_2\right)\frac{\mathrm{d}V_b}{\mathrm{d}t}$$

$$-\frac{2\pi r_{e2} l_2}{JK\beta_i(p_c - p_b + p_{i0})}\left(\frac{\left(\frac{\lambda}{D(\lambda+2G)}\right)_2\left(\frac{2\lambda G l}{(\lambda+2G)\pi r_e^2 D}\right)_2\left(\frac{r_e^2 - r_i^2}{r_e}\right)_2}{\sum_{i=1}^{3}\left(\frac{l}{\lambda+2G} + \frac{2l\lambda^2}{D(\lambda+2G)^2}\right)i}\right)\frac{\mathrm{d}V_b}{\mathrm{d}t}, \qquad (9-52)$$

where K is

$$K = \frac{1}{\beta_e(p_c + p_{e0})} + \frac{1}{\beta_i(p_c - p_b + p_{i0})} - \left(\frac{1}{J(\beta_i(p_c - p_b + p_{i0}))^2}\right) \qquad (9-53)$$

The circumferential strain of the pneumatic cuff outside wall is measured by fixing a piezoelectric sensor to the cuff. The sensor measures circumferential strains in the cuff external wall. Consequently, a relationship is required between the cuff pressure and the strain on the cuff external wall. Commercial cuffs are designed to be relatively stiff in the longitudinal direction [41]; therefore, it can be assumed that any change in the cuff volume is attributed to change in its radius. Using Eq. (9-21) gives

$$\frac{\mathrm{d}R}{\mathrm{d}t} = C_e\frac{\mathrm{d}p_c}{\mathrm{d}t} = \frac{1}{2\pi R l_2 \beta_e(p_c + p_{e0})}\frac{\mathrm{d}p_c}{\mathrm{d}t} \qquad (9-54)$$

Now the circumferential strain can be written as

$$\varepsilon = \frac{\mathrm{d}R}{R_0} \qquad (9-55)$$

To solve the differential equations numerically, the change in the volume of blood under the cuff $\frac{\mathrm{d}V_b}{\mathrm{d}t}$ will be derived in the next section.

Artery Model. A lumped parameter model is used to describe the blood volume changes under the cuff where all blood volume changes are assumed to be concentrated in a single representative section of the artery. Ursino and Cristalli [40, 41] modeled and defined expressions describing all pressure losses caused by the collapse of the brachial artery due to the cuff. These included losses from the conversion of pressure into kinetic energy, inertial and viscous losses, and expansion losses due to the reopening of the artery [40, 41]. These losses are simulated using MATLAB and are found to decrease the brachial artery pressure by a maximum of 4% near diastolic pressure. The computation power required to solve the differential equations describing the pressure losses outweigh the improvement in accuracy of the model and, therefore, were ignored.

As mentioned previously, it is assumed that the total change in blood volume under the pneumatic cuff is because of the change in blood volume in the brachial artery. This is obviously not entirely true because blood volume change is also present in the capillaries and veins of the upper arm. However, these segments are compressed during the measurement at relatively low cuff pressures, and their volume changes can be assumed to be insignificant [41]. Hence, the change in blood volume per unit length under the cuff is assumed to be equal to the change in the cross-sectional area of the brachial artery.

Now, the cuff and soft tissue model can be linked to the brachial artery model by finding an expression describing the compliance of the brachial artery under the cuff. This is given by a relationship between the area of the brachial artery under the pneumatic cuff A and the transmural pressure $p_a - p_e$. Two separate relations must be formed, one describing the artery compliance at positive transmural pressure and another describing the artery compliance at negative transmural pressure.

At positive transmural pressure, the artery can be assumed to maintain its circular cross section, and hence, the Laplace law can be used to describe the relationship:

$$(p_a - p_e)\sqrt{\frac{A}{\pi}} = \sigma h = (\sigma_e + \sigma_v)h \tag{9-56}$$

where

σ = circumferential stress
σ_e = elastic stress
σ_v = viscous stress
h = artery wall thickness

A bi-exponential relationship is used to describe the relationship between the brachial artery elastic circumferential stress and circumferential strain to reproduce experimental results reported by Learoyd and Taylor [45]:

$$\sigma_e(A) = \frac{E_0}{\beta'_a}(\exp(\beta'_a \varepsilon_a) + \exp(\beta''_a \varepsilon_a^2) - 2), \tag{9-57}$$

where

β'_a, β''_a, and E_0 = constants
ε_a = circumferential strain

ε_a is defined by

$$\varepsilon_a = \frac{\sqrt{A} - \sqrt{A_0}}{\sqrt{A_0}} \tag{9-58}$$

The wall thickness (h) at positive transmural pressure can be described by assuming the artery wall to be incompressible and that longitudinal strain is negligible:

$$h = -\sqrt{\frac{A}{\pi}} + \sqrt{\frac{A}{\pi} + 2h_0\sqrt{\frac{A_0}{\pi}} + h_0^2}, \tag{9-59}$$

where

h_0 and A_0 = wall thickness and cross-sectional area at zero transmural pressure, respectively

The viscous stress can be assumed to be linearly dependent on the rate of change of the artery wall strain and can be written as

$$\sigma_v = \eta \frac{d\varepsilon_a}{dt} = \frac{\eta}{2\pi\sqrt{\frac{A_0}{\pi}}\sqrt{\frac{A}{\pi}}}\frac{dA}{dt} \tag{9-60}$$

Finally, by using Eqs. (9-56) to (9-60), a relation for the rate of change of the brachial artery area at positive transmural pressure can be expressed as

$$\frac{dV_b}{dt} = \frac{dA}{dt} = \frac{2(p_a - p_e)\, A\sqrt{\frac{A_0}{\pi}}}{\eta h(A)} - 2\,\frac{\sqrt{\frac{A_0}{\pi}}\sqrt{\pi A}}{\eta}\,\sigma_e(A) \qquad (9-61)$$

As the transmural pressure become negative, the brachial artery collapses and loses its cylindrical shape. Hence, the Laplace law cannot be applied, and the tube law is used to describe the relationship between the transmural pressure and the artery cross-sectional area:

$$\frac{dV_b}{dt} = \frac{dA}{dt} = k\left(\frac{p_a - p_e}{k_p} - 1 + \left(\frac{A}{A_0}\right)^{-3/2}\right), \qquad (9-62)$$

where

k = proportionality factor
k_p = constant that depends on the geometry and material properties of the brachial artery

k_p is described by

$$k_p = \frac{h_0 E_0}{3 r_{a0}} \qquad (9-63)$$

The constant k_p determines the way the artery will collapse, whereas k is a constant parameter that is calculated by keeping $\frac{dA}{dt}$ continuous between positive and negative transmural pressures. Equation (9-62) can now be written as

$$\frac{dV_b}{dt} = \frac{dA}{dt} = \frac{2(p_a - p_e)\, A_0\sqrt{\frac{A_0}{\pi}}}{\eta h_0} - \frac{2k_p A_0\sqrt{\frac{A_0}{\pi}}}{\eta h_0}\left(1 - \left(\frac{A}{A_0}\right)^{-3/2}\right) \qquad (9-64)$$

Now that an expression linking the change in the blood volume under the cuff and the intravascular and extravascular brachial artery pressure has been established, the cuff-soft tissue-brachial artery model is linked, and the transmission of the pressure across the soft tissue and the pneumatic cuff can be found. The material properties and geometry for the cuff, arm soft tissue, and the brachial artery used in this model were adopted from the values reported by Ursino and Cristalli [39-41].

9.3 MODEL DEVELOPMENT AND SIMULATION

In this section, the acoustic model developed in Section 9.2.1 is used to build a physiologically based model of the central systemic arteries that can be used to simulate the brachial artery pressure. The brachial artery-soft tissue-cuff model developed in Section 9.2.2 is used to describe the transmission of this brachial artery pressure through the arm soft tissue and the pneumatic cuff. Finally, these two models are combined to obtain a complete model that can be used to non-invasively estimate the brachial artery pressure and strain contour created on the cuff outside wall.

9.3.1 Acoustic Model

In this section, the acoustic model developed in Section 9.2.1 is used to build a model that estimates the intravascular pressure at the brachial artery. First, the tapering and the variation of the material properties in the systemic arteries are examined and modeled. The input wave to the system from the heart is then determined and mathematically described. Also, the sites of wave reflection within the system are adopted from the available literature. Finally, the complete acoustic model is simulated by describing the propagation and reflection of pressure waves generated by the heart contraction and tracing these waves as they reach the brachial artery.

The central arteries take the shape of tapered tubes. The speed of wave propagations is influenced by the cross-sectional area, thickness, and modulus of elasticity of the artery; therefore, it is important to consider variations in these geometric and material properties when developing a model.

In this research, the arteries of interest are divided into small segments with constant geometric and mechanical properties. This assumption simplified the governing equations dramatically and reduced the complexity of the mathematical solution.

Westerhof et al [46] conducted experiments on human arteries and listed the thickness, radius, and the modulus of elasticity at ten positions along the aorta and 7 positions along the subclavian and brachial arteries. To find the minimum number of lumps leading to an accurate and acceptable solution, calculations are conducted on the aorta and brachial artery. A power function is found to be adequate to describe the variation in thickness and radius of the aorta, whereas a linear function is used to describe the thickness and radius variation of the subclavian-brachial arteries. The model is tested for different numbers of lumps using Eq. (9-12). Figure 9-7 shows the relation between the time it takes for the wave to travel through the aorta and brachial artery versus the number of lumps used. It is clearly indicated that convergence starts at 16 segments for the brachial artery and 24 segments for the aorta. Increasing the number of these segments shows insignificant changes in travel time, that is, doubling the number of segments after these values will only improve the accuracy of the travel time by less than 5%. Therefore, in this work, 16 and 24 segments are used for the brachial artery and aorta, respectively.

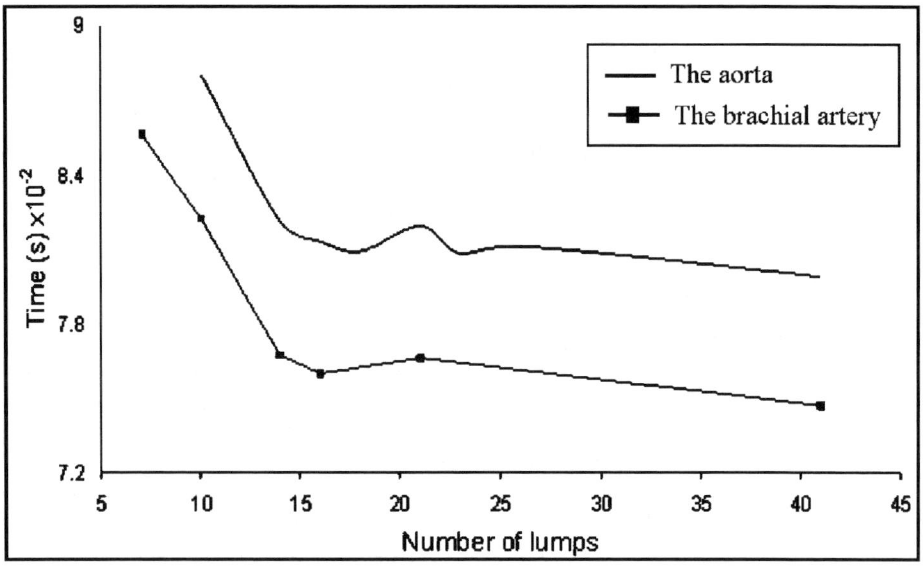

Figure 9-7. Pressure wave traveling time versus number of lumps

The precise location of the major reflection sites in the human arteries has been a controversial topic during the last century. Some claim [47] that the small arteries, 0.5 mm to 1 mm in diameter, are the main reflection sites, whereas others claim that reflections occur at all junctions [46]. Pollack et al [48] attributes the reflections to arteries ranging from 3.5 mm to 4 mm in diameter. O'Rourke and Taylor [49] support the idea of two reflection sites, one in the upper body and one in the lower body.

Many consider the iliac bifurcation as the main reflection site [50-52]; others consider an effective reflection site in the system that is located near the iliac bifurcation [53, 54]. These two hypotheses seem to have very strong support in the literature [50-52, 55-57]. Therefore, the iliac bifurcation is assumed to be the only significant source of reflection in the system in this work.

The geometry and material properties for the aorta and the iliac arteries are obtained from experimental data reported by Westerhof et al [46]. The compression of the upper arm by the pneumatic cuff causes a sudden decrease in the area of the brachial artery, which results in a large impedance mismatch that leads to a reflection at this site. Furthermore, from the data reported by Westerhof et al [46], the reflected brachial wave reflects negatively when it reaches the mouth of the subclavian artery because of the sudden increase in area [43, 58]. Hence, in this work, the following wave reflection model is proposed.

Any wave generated by the heart travels via the aorta into the brachial artery and iliac bifurcation. The aortic wave is reflected at the iliac bifurcation and then travels back to the brachial artery. The original wave, which traveled to the brachial artery, reflects back because of the cuff and travels to the mouth of the subclavian artery where it reflects back negatively to the brachial artery. Secondary reflections other than the one produced by the closure of the cuff are not considered in this work because their magnitude is too minute for the sensors on the cuff. The sum of the three waves reaches the brachial artery at different times as shown in Fig. 9-8.

The above model is simulated using the MATLAB software package. An invasive left-ventricle waveform obtained from Nichols and O'Rourke [3] is used to represent the periodic pressure wave generated from the heart. This wave is then represented using a Fourier series as input into the ascending aorta. The left ventricle supra-diastolic pressure

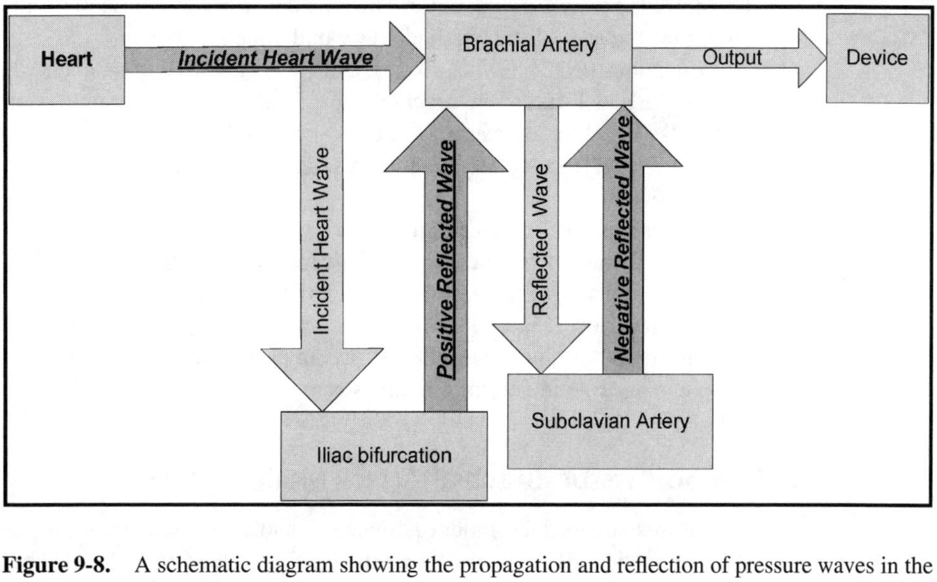

Figure 9-8. A schematic diagram showing the propagation and reflection of pressure waves in the system

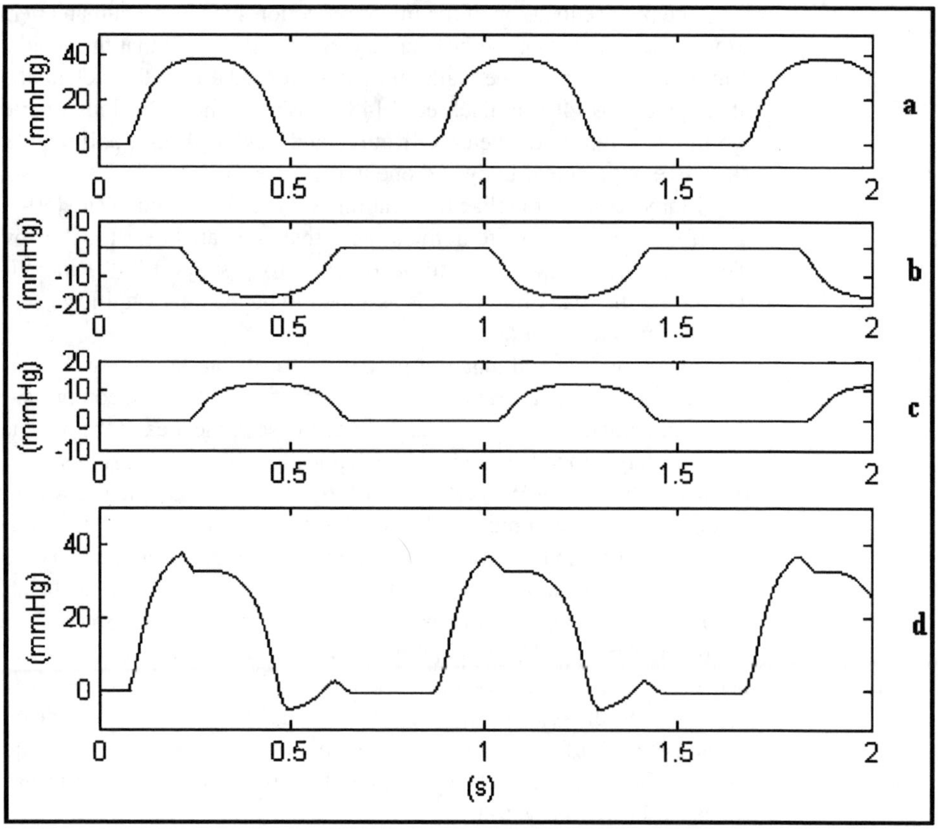

Figure 9-9. Pressure waves determined at the brachial artery: a) Heart wave, b) First reflected wave from the subclavian artery, c) Second reflected wave from the iliac bifurcation, d) Combined wave

is used because the aortic valve opens at this pressure, and the pressure of the ventricle is transferred to the ascending aorta. The brachial artery-soft tissue-cuff model is then used to determine the area of the brachial artery under the inflated cuff.

Equations (9-13), (9-14), (9-16), and (9-21) are used to calculate the reflection coefficient at the iliac bifurcation, brachial artery, and mouth of the subclavian artery. The radius and thickness of the iliac arteries are also obtained from the work by Westerhof et al [46]. These are used to calculate the area mismatch between the descending aorta and the iliac bifurcation.

The travel time is then calculated using Eq. (9-12) for each element. The total time for the waves to travel from the heart, and the respective reflection sites to the brachial artery is determined by adding the time for each element between the heart and the reflection sites to the brachial artery. Figure 9-9 shows the three waves that make up the brachial artery wave and the combination of these waves. It is important to note that the timing of these waves plays a major role in determining the shape of the brachial artery wave.

9.3.2 Cuff-Soft Tissue-Brachial Artery Model

In this section, the model parameters and input quantities used in this research for the pneumatic cuff, soft tissue, and the brachial artery are defined. The lumped parameter models are then linked together and simulated to describe the transmission of pressure from the brachial artery through the arm soft tissue to the pneumatic cuff. The ODEs are then numerically solved using the MATLAB software package.

The parameters of the cuff β_e, p_{e0}, β_i, and p_{i0} were adopted from experiments conducted by Ursino and Cristalli [39, 41] on the inside and the outside the cuff to characterize their compliance. The parameters of the arm soft tissue were also adopted from experiments conducted by Ursino and Cristalli [41]. Average values for the upper arm's initial radius and length were used. The arm tissue elastic properties were estimated in an experimental study performed on healthy subjects. The Young's modulus E_s is assumed to increase linearly with the pressure on the outside surface of the arm p_b.

The geometry for the brachial artery A_0 and h_0 have been adopted from data reported in previous research [41]. The length of the brachial artery L is assumed to be equal to the cuff length L_2 because this is where all the interactions among the three models take place. Also, the parameters that describe the mechanics of the brachial artery wall E_0, β'_a, β''_a, and η have been adopted from the work reported by Ursino and Cristalli [40, 41].

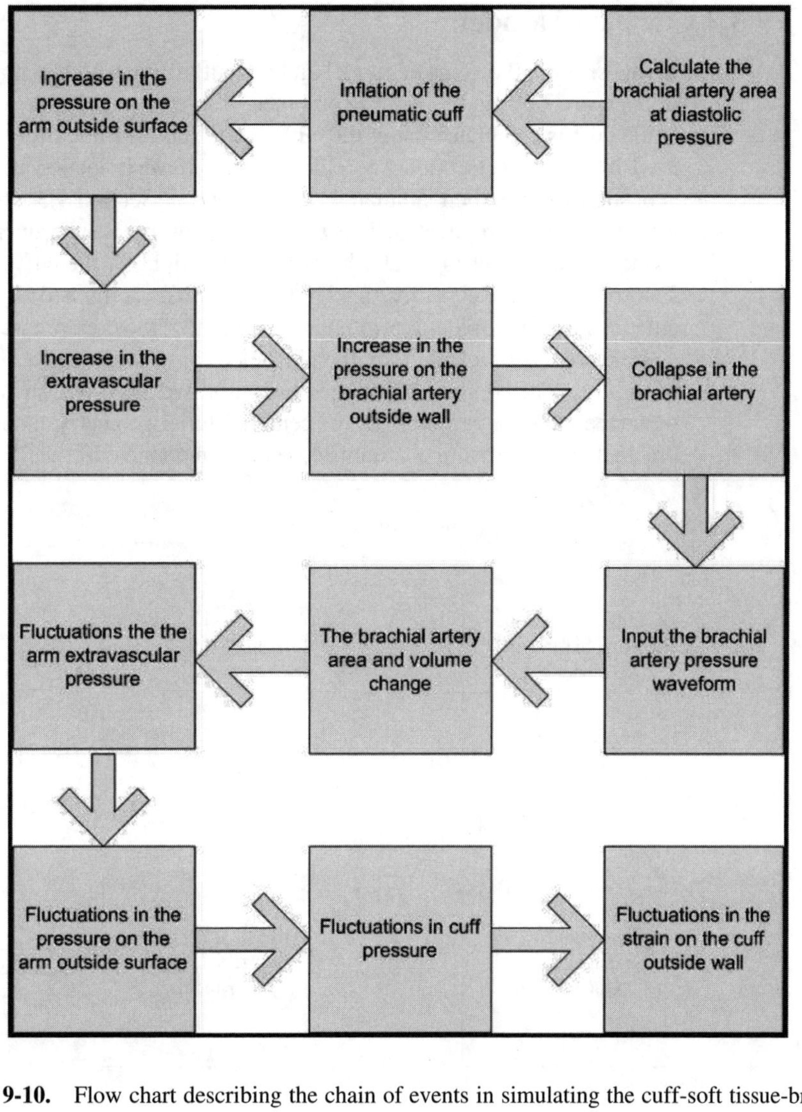

Figure 9-10. Flow chart describing the chain of events in simulating the cuff-soft tissue-brachial artery model

The three lumped parameter models are linked and simulated on a computer. Before inflating the cuff, the area of the brachial artery at diastolic pressure is estimated by using the brachial artery model described in Section 9.2.2. The intravascular pressure is iteratively increased from 0 mmHg to 80 mmHg while keeping the extravascular pressure at 0 mmHg. The brachial artery area is subsequently calculated using the brachial artery model. The cuff pressure in the cuff model is iteratively increased from zero to the required pressure for the simulation and change in the upper arm soft tissue and the brachial artery state variables is calculated using the cuff, soft tissue, and the brachial artery models. The brachial artery pressure is then fed into the brachial artery model to determine the rate of change in the brachial artery volume. This is then input into the soft tissue and cuff model where changes in the extravascular pressure, pressure on the arm outside surface, and other state variables were calculated. The change in the pressure on the outside surface of the upper arm is fed into the cuff model, where changes in the cuff pressure and the strain on the cuff outside wall were determined. Figure 9-10 shows the sequence of events described above.

9.3.3 Combined Model

In this section, the acoustic model and the cuff-soft tissue-brachial artery model described in Sections 9.2.1 and 9.2.2 are combined. The area of the brachial artery under the pneumatic cuff is calculated using the cuff-soft tissue-brachial artery model. This area is then used by the acoustic model to calculate the wave reflection at the brachial artery. The acoustic model is then simulated as described in Section 9.3 where brachial artery pressure waveform is replicated. This pressure waveform is subsequently fed into the cuff-soft tissue-brachial artery model where the fluctuations in the cuff outer wall are simulated. The combined model is illustrated in Fig. 9-11. The combined physiologically based mathematical model provides a computer simulation of the pressure waves in the central arteries and the contours picked up by the piezoelectric strain sensor attached to the external wall of the cuff. The computer simulated model provides a quick and easy-to-use tool to investigate the effect of variations in the central arterial geometry and material properties on the brachial artery pressure contours and the strain on the cuff outside wall. It can also be used

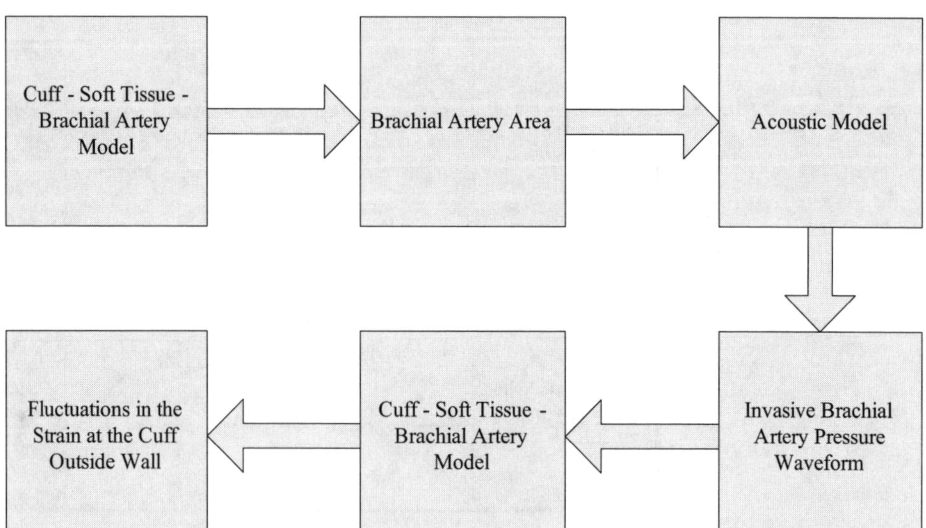

Figure 9-11. A flow diagram showing the connections among the combined model

to investigate the effect of variations in the pneumatic cuff and soft tissue properties on the acquired pressure and strain contours. This model has been adopted to help investigate the sensitivity of oscillometric cuff-based measurements to factors such as artery geometry, material properties, and cuff pressure.

9.4 MODEL RESULTS AND APPLICATIONS

In this section, the simulation results for the combined model discussed in Section 9.3 are presented. The brachial artery pressure and strain on the pneumatic cuff outer wall are first simulated using the model where important features in the simulated waveforms are then defined and extracted. These feature points are used to calculate the brachial AI and the time lag. The effects of variations in a number of model parameters on the AI and time lag are then investigated by simulating the model. The model results are compared with the results obtained from experimental work and independent conclusions are drawn. The significance of the model developed in this work and the fact that it can be used to investigate the effect of variations in a number of cardiovascular, arm, and cuff parameters is also discussed.

9.4.1 Simulation and Feature Extraction

The combined model discussed in Section 9.3 is used to simulate the brachial artery pressure and pneumatic cuff external wall strain contours as shown in Fig. 9-12. Average model parameters are adopted from the literature and used to simulate the contours for a young, healthy adult male [46]. A cuff pressure of 120 mmHg (being above systolic pressure) is used to simulate these plots.

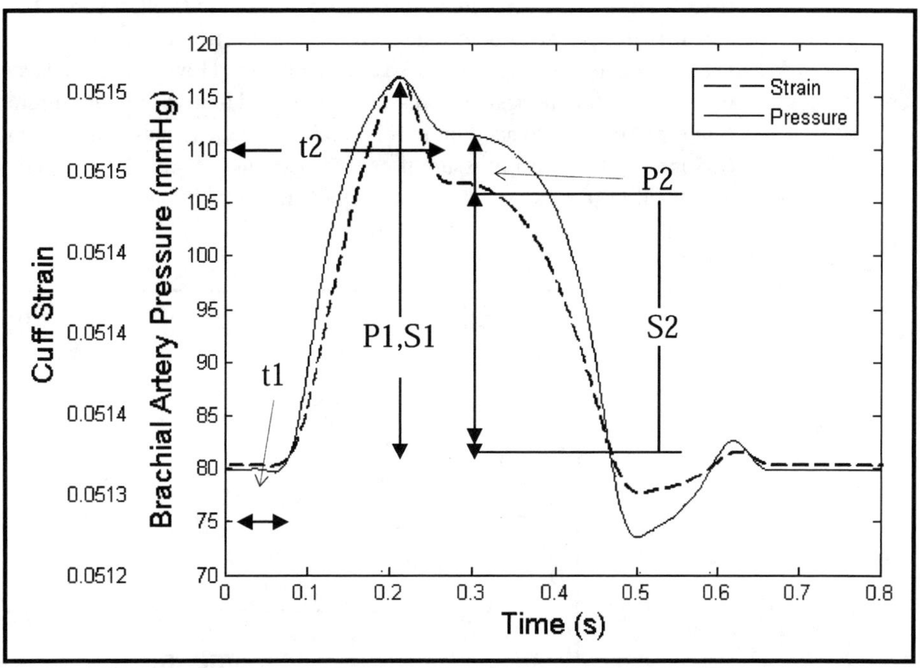

Figure 9-12. Pressure and strain contours simulated by the model for normal healthy adult input parameters

Four feature points are extracted from the pressure contours. These feature points are the time duration before the deflection marking the arrival of the incident wave ($t1$), the time duration before the deflection marking the arrival of the reflected wave ($t2$), the peak of the first pressure deflection minus the diastolic pressure ($P1$), and the peak of the second pressure deflection minus the diastolic pressure ($P2$).

The same feature points are extracted from the strain contours where $S1$ is the peak of the first strain deflection minus the strain at diastolic pressure and $S2$ is the peak of the second strain deflection minus the strain at diastolic pressure.

These features points are then used to calculate the brachial AI_p and AI_s and the time lag, which are used as a measure of arterial stiffness. The brachial AI is seldom mentioned in the open literature because of the unavailability of non-invasive brachial blood pressure waveform monitoring devices. The brachial artery pressure AI (AI_p) can be written as

$$AI_p = \frac{P2}{P1}, \quad (9-65)$$

where the brachial artery strain AI_s can be written as

$$AI_s = \frac{S2}{S1}, \quad (9-66)$$

and the time lag is written as

$$\text{time lag} = t2 - t1 \quad (9-67)$$

The pressure contour obtained from the model has a similar shape and contains the feature points found in invasively measured brachial artery contours reported in the literature as shown in Fig. 9-13 [3, 59]. In both this model and invasive data, the reflected pressure wave arrives at the brachial artery during diastole in healthy subjects.

It is important to note that there is no reflection at the measurement site in invasive measurements because there is no cuff occlusion. However, there is a significant reflection that comes from the resistive vessels of the hand in invasive measurements [3]. This reflection is not taken into account in this model because it is assumed that the occlusion caused by supra-systolic cuff pressure prevents the transmission of pressure beyond the occlusion. The validity of this assumption becomes questionable as the cuff pressure is reduced be-

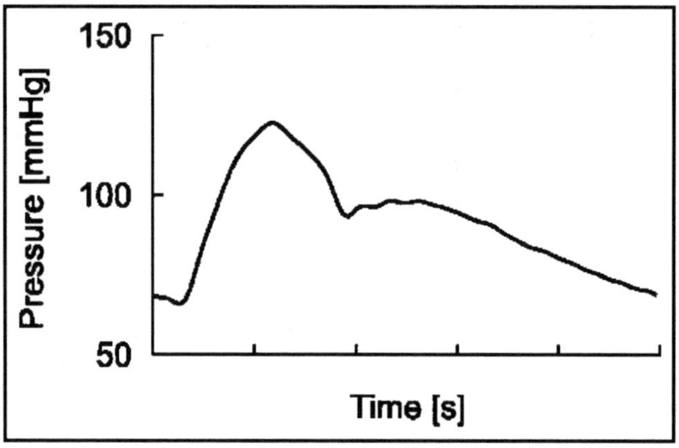

Figure 9-13. Brachial artery pressure pulse for healthy young subject [48]

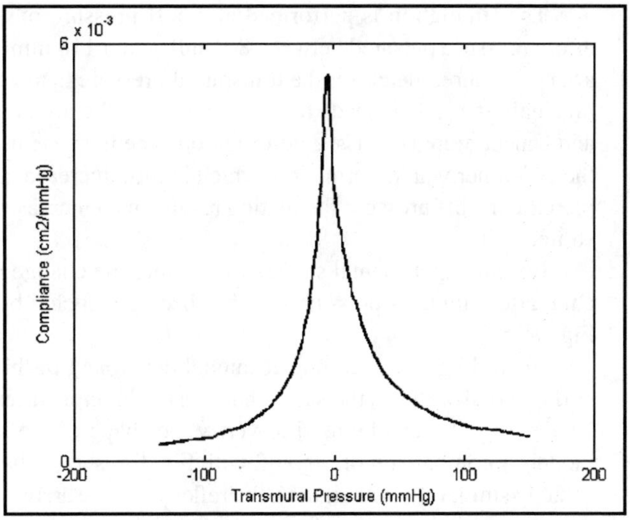

Figure 9-14. The compliance of the brachial artery versus transmural pressure

cause more pressure will be transmitted beyond the cuff where the reflection from the hand vessels becomes significant.

The cuff strain contour appears to be very similar to the brachial pressure contour. However, if examined more closely, it is clear that the relative amplitude between the first and second peaks is different. It appears that the first peak is amplified, whereas the second peak is diminished when compared with the pressure contour.

The change can be explained by reviewing the model of brachial hemodynamics described in Section 9.2.2. Figure 9-14 shows the relationship between the compliance of the brachial artery versus the transmural pressure as simulated by the model. This relationship is also confirmed by experimental studies conducted by Linden and Drzewiecki [60]. Figure 9-14 shows that the artery compliance is highest at zero transmural pressure. There is a very small change in the compliance of the brachial artery with age; therefore, this relationship should not be affected with aging or central arterial stiffening [3].

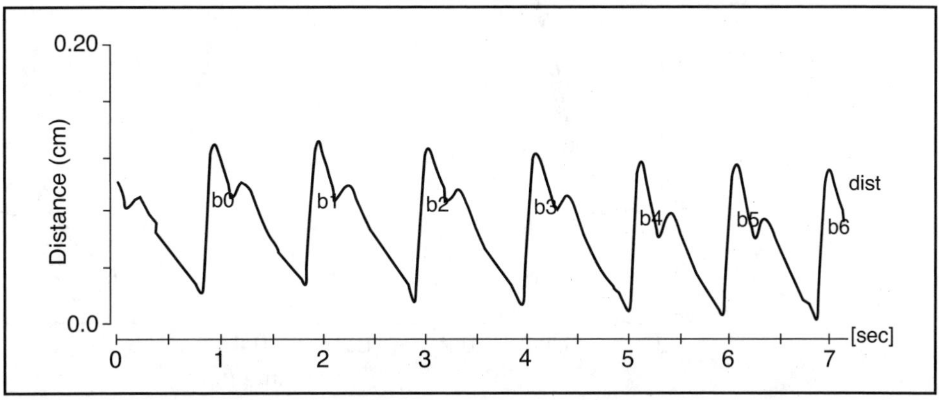

Figure 9-15. Brachial artery cross-sectional area under a pressure cuff versus time for a healthy adult man [50]

The simulation is performed at a cuff pressure of 120 mmHg, whereas the brachial artery pressure pulsated between 80 mmHg and 118 mmHg. Hence, the higher the brachial artery pressure, the closer the transmural pressure is to zero, resulting in a more compliant brachial artery. This leads to more change in the artery area for a given pressure change, and hence, more soft tissue deformation. The increase in soft tissue deformation increases the cuff inner wall deformation, which in turn, increases the cuff pressure fluctuations. The increase in cuff pressure fluctuation results in an increase in the outer wall deformation and strain.

Recent experimental studies to measure the change in cross-sectional area of the brachial artery under a pressure cuff has been conducted by Bank et al [61] and is shown in Fig. 9-15.

According to the combined model developed in this work, the strain contour on the cuff external wall has the same shape as the brachial artery cross-sectional area variations. Hence, Fig. 9-6 can be used to verify the shape of the strain contours obtained from the model. A comparison of Fig. 9-2 with Fig. 9-6 shows that the shape of the contours is reasonably similar. In both figures, the reflected wave arrives at the brachial artery after around 0.2 sec, which shows that the acoustic model used in this work is adequate in predicting the propagation velocity of the pressure waves in the cardiovascular system.

The AI from the model simulation and experimental results are in close agreement at around 80% correlation. The primary difference between the two contours is that the simulated contour returns to the diastolic level immediately, whereas the measured contour slowly settles down. This can be attributed to the fact that the model does not consider secondary wave reflections that come later in diastole.

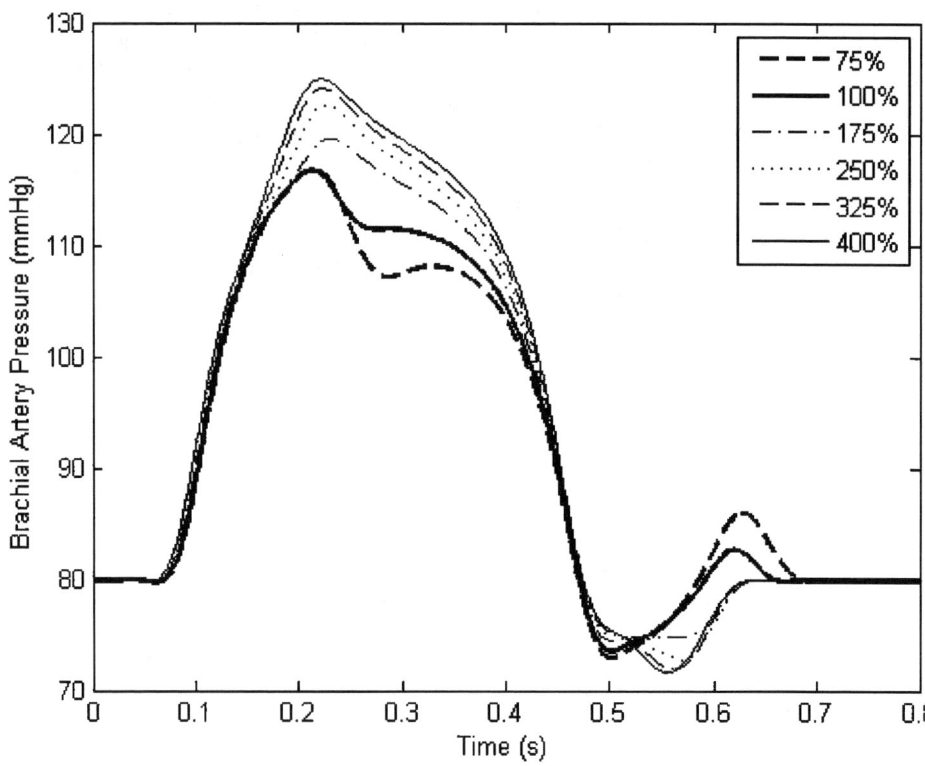

Figure 9-16. Brachial artery pressure waveforms at different aortic stiffness values ranging from healthy (75%) to diseased (400%)

9.4.2 Effect of Artery Stiffness

The effect of the stiffness of the aorta on the brachial artery pressure waveform and the strain on the external wall of the pneumatic cuff is investigated by simulating the combined model at different stiffness values and keeping all other parameters constant at their normal healthy level. A global stiffness increment is assumed in the simulation.

Figure 9-16 shows the brachial artery pressure waveforms at different aortic stiffnesses, ranging from a healthy aortic stiffness (75%) to a diseased/aged aortic stiffness (400%), and keeping all other parameters constant. Figure 9-17 shows the strain contours on the outer wall of the pneumatic cuff over this range of stiffnesses. The simulation uses a cuff pressure of 120 mmHg, and all other parameters are kept constant.

Recent in vivo human studies by Pearson et al [62] and other published research [3] confirm that aortic stiffness increases by an average of 300% in humans between the ages of 10 years and 80 years. The effects of variations in aortic stiffness on the AI of the brachial artery pressure waveforms and the strain contours on the pneumatic cuff outer wall are shown graphically in Fig. 9-18.

The figure shows a prediction that an increase in aortic stiffness results in an increase in both the brachial pressure and strain AIs. The pressure AI increased from 76% at 75% of the normal aortic stiffness for a young, healthy adult male to 147% at 400% of the normal aortic stiffness, whereas the strain AI increased from 61% to 218%. The increase in pressure augmentation is caused by the increase in propagation velocity of the pressure waves in the aorta because of the increased stiffness, which causes the reflected wave from the lower body to return to the brachial artery early where it augments the incident wave. This can also be seen from the decreased time lag between the upstroke of the incident wave and the arrival of the reflected wave from 0.206 sec to 0.083 sec, as shown in Fig. 9-19.

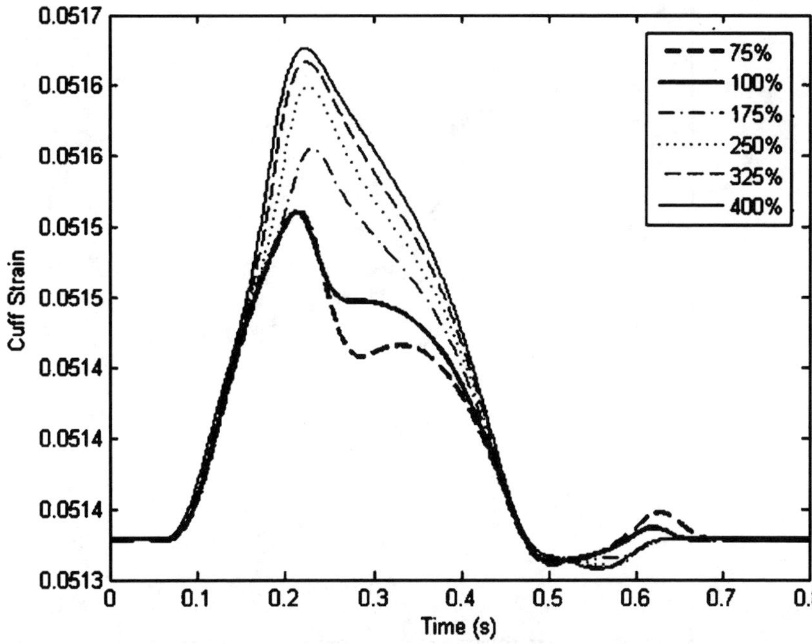

Figure 9-17. Strain on the pneumatic cuff outer wall contours at different aortic stiffness values ranging from healthy (75%) to diseased (400%)

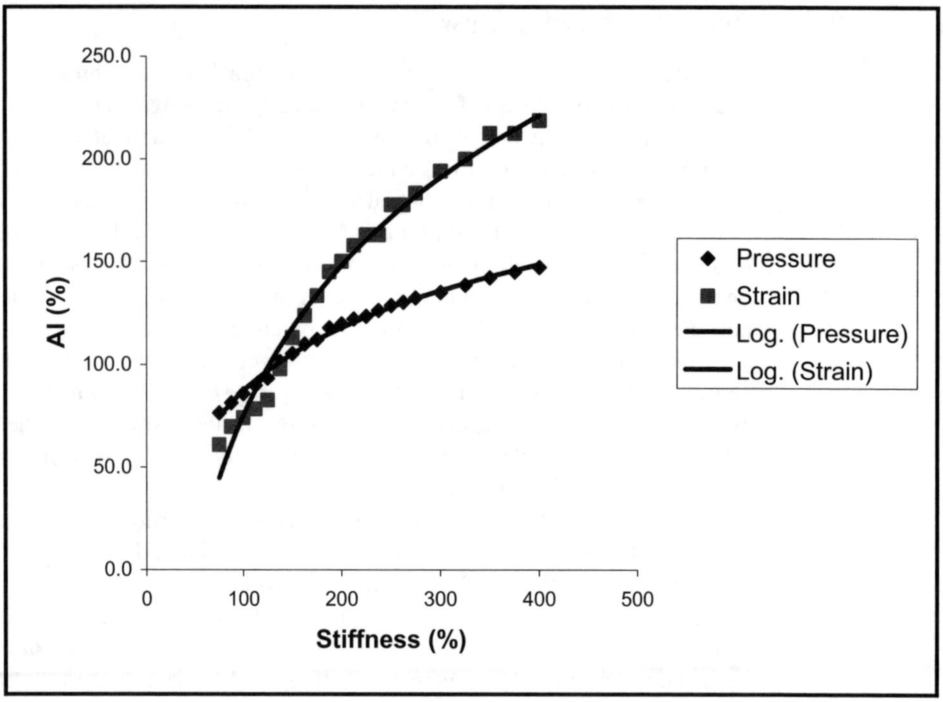

Figure 9-18. The effect of variations in aortic stiffness on the brachial AI

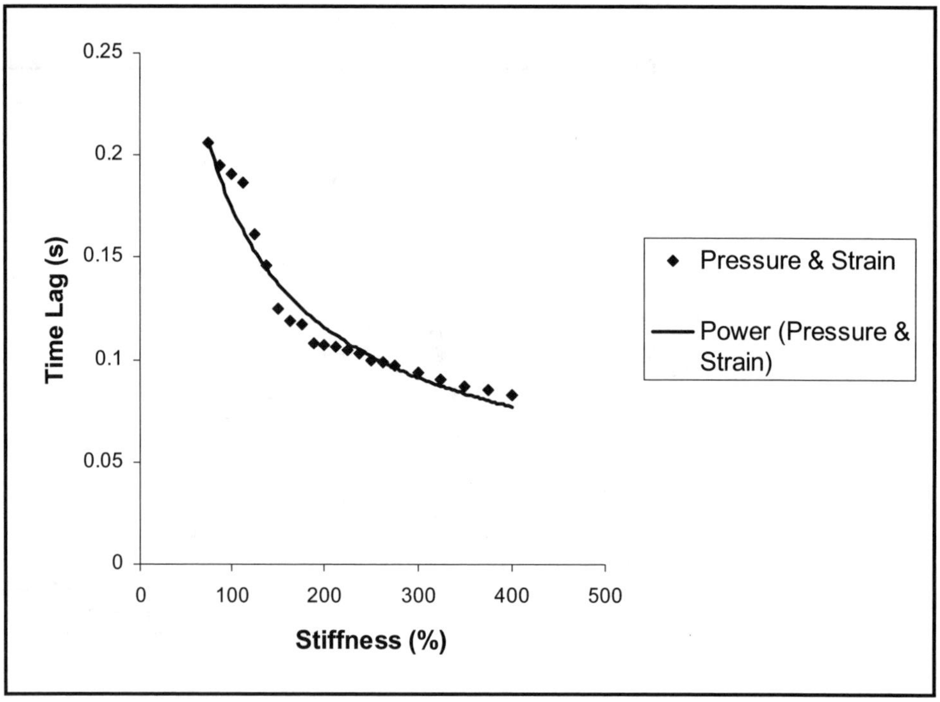

Figure 9-19. The effect of variations in aortic stiffness on the time lag of the arrival of the reflected wave in both the pressure and strain contours

It is evident from Fig. 9-18 that the strain AI (AI_s) is less than the pressure AI (AI_p) for indices less than 100%. The reason for this is that the incident wave peak is higher than the reflected wave peak, and hence, the transmural pressure is closer to zero where there is increased arterial and cuff deflection.

9.4.3 Effect of Artery Radius

The effect of the aortic radius on the brachial artery pressure waveform and the strain on the external wall of the pneumatic cuff is investigated by simulating the combined model at different aortic internal radii and keeping all other parameters constant at their normal healthy level. The aortic radius elevation is assumed to act globally across the entire aorta.

Figure 9-20 shows the brachial artery pressure waveforms at different internal radii ranging from a normal radius (100%) to an elevated aortic radius that could be expected from degeneration of the arteries (200%). Figure 9-21 shows the strain contours on the outer wall of the pneumatic cuff at different aortic radii ranging from a normal radius (100%) to an elevated aortic radius (200%).

In vivo studies conducted by Pearson et al [62] show that the radius of the aorta increases by an average of 100% in human subjects from the age of 10 to 80 years. Figure 9-22 shows the relationship between aortic radius and the pressure and cuff strain AIs. The pressure AI (AI_p) decreased from 86% at normal aortic radius to 51% at 200% aortic stiffness, whereas the cuff strain AI (AI_s) decreased from 74% at normal aortic radius to 30% at 200% the normal aortic radius. This is because of the decrease in the pressure wave propagation velocity as the radius of the aorta increases.

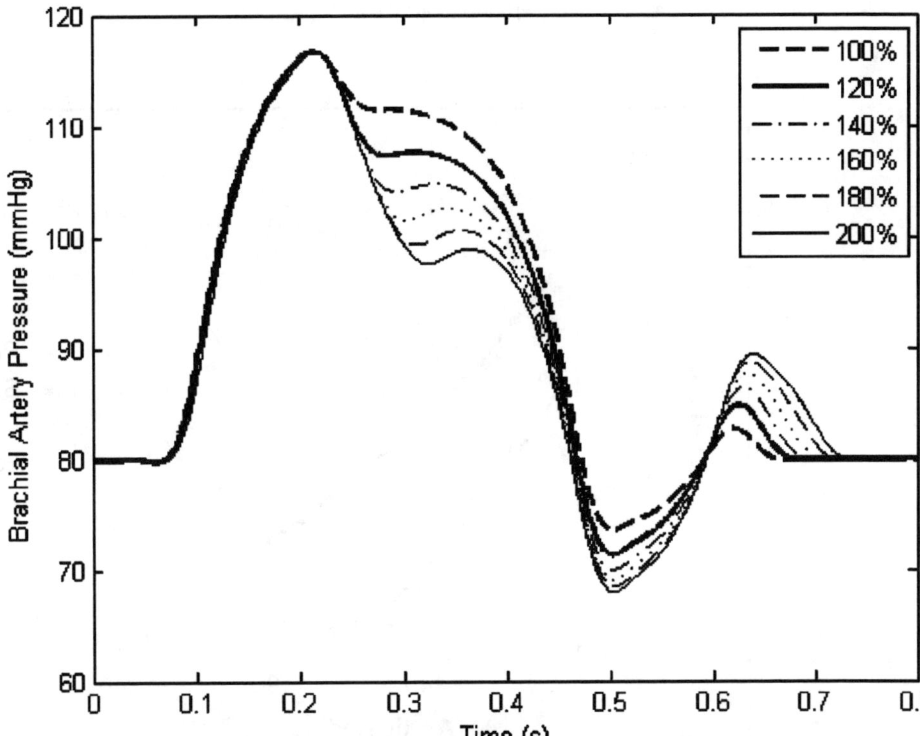

Figure 9-20. Brachial artery pressure waveforms at different aortic radius values ranging from normal (100%) to elevated (200%)

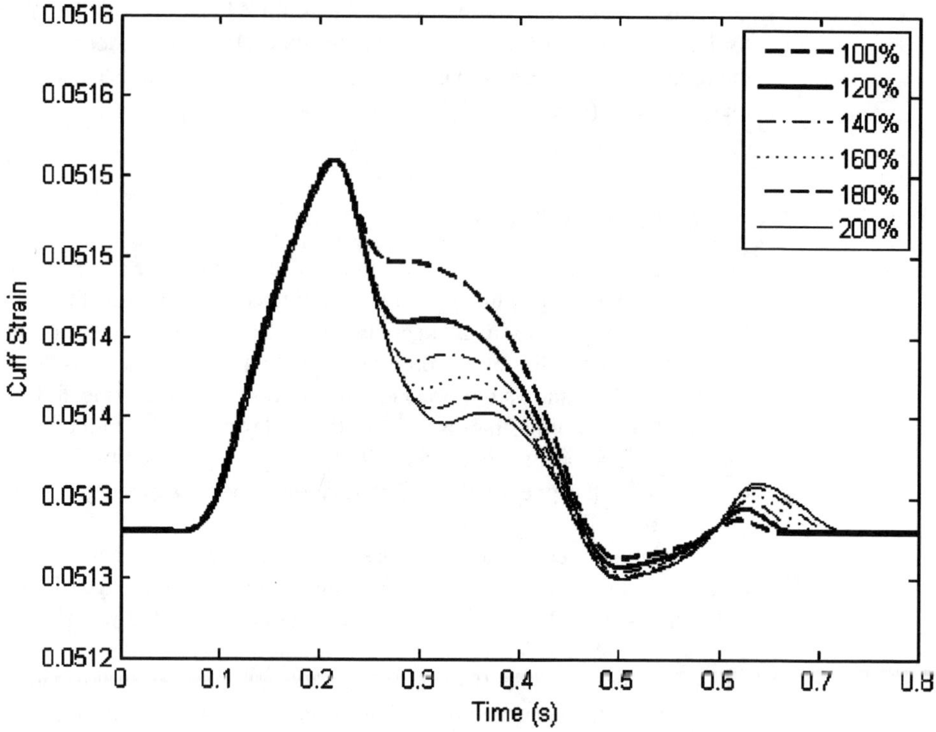

Figure 9-21. Strain on the pneumatic cuff outer wall contours at different aortic radii ranging from healthy (100%) to diseased (200%)

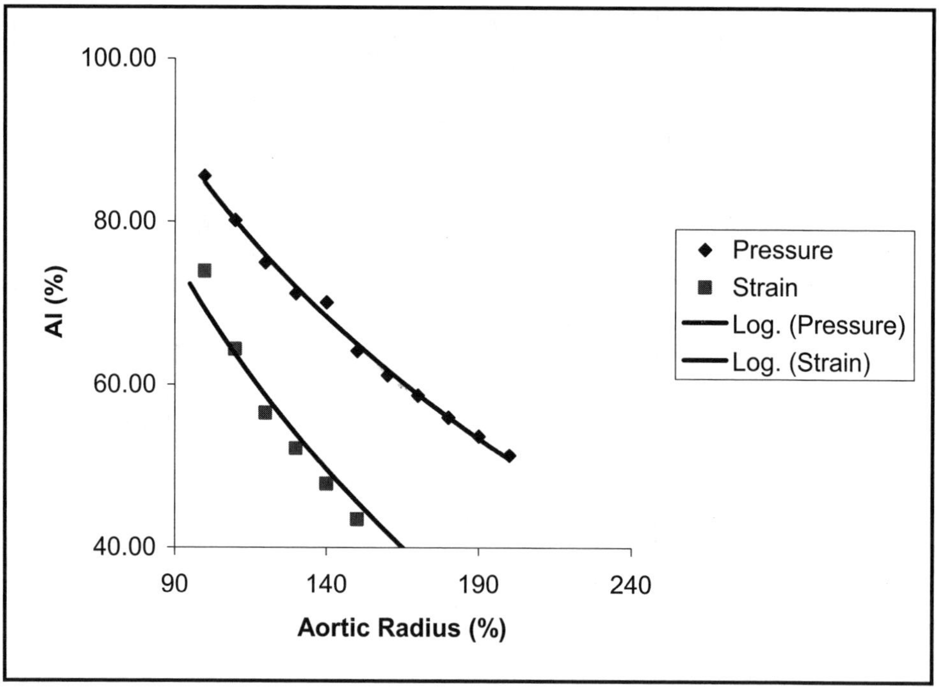

Figure 9-22. The effect of aortic radius variations on the brachial pressure and cuff strain AIs

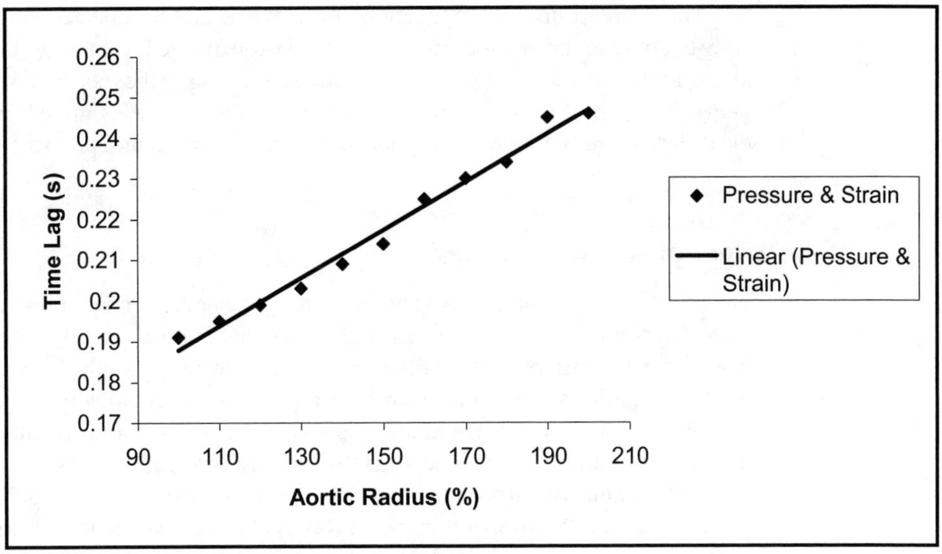

Figure 9-23. The effect of aortic radius variations on the time lag of the reflected wave

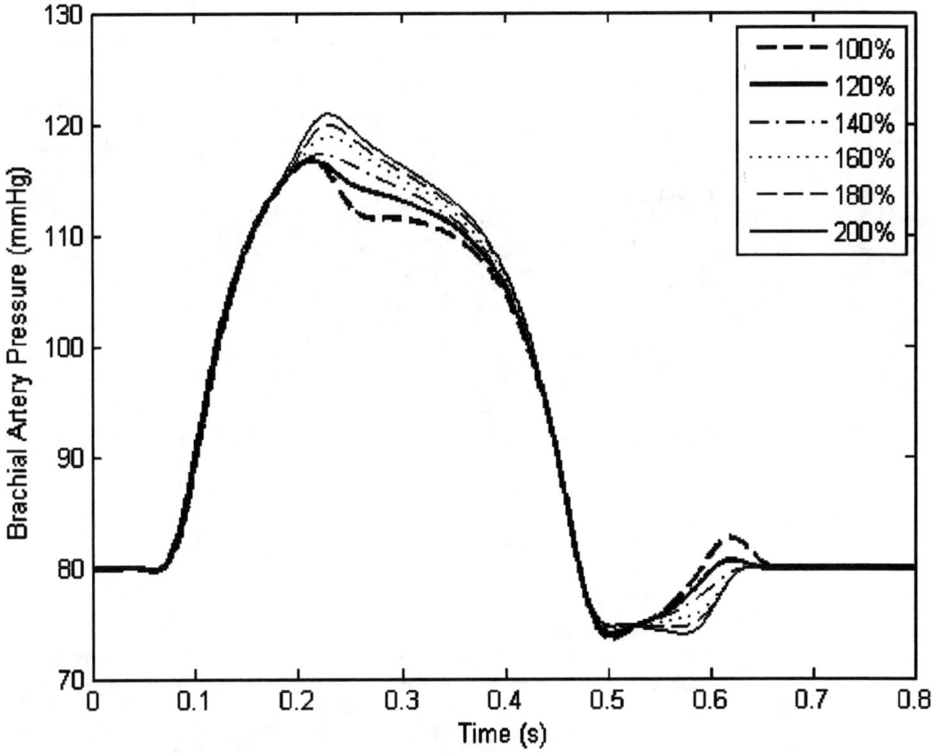

Figure 9-24. Brachial artery pressure waveforms at different aortic thicknesses ranging from healthy (100%) to diseased (200%)

The decrease in the propagation velocity caused the reflected wave from the lower body to arrive to the brachial artery later in diastole where less pressure augmentation takes place. This is also shown by the increase in the time lag of the reflected wave from as shown in Fig. 9-23. The difference between the pressure and the cuff strain AIs is also attributed to the different brachial artery compliance at different transmural pressures.

9.4.4 Effect of Aortic Thickness

The effect of the global thickness of the aorta on the brachial artery pressure waveforms and the strain contours of the pneumatic cuff outer wall is investigated by simulating the combined model at different aortic thicknesses while keeping all other model parameters constant. The aortic thickness elevation is also assumed to act globally across the entire aorta. Figure 9-24 shows the brachial artery pressure waveforms at different aortic thicknesses ranging from a normal thickness (100%) to an elevated thickness (200%). Figure 9-25 shows the strain contours on the outer wall of the pneumatic cuff at different aortic thicknesses ranging from normal thickness (100%) to an elevated aortic thickness (200%).

In vivo experimental results from Pearson et al [62] confirm that the thickness of the human aorta doubles between the ages of 10 years and 80 years. The relationship among the thickness of the aorta, the brachial artery pressure, and cuff strain AIs is plotted in Fig. 9-26.

The brachial artery pressure AI increased from 85.60% at normal aortic thickness to 116.19% at double the thickness. The increase in pressure augmentation is a result of the increase in the propagation velocity of the pressure waveform caused by the increased artery

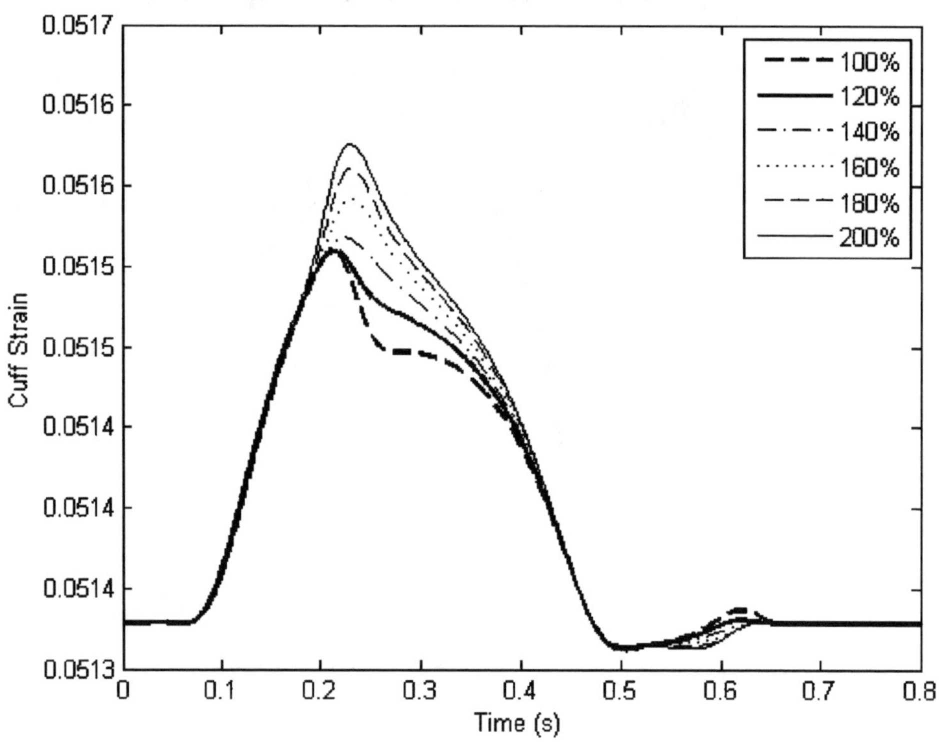

Figure 9-25. Pneumatic cuff outer wall strain contours at different aortic thicknesses ranging from healthy (100%) to diseased (200%)

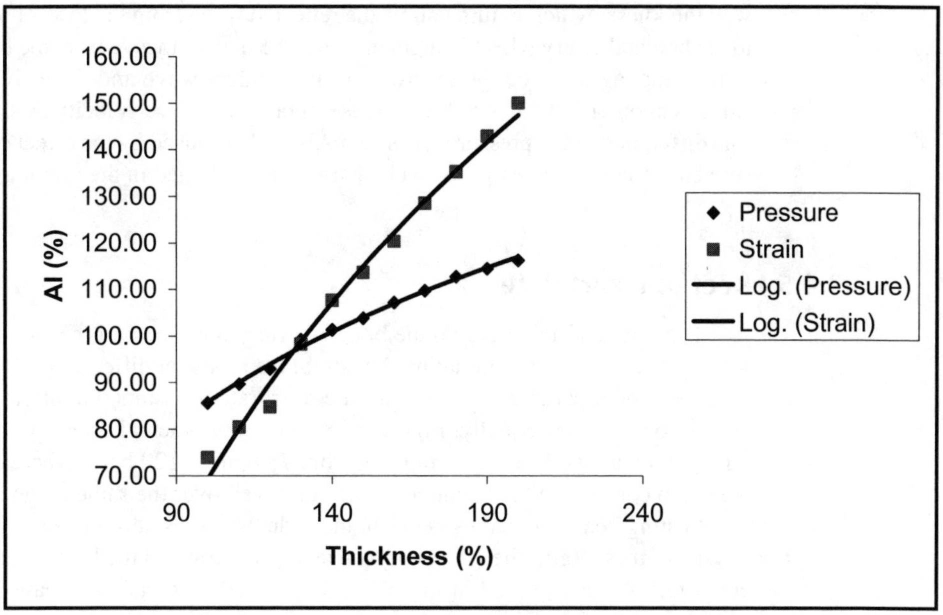

Figure 9-26. The effect of variations in aortic thickness on the brachial artery and the cuff strain AIs

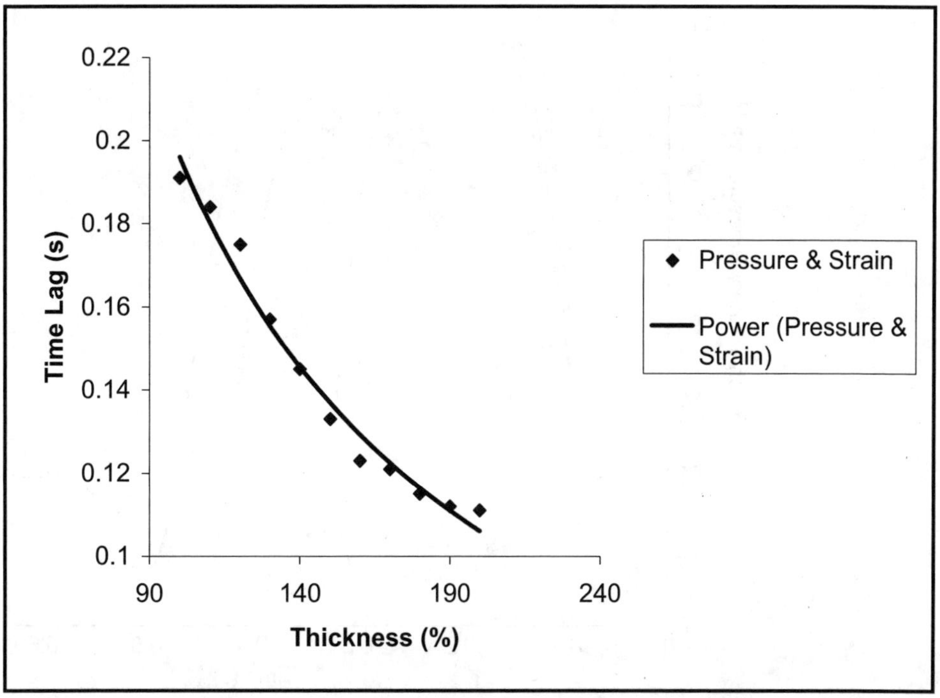

Figure 9-27. The effect of variations in aortic thickness on the time lag between the upstroke of the incident wave and the arrival of the reflected wave to the brachial artery

wall thickness, which in turn causes the reflected wave from the lower body to return faster to the brachial artery where it augments with the incident wave from the heart. The decrease in the time lag between the upstroke of the incident wave and the arrival of the reflected wave is another indicator of the increase in pressure wave velocity as shown in Fig. 9-27. The difference in the pressure and strain AIs is also caused by the effect of transmural pressure on the compliance of the brachial artery as explained in the previous sections.

9.4.5 Effect of Heart Rate

The effect of heart rate on the brachial artery pressure and the strain on the cuff outer wall is investigated by simulating the combined model at different heart rates and keeping all other model parameters constant. The heart rate is assumed to affect both the diastolic and systolic periods equally. Figure 9-28 shows the brachial artery pressure contours at a range of simulated heart rates ranging from 75 bpm to 120 bpm, whereas Fig. 9-29 shows the strain contour on the pneumatic cuff outer wall over the same range of heart rates.

Resting heart rate varies between individuals and usually increases with age and poor physical fitness [63]. The effects of increasing heart rate on the brachial artery pressure and cuff strain AIs are plotted in Fig. 9-30. The model predicts a decrease in the pressure AI (AI_p) from 85.6% at 75 bpm to 64.96 % at 120 bpm. The cuff strain AI (AI_s) also decreases

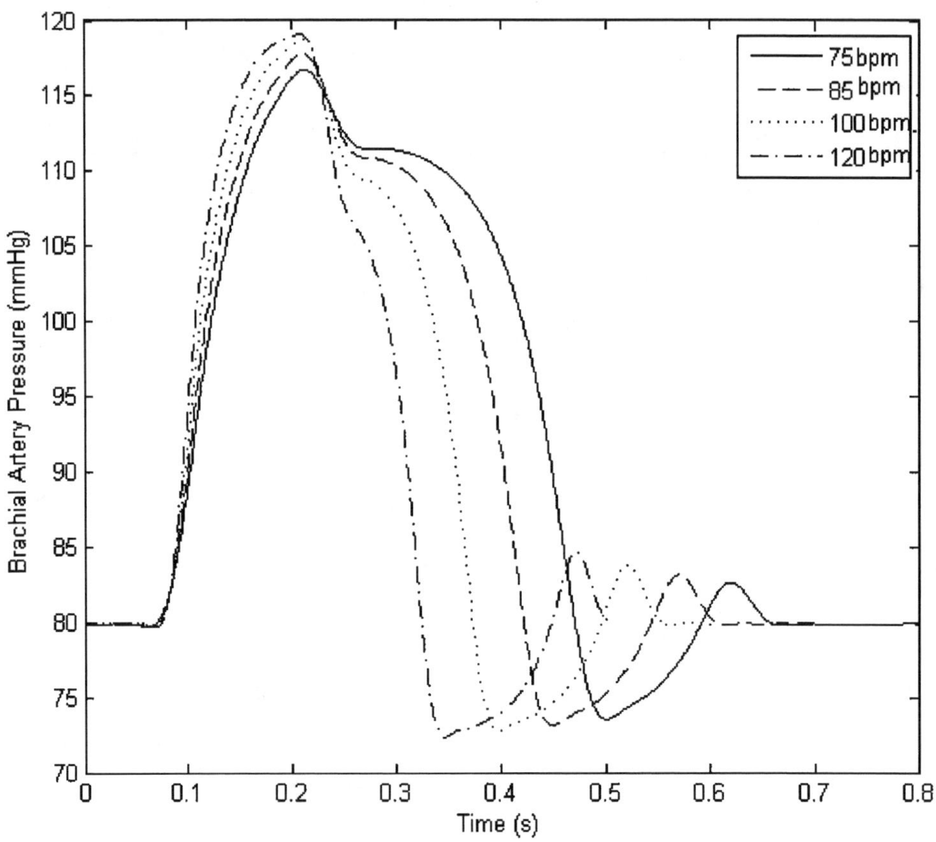

Figure 9-28. Brachial artery pressure simulation at different heart rates ranging from 75 bpm to 120 bpm

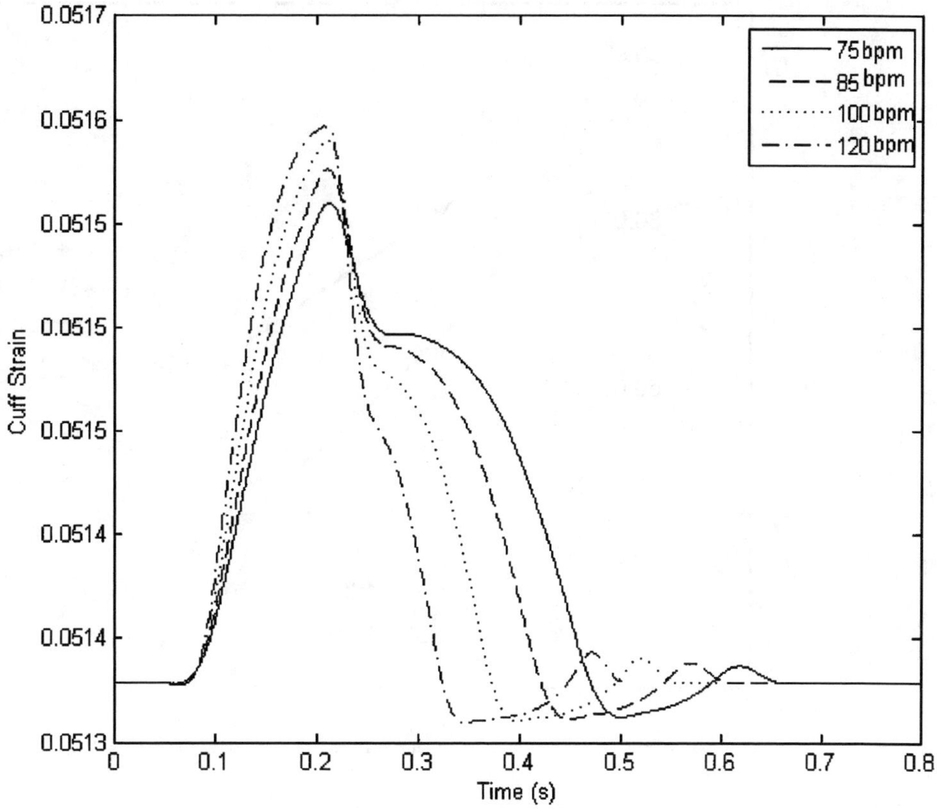

Figure 9-29. Strain contour on the pneumatic cuff outer wall at a number of heart rates ranging from 75 bpm to 120 bpm

from 74% at 75 bpm to 44% at 125 bpm. The decrease in pressure augmentation is caused by the decrease in the systolic period. The pressure wave velocity is independent of the heart rate; hence, the reflected pressure wave still arrives at the brachial artery at the same time, which is confirmed by the unchanged time lag value. However, the reflected pressure wave now arrives in diastole because of the shortened systolic period, and hence, less pressure augmentation takes place. This phenomenon is also supported by an increasing number of recent studies [64]. These results demonstrate that it is essential to adjust for the heart rate when developing any stiffness detection algorithms based on AI.

9.4.6 Effect of Cuff Pressure

The effect of cuff pressure on the brachial artery pressure waveform and the strain contour on the cuff outer wall is investigated by simulating the combined model at different cuff pressures. Figure 9-31 shows the brachial artery pressure contours at a range of cuff pressures ranging from 90 mmHg to a significantly high cuff pressure of 175 mmHg. This figure is then used to extract the four feature points; the AI and the time lag were then calculated.

Figure 9-32 shows the strain contours on the pneumatic cuff outer wall at a range of cuff pressures ranging from 90 mmHg to a significantly high cuff pressure of 170 mmHg. This figure is then used to extract the four feature points, the strain AI, (AI_s) and the time lag.

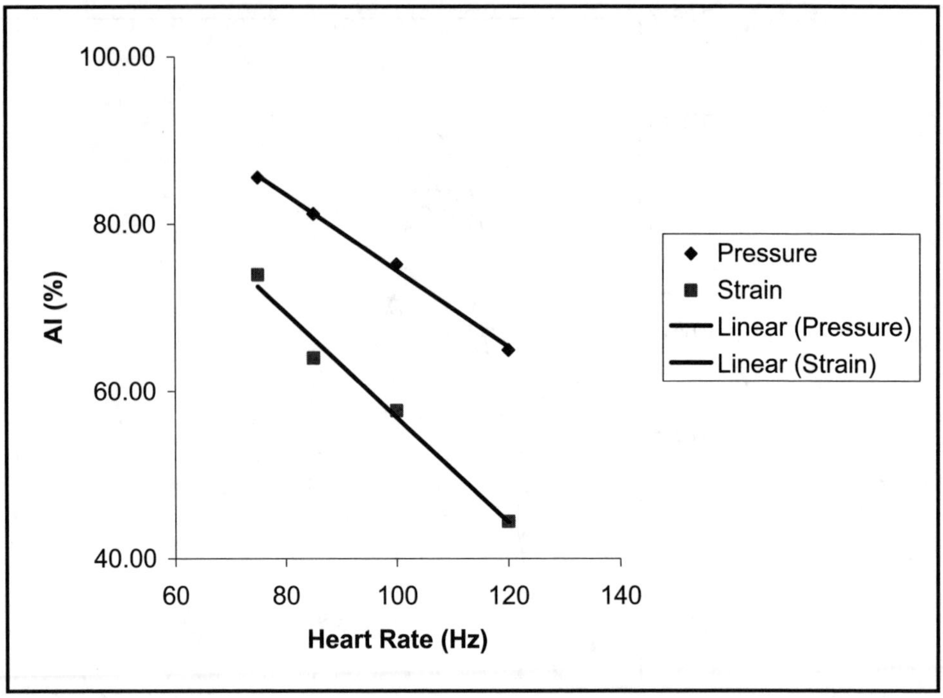

Figure 9-30. The effect of heart rate on the brachial artery pressure and cuff strain AIs

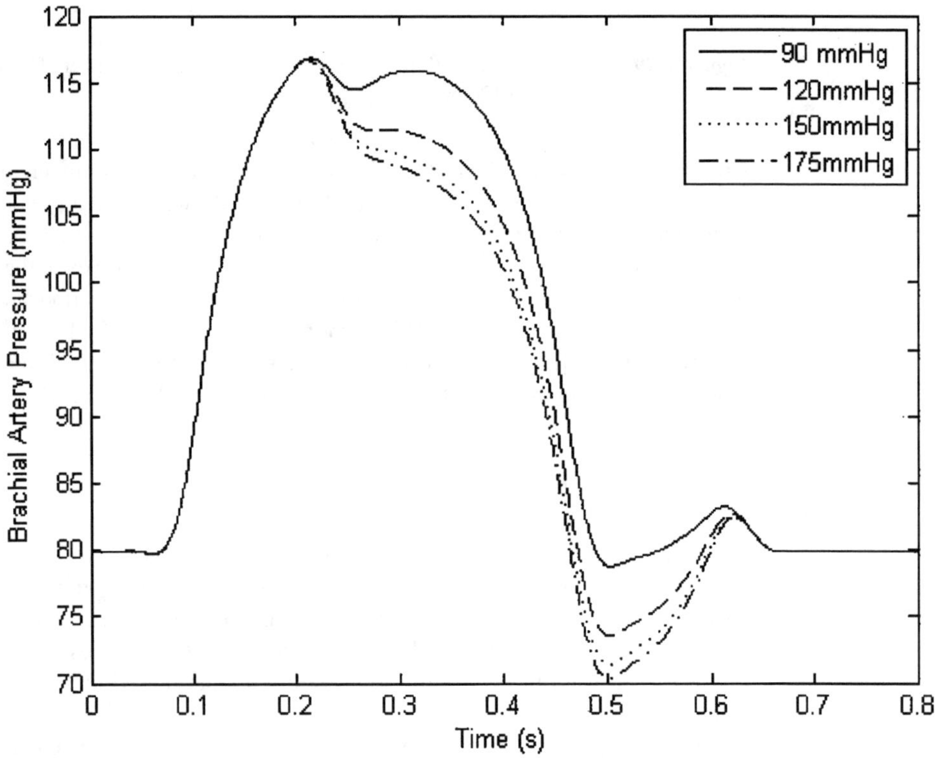

Figure 9-31. Brachial artery pressure simulation at a number of cuff pressures ranging from 90 mmHg to 175 mmHg

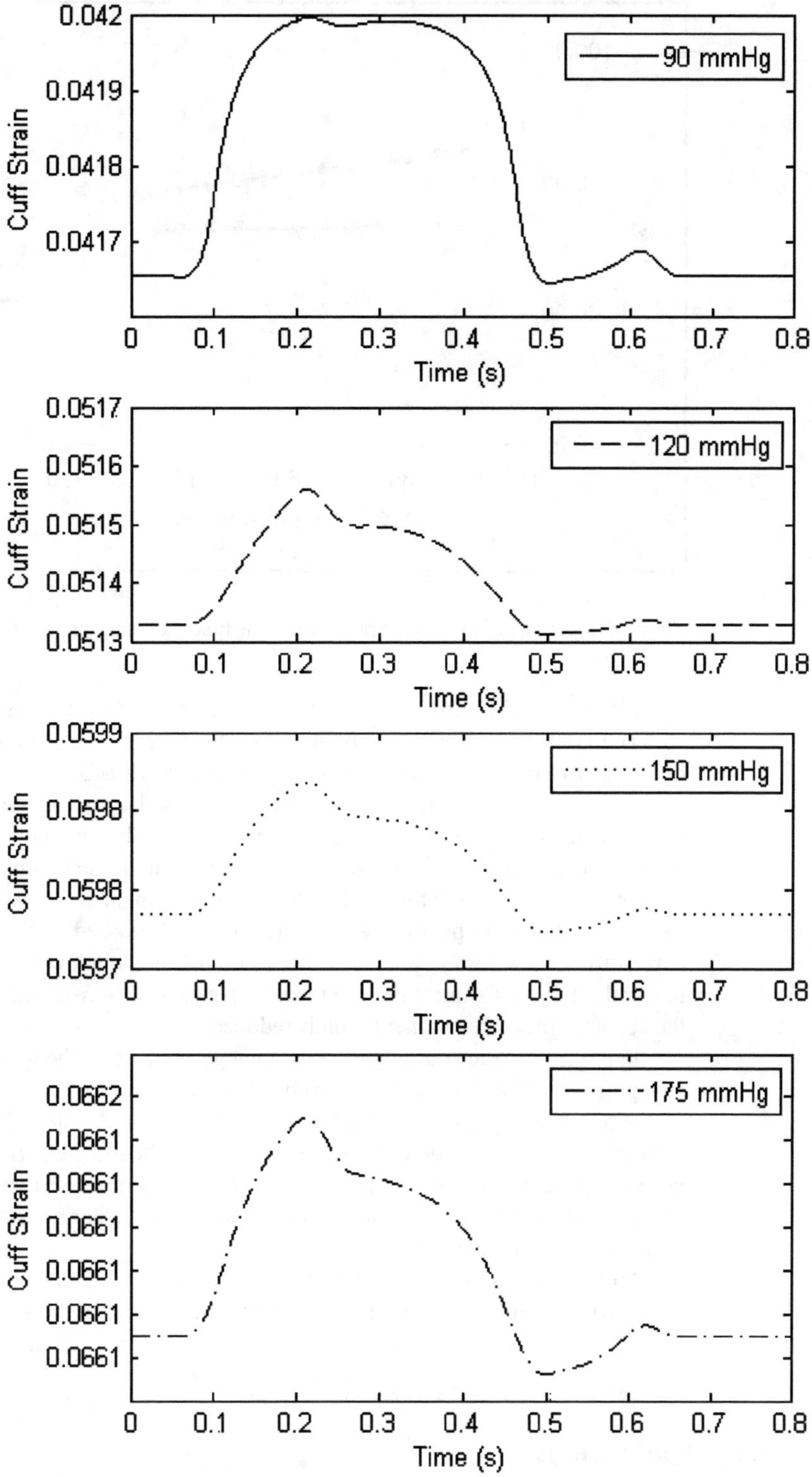

Figure 9-32. Strain contours on the pneumatic cuff outer wall at a number of cuff pressures ranging from 90 mmHg to 175 mmHg

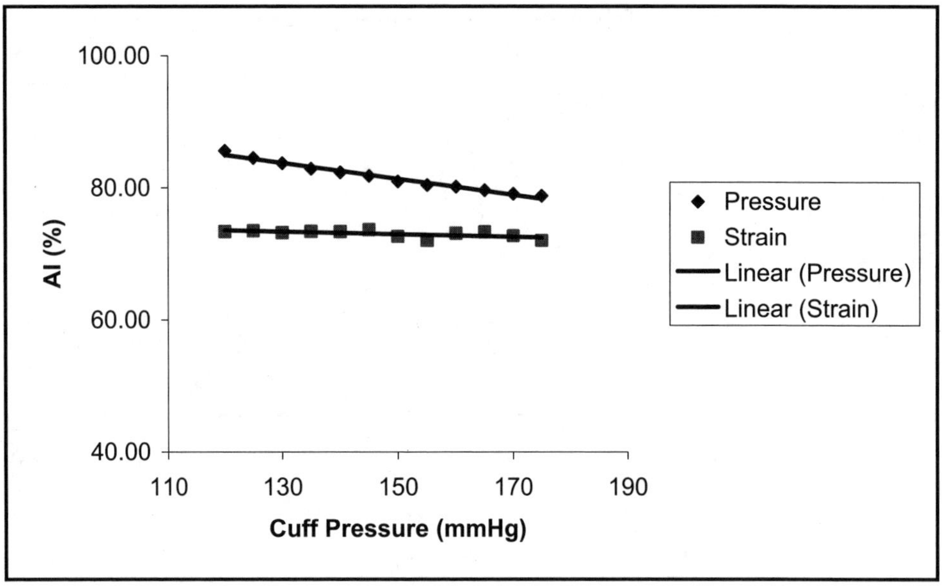

Figure 9-33. The effect of cuff pressure on the brachial artery pressure and the cuff strain AIs

It is essential to observe the effect of the value of cuff pressure on the pressure and strain contours and their feature points. The model predicts that pressure and strain AIs will decrease with increasing cuff pressure as shown in Fig. 9-33. This decrease is attributed to the increase in the magnitude of the negative reflected wave. As the cuff pressure increases, the cross-sectional area of the brachial artery under the cuff decreases, which in turn increases the impedance mismatch between the section of the brachial artery proximal to the cuff and the section under the cuff. The impedance mismatch increase, therefore, enlarges the amplitude of the positive wave reflection as discussed in Section 9.3. This reflected wave further negatively reflects at the mouth of the subclavian artery and returns to the brachial artery at a similar time to the positive reflection from the lower body, and hence, the apparent pressure augmentation is reduced.

Furthermore, the strain AI does not change as much as the pressure AI with increasing cuff pressures. This can be explained by the dependence of the brachial artery compliance on the transmural pressure as explained in Section 9.4. As the cuff pressure increases, the transmural pressure becomes more negative where the brachial artery is less compliant and not as dependent on the transmural pressure as it is near zero transmural pressure. This decreases the variation in the artery response between the first and second pressure peaks as discussed in Section 9.4.1.

From a modeling point of view, it is recommended that the device readings are taken at above systolic cuff pressure. This is to ensure the validity of the assumption imposed in this work that there be no reflections from sites downstream of the measurement site.

9.4.7 Applications

The results of the model simulation demonstrated that the model developed in this work is able to simulate the brachial artery pressure and cuff outer wall strain contours. The results also show that the model is able to predict the effect of changes in aortic stiffness, aortic thickness, aortic cross-sectional area, heart rate, and cuff pressure on the simulated pressure and strain waveforms.

All model results obtained in this work are qualitatively in agreement with published results. Quantitative validation of the model developed in this work is extremely difficult for a number of reasons. This is the first published investigation that looks at the brachial AI and time lag as a measure of arterial stiffness. Hence, there is no evidence of previous experimental studies that report any model or experimental results that show the effect of any of the above parameters on the brachial pressure and strain contours. Furthermore, conducting an experimental investigation to investigate the effect of aortic material and geometric properties on the measured contours will not produce conclusive results because it is impossible to isolate parameters in the system. That is, there are a large number of factors that affect the propagation and reflection of pressure waves in the system as demonstrated earlier, and measuring the contours for different subjects would only show the combined effect of an enormous range of factors.

Nevertheless, some measurements using a strain gauge attached to the external cuff wall can be used to qualitatively compare the waveforms obtained from the model presented here and readings from the device as shown in Fig. 9-34. The shape of the strain contours obtained from the device compared very well with the model results. Some of the strain contours showed two pressure peaks, whereas some contours showed increased pressure augmentation. These could be a result of increased aortic stiffness, but such a hypothesis cannot be verified.

In both reality and modeling, aortic PWV not only depends on the aortic stiffness but also on the radius and thickness of the aorta as shown in the model simulation. However,

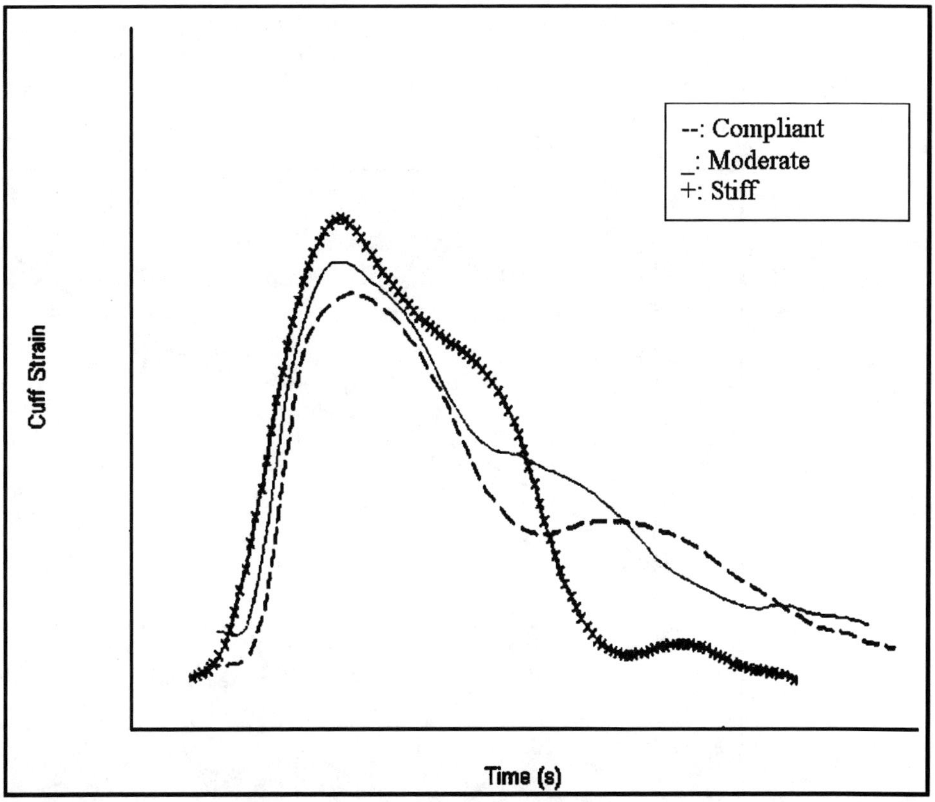

Figure 9-34. Strain contours obtained from a piezoelectric strain gauge attached to the outer cuff wall for three individuals. The legend shows a speculative artery stiffness state

recent in vivo human studies by Pearson et al [62] confirm that aortic stiffness increases by an average of 300% in humans between the ages of 10 years and 80 years, with little or no change in the thickness-to-radius ratio. Furthermore, there is much less geometric variation within the same age group when compared with aortic stiffness [62]. Hence, from Eq. (9-12), it can be seen that aortic PWV is mainly dependent on aortic stiffness.

Aortic PWV is simulated using age-dependent aortic stiffness values obtained from Pearson et al [62] as shown in Fig. 9-35. These simulated PWV curves compared well with in vivo age-dependent aortic PWV measurements reported in the literature as shown in Fig. 9-35 [3, 65]. This further demonstrates that the model is able to accurately predict arterial PWV and that aortic stiffness is the major determinant of aortic PWV, which can be estimated from the feature points on the strain contours.

The results of the mathematical model also show that heart rate and cuff pressure affect the pressure and the cuff strain AIs. Hence, it is essential to adjust for these two factors when developing any stiffness detection algorithms based on the AIs. Also from a modeling point of view, it is recommended that the readings are taken at above systolic cuff pressure. This is to ensure the validity of the "no hand reflection" assumption imposed in this model.

The model can be used to develop the stiffness detection algorithm for the Pulsecor WEP monitor to serve two purposes. The device can be used to detect elevated arterial stiffness by using average age and height-specific arterial geometry data obtained from the open literature. Otherwise, the exact severity of arterial stiffness can be found by performing an upper body magnetic resonance imaging scan on the patient, where it can be used to derive the patient's arterial geometry information required by the model.

The model is also designed in a way that makes it capable of simulating the effect of more parameters such as arm soft tissue, pneumatic cuff material, and geometric properties on the contours. This can be done by simply changing the model input parameters in the MATLAB program. This can be very useful in the cuff design process and in investigating the effect of arm soft tissue geometry and material properties on the acquired waveforms.

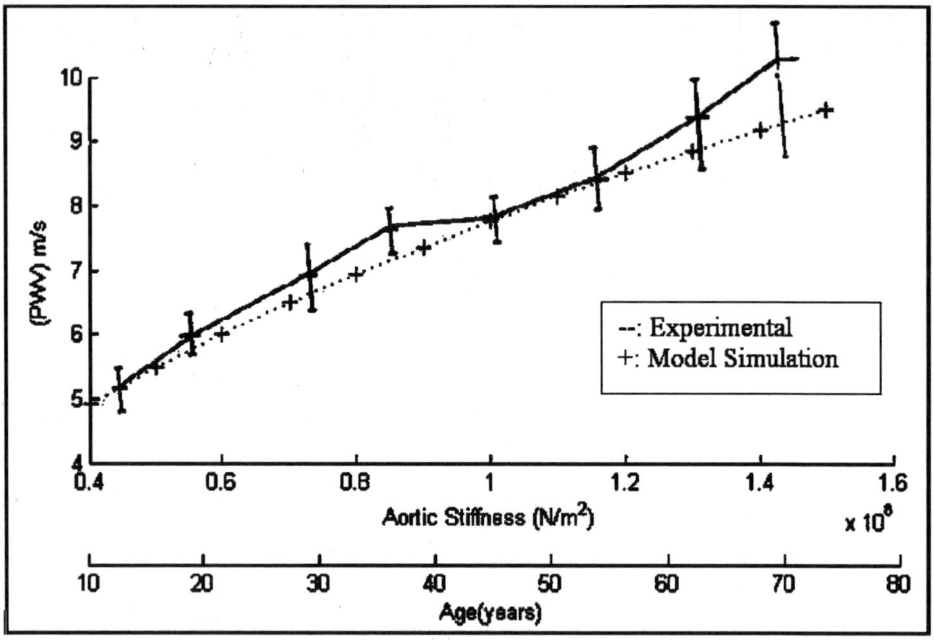

Figure 9-35. A comparison between experimental and model (PWV) versus aortic stiffness and patient's age

REFERENCES

[1] NZ Guidelines Group, 2003, *Assessment and Management of Cardiovascular Risk*, NZGG, Wellington, NZ.

[2] National Heart, Lung and Blood Institute, 2004, *Chartbook on Cardiovascular, Lung and Blood Disease*, NHLBI, Bethesda, MD.

[3] Nichols, W., and O'Rourke, M., 1998, *McDonald's Blood Flow in Arteries Theoretical, Experimental and Clinical Principles*, 4th ed., Hodder Arnold, London.

[4] Dewood, A., Spores, J., Notske, R., Mouser, L. T., Burroughs, R., Golden, M. S., and Lang, H. T., 1980, "Prevalence of Total Coronary Occlusion During the Early Hours of Transmural Myocardial Infraction," *New England Journal of Medicine*, **303**(16), pp. 897-902.

[5] Martin, P., 1978, "On Abdominal Aortic Aneurysms," *Journal of Cardiovascular Surgery*, **19**, pp. 597-598.

[6] Liu, H., Fukasaku, K., Iwase, H., Matsunaga, N., He, Y., Yokoi, K., and Himeno, R., 2003, "Influences of Non-Planarity, Bifurcation, Dynamics, Inflow and Outflows on Blood Flow Patterns in the Aortic Arch: A Multi-Scale Computational Study," *Summer Bioengineering Conference*, Sonesta Beach Resort, Key Biscayne, FL.

[7] Atherosclerosis, February 27, 2007, Wikipedia, retrieved February 27, 2007, http://en.wikipedia.org/wiki/Atherosclerosis.

[8] Tortora, G. J., and Grabowski, S. R., 1999, *Principles of Anatomy and Physiology*, 9th ed., John Wiley and Sons, Chichester, New York.

[9] Pannier, B., Avolio, P., Hoeks, A., Mancia, G., and Takazawa, K., 2002, "Methods and Devices for Measuring Arterial Compliance in Humans," *American Journal of Hypertension*, **15**, pp. 743-753.

[10] Laurent, S., Caviezel, B., Beck, L., Girerd, X., Billaud, E., Boutouyrie, P., Hoeks, A., and Safar, M., 1994, "Carotid Artery Distensibility and Distending Pressure in Hypertensive Humans," *Hypertension*, **23**, pp. 878-883.

[11] Verbeke, F., Segers, P., Heireman, S., Vanholder, R., Verdonck, P., and Van Bortel, L. M., 2005, "Noninvasive Assessment of Local Pulse Pressure, Importance of Brachial-to-Radial Pressure Amplification," *Hypertension*, **46**, pp. 244-248.

[12] Latham, D., Westerhof, N., Sipkema, P., Rubal, B. J., Reuderink, P., and Murgo, J. P., 1985, "Regional Wave Travel and Reflections Along the Human Aorta: A Study With Six Simultaneous Micromanometric Pressures," *Circulation*, **72**, pp. 1257-1269.

[13] Laurent, S., Hayoz, D., Trazzi, S., Boutouyrie, P., Waeber, B., Omboni, S., Brunner, H. R., Mancia, G., and Safar, M., 1993, "Isobaric Compliance of the Radial Artery is Increased in Patients With Essential Hypertension," *Journal of Hypertension*, **11**, pp. 89-98.

[14] Blacher, J., Guerin, P., Pannier, B., Marchais, S. J., Safar, M. E., and London, G. M., 1999, "Impact of Aortic Stiffness on Survival in End-Stage Renal Disease," *Circulation*, **99**, pp. 2434-2439.

[15] Willum-Hansen, T., Staessen, A., Torp-Pedersen, C., Torp-Pedersen, C., Rasmussen, S., Thijs, L., Ibsen, H., and Jeppesen, J., 2006, "Prognostic Value of Aortic Pulse Wave Velocity as Index of Arterial Stiffness in the General Population," *Circulation*, **113**, pp. 664-670.

[16] Van Bortel, M., Duprez, D., Starmans-Kool, J., Safar, M. E., Giannattasio, C., Cockcroft, J., Kaiser, D. R., and Thuillez, C., 2002, "Clinical Applications of Arterial Stiffness, Task Force III: Recommendations for User Procedures," *American Journal of Hypertension*, **15**, pp. 445-452.

[17] Mackenzie, S., Wilkinson, B., and Cockcroft, J. R., 2002, "Assessment of Arterial Stiffness in Clinical Practice," *QJM*, **95**, pp. 67-74.

[18] O'Rourke, F., 1982, *Arterial Function in Health and Disease*, Churchill Livingstone, Edinburgh.

[19] London, G., Guerin, A., Pannier, B., Marchais, S., Benetos, A., and Safar, M., 1992, "Increased Systolic Pressure in Chronic Uremia. Role of Arterial Wave Reflections," *Hypertension*, **2**, pp. 10-19.

[20] Chen, C.-H., Nevo, E., Fetics, B., Pak, P. H., Yin, F. C., Maughan, W. L., and Kass, D. A., 1997, "Estimation of Central Aortic Pressure Waveform by Mathematical Transformation of Radial Tonometry Pressure: Validation of Generalized Transfer Function," *Circulation*, **95**, pp. 1827-1836.

[21] Pauca, A., O'Rourke, M., and Kon, N., 2001, "Prospective Evaluation of a Method for Estimating Ascending Aortic Pressure From the Radial Artery Pressure Waveform," *Hypertension*, **38**, pp. 932-937.

[22] Nichols, W., 2005, "Clinical Measurement of Arterial Stiffness Obtained From Non-Invasive Pressure Waveforms," *American Journal of Hypertension*, **18**, pp. S3-S10.

[23] Takazawa, K., Tanaka, N., Fujita, M., Matsuoka, O., Saiki, T., Aikawa, M., Tamura, S., and Ibukiyama, C., 1998, "Assessment of Vasoactive Agents and Vascular Ageing by the Second Derivative of Photoplethysmogram Waveform," *Hypertension*, **32**, pp. 365-370.

[24] Chowienczyk, P., Kelly, R., MacCallum, H., Millasseau, S. C., Andersson, T. L., Gosling, R. G., Ritter, J. M., and Anggård, E. E., 1999, "Photoplethysmographic Assessment of Pulse Wave Reflection: Blunted Response to Endothelium-Dependent beta2-Adrenergic Vasodilation in Type II Diabetes Mellitus," *Journal of the American College of Cardiology*, **34**, pp. 2007-2014.

[25] Pontrelli, G., and Rossoni, E., 2003, "Numeric Modelling of the Pressure Wave Propagation in the Arterial Flow," *International Journal for Numerical Methods in Fluids*, **43**, pp. 651-671.

[26] Stergiopulos, N., Westerhof, B., and Westerhof, N., 1999, "Total Arterial Inertance as the Fourth Element of the Windkessel Model," *American Journal of Physiology*, **276**, pp. 81-88.

[27] Olfusen, M., and Nadim, A., 2004, "On Deriving Lumped Models for Blood Flow and Pressure in the Systemic Arteries," *Mathematical Biosciences and Engineering*, **1**, pp. 61-80.

[28] Formaggia, L., Lamponi, D., and Quarteroni, A., 2003, "One Dimensional Models for Blood Flow in Arteries," *Journal of Engineering Mathematics*, **47**, pp. 251-276.

[29] Olfusen, M., 1999, "Structured Tree Outflow Condition for Blood Flow in Larger Systemic Arteries," *American Journal of Physiology*, **276**, pp. H257-H268.

[30] Steele, N., and Taylor, C., 2003, "Simulation of Blood Flow in the Abdominal Aorta at Rest and During Exercise Using a 1-D Finite Element Method With Impedance Boundary Conditions Derived From a Fractal Tree," *Summer Bioengineering Conference*, Sonesta Beach Resort, Key Biscayne, FL.

[31] Tang, D., Yang, C., and Huang, Y., 1999, "Wall Stress and Strain Analysis Using a Three-Dimensional Thick-Wall Model With Fluid-Structure Interactions for Blood Flow in Carotid Arteries With Stenoses," *Computers and Structures*, **72**, pp. 341-356.

[32] O'Rourke, M., 1982, "Vascular Impedance in Studies of Arterial and Cardiac Function," *Physiological Reviews*, **62**, pp. 570-623.

[33] Hamilton, W., and Dow, P., 1939, "An Experimental Study of the Standing Waves in the Pulse Propagated Through the Aorta," *American Journal of Physiology*, **125**, 48-59.

[34] Alexander, R., 1952, "The Genesis of the Aortic Standing Waves," *Circulation Research*, **1**, pp. 145-151.

[35] Remington, J., 1963, "The Physiology of the Aorta and the Major Arteries," *American Physiological Society Handbook of Physiology*, American Physiology Society, Bethesda, MD, pp. 799-838.

[36] Mauck, G., Smith, C., Geddes, L., and Bourland, J. D., 1980, "The Meaning of the Point of Maximum Oscillations in Cuff Pressure in the Indirect Measurement of Blood Pressure: Part II," *ASME Journal of Biomechanical Engineering*, **102**, pp. 28-33.

[37] Forster, F., and Turney, D., 1986, "Oscillometric Determination of Diastolic, Mean, and Systolic Blood Pressure. A Numerical Model," *ASME Journal of Biomechanical Engineering*, **108**, pp. 359-364.

[38] Drzewiecki, G., Bansal, V., Karam, E., Hood, R., and Apple, H., 1993, "Mechanics of the Occlusive Arm Cuff and Its Application as a Volume Sensor," *IEEE Transactions of Biomedical Engineering*, **40**, pp. 704-708.

[39] Ursino, M., and Cristalli, C., 1995, "Mathematical Modeling of Noninvasive Blood Pressure Estimation Techniques — Part I — Pressure Transmission Across the Arm Tissue," *ASME Journal of Biomechanical Engineering*, **117**, pp. 107-116.

[40] Ursino, M., and Cristalli, C., 1995, "Mathematical Modeling of Noninvasive Blood Pressure Estimation Techniques — Part II — Brachial Hemodynamics," *ASME Journal of Biomechanical Engineering*, **117**, pp. 117-126.

[41] Ursino, M., and Cristalli, C., 1996, "A Mathematical Study of Some Biomechanical Factors Affecting the Oscillometric Blood Pressure Measurement," *IEEE Transactions on Biomedical Engineering*, **43**, pp. 761-778.

[42] Taylor, M., 1959, "The Influence of the Anomalous Viscosity of Blood Upon Its Oscillatory Flow," *Physics of Medicine and Biology*, **3**, pp. 273-290.

[43] Fung, Y., 1997, *Biomechanics Circulation*, 2nd ed., Springer, New York.

[44] Chang, C., and Atabek, H., 1961, "The Inlet Length for Oscillatory Flow and Its Effects on the Determination of the Rate of Flow in the Arteries," *Physics of Medicine and Biology*, **6**, pp. 303-317.

[45] Learoyd, B., and Taylor, M., 1966, "Alterations With Age in the Viscoelastic Properties of Human Arterial Walls," *Circulation Research*, **18**, pp. 278-292.

[46] Westerhof, N., Bosman, F., De Vries, C. J., and Noordergraaf, A., 1969, "Analog Studies of Human Systemic Arterial Tree," *Journal of Biomechanics*, **2**, pp. 121-143.

[47] Bergel, D. H., and Milnor, W. R., 1965, "Pulmonary Vascular Impedance in the Dog," *Circulation Research*, **26**, pp. 401-415.

[48] Pollack, G. H., Reddy, R. V., and Noordergraaf, A., 1968, "Input Impedance, Wave Travel and Reflections in the Human Pulmonary Arterial Tree: Studies Using an Electrical Analogue," *IEEE Transactions of Biomedical Engineering*, **15**, pp. 151-164.

[49] O'Rourke, E. F. M., and Taylor, M. G., 1967, "Input Impedance of the Systemic Circulation," *Circulation Research*, **20**, pp. 365-380.

[50] Mills, C. J., Gabe, I. T., Gault, J. H., Mason, D. T., Ross, J. Jr., Braunwald, E., and Shillingford, J. P., 1970, "Pressure-Flow Relationships and Vascular Impedance in Man," *Cardiovascular Research*, **4**, pp. 405-417.

[51] Mills, C. J., and Gabe, I. T., 1972, "Pulsatile Blood Velocity and Pressure and the Computer Analysis of Cardiovascular Data," *Quantitation in Cardiology*, Snellen, H. A., Hemker, H. C., Hugenholtz, P. G., and Van Bemmel, J. H., eds., Leiden University Press, Leiden, pp. 99-113.

[52] Newman, D., King, D., Gosling, R., and Wille, S., 1972, "Reflections at Aortic-Iliac Junctions." *Blood Flow Measurement*, V. C. Roberts, ed., Sector Publishing, London, pp. 44-47.

[53] Murgo, J., Westerhof, N., Giolma, J., and Altobelli, S. A., 1980, "Aortic Input Impedance in Normal Man: Relationship to Pressure Wave Shapes," *Circulation*, **62**, pp. 105-116.

[54] Nichols, W., O'Rourke, M., Avolio, A., Yaginuma, T., Murgo, J. P., Pepine, C. J., and Conti, C. R., 1987, "Age-Related Changes in Left Ventricular-Arterial Coupling," *Vascular/Ventricular Coupling*, F. C. P. Yin, ed., Springer-Verlag, New York, pp. 79-114.

[55] Hamilton, W., and Dow, P., 1939, "An Experimental Study of the Standing Waves in the Pulse Propagated Through the Aorta," *American Journal of Physiology*, **125**, pp. 48-59.

[56] Alexander, R., 1952, "The Genesis of the Aortic Standing Waves," *Circulation Research*, **1**, pp. 145-151.

[57] Remington, J., 1963, "The Physiology of the Aorta and the Major Arteries," *American Physiological Society Handbook of Physiology*, American Physiology Society, Bethesda, MD, pp. 799-838.

[58] Remington, J., 1965, "Quantitative Synthesis of Head and Foreleg Arterial Pulsed in the Dog," *American Journal of Physiology*, **208**, pp. 968-983.

[59] Westerhof, B., Guelen, I., Stok, W., Wesseling, K. H., Spaan, J. A. E., Westerhof, N., Bos, W. J., and Stergiopulos, N., 2006, "Arterial Pressure Transfer Characteristics: Effects of Travel Time," *American Journal of Physiology: Heart Circulation Physiology*, **292**, pp. 800-807.

[60] Linden, R., and Drzewiecki, G., 2000, "Noninvasive Measurement of Brachial Artery Compliance Variability," *Proceedings of the IEEE 26th Annual Northeast Bioengineering Conference*, April 8-9, 2000, pp. 39-40.

[61] Bank, A., Kaiser, D., Rajala, S., and Cheng, A., 1999, "In Vivo Human Brachial Artery Elastic Mechanics: Effects of Smooth Muscle Relaxation," *Circulation*, **100**, pp. 41-47.

[62] Pearson, A., Guo, R., Orsinelli, D. A., Binkley, P. F., and Pasierski, T. J., 1994, "Transesophageal Echocardiographic Assessment of the Effects of Age, Gender, and Hypertension on Thoracic Aortic Wall Size, Thickness, and Stiffness," *American Heart Journal*, **128**, pp. 344-351.

[63] Nishime, E., Cole, C., Blackstone, E., Pashkow, F. J., and Lauer, M. S., 2000, "Heart Rate Recovery and Treadmill Exercise Score as Predictors of Mortality in Patients Referred for Exercise ECG," *JAMA*, **284**, pp. 1392-1398.

[64] Wilkinson, I. B., MacCallum, H., Flint, L., Cockcroft, J. R., Newby, D. E., and Webb, D. J., 2000, "The Influence of Heart Rate on Augmentation Index and Central Arterial Pressure in Humans," *Journal of Physiology*, **525**, pp. 263-270.

[65] Ho, K., 1982, "Effects of Aging on Arterial Distensibility and Left Ventricular Load in an Australian Population," thesis, University of New South Wales, Australia.

Chapter 10

NON-INVASIVE METHOD FOR MEASURING LOCAL VISCOELASTICITY OF ARTERIAL VESSELS

Xiaoming Zhang and James F. Greenleaf*
Department of Physiology and Biomedical Engineering, Mayo Clinic College of Medicine, Rochester, MN 55905, USA

Corresponding author:
Email: zhang.xiaoming@mayo.edu
Fax: (507)266-0361
Tel: (507)538-1951

ABSTRACT

Pulse wave velocity (PWV) is widely used for estimating the stiffness of an artery. It is well known that a stiffened artery can be associated with various diseases and with aging. Usually, PWV is measured using the "foot-to-foot" method in which the travel time of the blood pressure wave is measured over a distance. The "foot" of the pressure wave is not clear due to reflected waves. Also, PWV is an average indicator of artery stiffness between the two measuring points, and therefore does not identify local stiffness. The authors have developed a wave method in which the localized ultrasound radiation force is non-invasively generated in the arterial wall. The velocity of wave propagation in the artery can be measured accurately over a few millimeters. A mathematical model for wave propagation along the artery is developed with which the viscoelasticity of the artery can be evaluated non-invasively and locally.

Keywords: Arterial vessel, pulse wave velocity, viscoelasticity, ultrasound.

10.1 INTRODUCTION

According to recent statistics by the American Heart Association, cardiovascular disease (CVD) has been the number one killer in the United States [1]. CVD claims more lives each year than the next five leading causes of death combined, which are cancer, chronic lower respiratory diseases, accidents, diabetes mellitus, and influenza and pneumonia. It has long been recognized that a high percentage of all cardiovascular disease is associated with a hardening of the arteries or arteriosclerosis [2]. Arteriosclerosis involves the buildup of plaques on the inside of the artery walls, which causes thickening and hardening of the arteries. If not treated, it can ultimately lead to high blood pressure and to angina (chest pain), heart attack, stroke, and/or sudden cardiac death.

The mechanical properties of the blood vessels are essential elements in cardiovascular behaviour [3-6]. It is well known that a stiffened artery can be associated with various diseases and with aging. Increased stiffness of the arteries has recently gained acceptance as a potential risk factor for cardiovascular and many other diseases [7, 8].

Pulse wave velocity (PWV) is widely used for estimating the stiffness of an artery [9-16]. PWV is defined as the distance between two measurement points divided by the pulse wave transit time from the two points. PWV is directly related to the elastic modulus of the artery by the well-known Moens-Korteweg equation [3]

$$c_p = \sqrt{\frac{Eh}{2R\rho}} \tag{10 – 1}$$

where c_p is the PWV, R is radius, E is Young's modulus, h is thickness of the artery, and ρ is mass density of the blood. This equation was modified by Bramwell and Hill [17]

$$c_p = 3.57/\sqrt{D_p} \tag{10 – 2}$$

where $D_p = (dV/V)/dp$ is distensibility defined as the relative change in volume (dV/V) for a given change in pressure (dp).

Despite the simple definition of PWV, some problems still remain that limit the interpretation of the available data and the general applicability of PWV measurement. Usually, PWV is measured using the "foot-to-foot" method. The "foot" of the pressure wave is not clear due to reflected waves. In addition, the pressure pulse, produced by the pumping of the heart, is predominately a low-frequency wave. Therefore, it is difficult to measure small time changes from the pulse waveforms. A relatively long distance is needed for reliable time change measurement. Measurement over a long distance means PWV produces an average value of artery stiffness between the two measuring points. It is therefore difficult to identify local stiffness variation of a few millimeters length. For early diagnosis of arteriosclerosis, a measurement of local stiffness in a few millimeters length is needed [18] because in the early stage of arteriosclerosis, a fibrous spot is several millimeters in diameter and becomes homogeneously hard in the final stage.

Recently, the authors have developed a novel method for non-invasively estimating of elastic properties of arteries [19-22]. In this method, a bending wave is generated in an arterial vessel by the radiation force of ultrasound and the wave velocity along the longitudinal direction of the artery is measured. A new wave propagation theory in arteries is developed from which the viscoelastic modulus of the artery is estimated locally in a few millimeters.

10.2 THEORY

A new theory for studying wave propagation in an artery is proposed in which the artery is considered as a three-dimensional elastic cylindrical tube. The cylindrical tube has constant thickness h and median radius R. The x coordinate is taken in the axial direction of the tube, where the θ and z coordinates are in the circumferential and radial directions of the tube, respectively, as shown in Fig. 10-1. The three-dimensional displacements of the tube are defined by u, v, and w in the x, θ, and z directions, respectively.

The equations of motion for a cylindrical tube can be written by Love's theory as [19]

$$\begin{bmatrix} L_{11} & L_{12} & L_{13} \\ L_{21} & L_{22} & L_{23} \\ L_{31} & L_{32} & L_{33} \end{bmatrix} \begin{Bmatrix} u \\ v \\ w \end{Bmatrix} = \begin{Bmatrix} 0 \\ 0 \\ 0 \end{Bmatrix} \tag{10-3}$$

where

$$L_{11} = R^2\frac{\partial^2}{\partial x^2} + \frac{1-v}{2}\frac{\partial^2}{\partial \theta^2} - \gamma\frac{\partial^2}{\partial t^2},\ L_{12} = R\frac{1+v}{2}\frac{\partial^2}{\partial x \partial \theta},\ L_{13} = Rv\frac{\partial}{\partial x},$$

$$L_{21} = R\frac{1+v}{2}\frac{\partial^2}{\partial x \partial \theta},\ L_{22} = R^2\frac{1-v}{2}\frac{\partial^2}{\partial x^2} + \frac{\partial^2}{\partial \theta^2} + \beta\left\{(1-v)R^2\frac{\partial^2}{\partial x^2} + \frac{\partial^2}{\partial \theta^2}\right\}$$

$$-\gamma\frac{\partial^2}{\partial t^2},\ L_{23} = \frac{\partial}{\partial \theta} - \beta\left(R^2\frac{\partial^3}{\partial x^2 \partial \theta} + \frac{\partial^3}{\partial \theta^3}\right),\ L_{31} = -Rv\frac{\partial}{\partial x},\ L_{32} = -\frac{\partial}{\partial \theta}$$

$$-\beta\left\{(2-v)R^2\frac{\partial^3}{\partial x^2 \partial \theta} + \frac{\partial^3}{\partial \theta^3}\right\},\ L_{33} = -1 - \beta\left(R^4\frac{\partial^4}{\partial x^4} + 2R^2\frac{\partial^4}{\partial x^2 \partial \theta^2} + \frac{\partial^4}{\partial \theta^4}\right)$$

$$-\gamma\frac{\partial^2}{\partial t^2}$$

where t is time,

$$\beta = \frac{h^2}{12R^2},\ \gamma = \frac{\rho R^2(1-\sigma^2)}{E}.$$

The displacements of the tube can be expressed in the format of wave propagation, associated with an axial wave number k_x and circumferential mode parameter n, and defined by

$$\begin{aligned} u &= U_{\mathrm{m}}\cos(n\theta)\exp(i\omega t - ik_x x) \\ v &= V_{\mathrm{m}}\sin(n\theta)\exp(i\omega t - ik_x x) \\ w &= W_{\mathrm{m}}\cos(n\theta)\exp(i\omega t - ik_x x) \end{aligned} \tag{10-4}$$

where U_{m}, V_{m}, and W_{m} are the wave amplitudes in the axial x, circumferential θ, and radial z directions, respectively, and ω is the circular driving frequency.

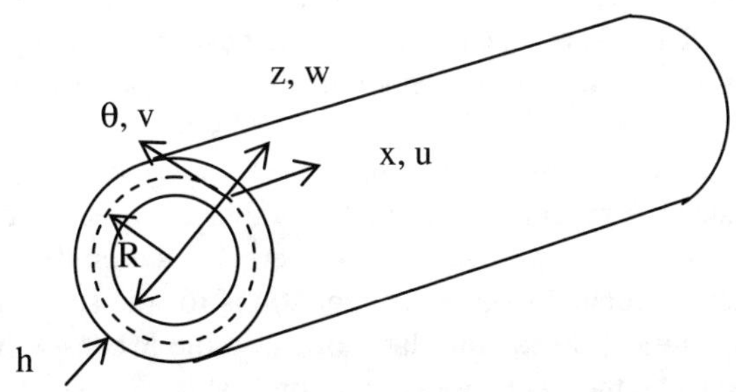

Figure 10-1. Coordinate system for an arterial tube in which z is the coordinate in the radial direction of the tube and w is the displacement of the tube in the z coordinate

Substituting Eq. (10-4) into Eq. (10-3), results in the equation of wave number form

$$a_8k_x^8 + a_6k_x^6 + a_4k_x^4 + a_2k_x^2 + a_0 = 0 \qquad (10-5)$$

where a_8, a_6, a_4, a_2, and a_0 are coefficients depending on shell's geometric and material properties. Solving the above equation results in four pairs of roots. These wave numbers can be separated into two groups. Each group consists of four waves. The first group contains backward waves associated with a semi-infinite tube, $-\infty \le x \le 0$ (left side), excited at the edge at $x = 0$. The second group describes forward waves associated with a semi-infinite tube, $\infty \ge x \ge 0$ (right side), excited at the edge at $x = 0$. If k_x is pure real or pure imaginary, one obtains a propagating wave or a near-field wave, respectively. If k_x is complex, which must be in conjugate pairs, one obtains an attenuated standing wave, which means that the wave amplitudes decay in one direction but the waves propagate in both directions. Solving the above equation for each frequency realizes the related wave number k_x. The dispersion curve of wave velocity with frequency can then be calculated. Only the propagating waves are of interest.

10.3 EXPERIMENTS

The radiation force of ultrasound is used to non-invasively generate a localized force on the artery. The excitation of this force produces a bending wave propagating along the artery. In this technique, a confocal ultrasound transducer is used to generate either a short impulse wave or a low-frequency harmonic wave in the artery. The focal size of the ultrasound beams is about 0.7 mm in diameter which generates almost a point force on the artery. The center of the artery is placed at the focal plane of the transducer. The wave propagation in the artery can be measured by either laser or ultrasound methods. Viscoelasticity of the vessel can be inversely obtained by the wave measurements. A typical experiment setup is shown in Fig. 10-2.

10.4 RESULTS AND DISCUSSION

The wave velocity can be calculated by measuring the phase change at a known distance for a selected set of frequencies. After measuring the wave velocities for different frequencies, the dispersion curve, which is the wave velocity variation with frequency, can be obtained. Figure 10-3 shows the dispersion curve for an artery embedded in a tissue mimicking gelatin phantom [19]. The frequency ranges from 100 Hz to 500 Hz. The measured wave velocity is about 3 m/sec at 100 Hz and 6.5 m/sec at 500 Hz. Also shown in the figure are the dispersion curves calculated using the theory in this paper for a propagation mode of $n = 1$ with Young's modulus of $E = 300$ kPa. The other parameters used in the calculation are the mass density $\rho = 1180$ kg/m^3, Poisson's ratio $\sigma = 0.49$, thickness h 0.8 mm, and outer diameter 4.8 mm. The value of $E = 300$ kPa fits well with the experimental data. The modulus value of $E = 300$ kPa is used for calculation of the pulse wave velocity with the Moens-Korteweg and Bergel equations. Figure 10-3 also shows the wave velocities calculated using these equations for the same modulus with both experimental and theoretical dispersion data. Because the Moens-Korteweg and Bergel equations do not consider the frequency dispersion, a single constant value is shown for the all frequencies with each equation.

The decay rate of wave amplitude is related to the viscous part of the elastic modulus. The decay rate is calculated from the vibration magnitudes at two locations by

Figure 10-2. Typical experimental setup in which a confocal ultrasound transducer remotely produces a localized force at the artery and an ultrasound Doppler transceiver or a laser vibrometer is used to measure the resulting vibration of the artery

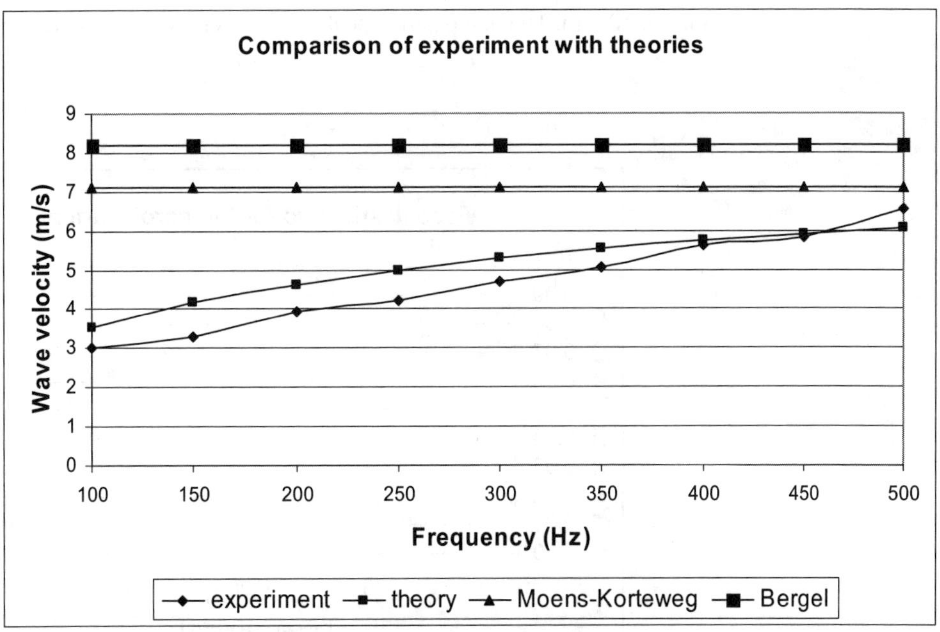

Figure 10-3. Dispersion of wave propagation in an excised artery embedded in a tissue mimicking gelatin phantom. The wave velocity is compared between the authors' theory and experiment as well as the Moens-Korteweg and Bergel equation

$$\mu = -\log(W_{m_2}/W_{m_1})/\Delta x \qquad (10-6)$$

where W_{m_2} and W_{m_1} are magnitudes of vibration at the two points, μ is decay rate, and Δx is distance between the two points.

Due to the viscoelastic behavior of the vessel, the elastic modulus is a complex quantity that can be written as

$$E^* = E(1 + i\eta) \qquad (10-7)$$

where E^* is complex modulus, E is real modulus without viscoelastic effect, and η is viscoelastic ratio that is related to the decay rate by $\eta = 2\mu c/\omega$. The magnitude of the complex modulus is $|E^*| = E \times \sqrt{1+\eta^2}$.

Due to the viscoelastic properties of the artery and its surrounding gelatin, the amplitude of the wave decays as it propagates. This decay reveals the viscoelastic characteristics of the artery and surrounding tissue. By measuring the decay rate over distance for each frequency, the viscoelastic ratio is calculated and shown in Fig. 10-4. It is interesting to note that the decay rate increases with the frequency, which is expected, however the imaginary part of modulus or the viscoelastic ratio decreases with the frequency, and then remains almost constant for high frequencies.

A scanning technique is developed to measure the wave velocity in the artery in which the ultrasound Doppler transceiver is fixed and the ultrasound force transducer is moved [20]. An image of the wave propagation in a rubber tube embedded in tissue mimicking gelatin is shown in Fig. 10-5a for frequency of 300 Hz. The time resolution was 20 microseconds, while the distance resolution was 0.1 mm. The intensity of the image is the wave amplitude that was digitized between ±2048. To obtain this image, the force transducer was scanned at points with a distance of 0.1 mm. For each point, the tube vibration velocity signal was sampled with the continuous wave CW Doppler transceiver at 50,000 samples/sec. The wave velocity of the 300-Hz wave can be calculated by measuring the transit time of the wave propagation at a known distance. Figure 10-5b shows two wave shapes at a distance of 10 mm. The first positive peak was chosen as the reference point to measure the

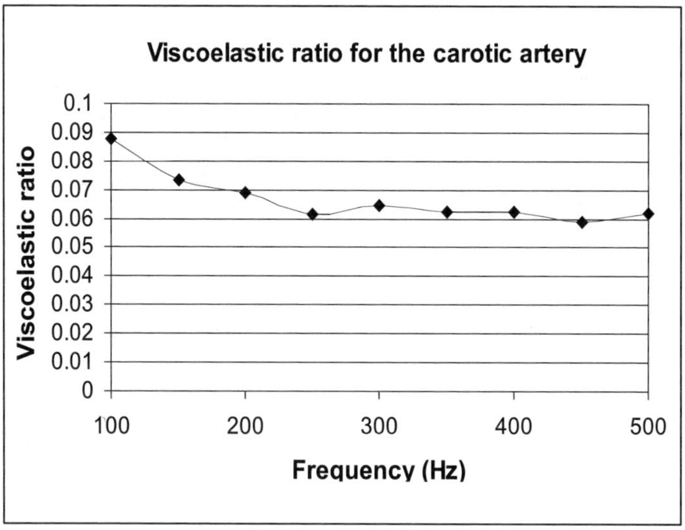

Figure 10-4. Viscoelastic ratio η vs. frequency for the excised artery in Fig. 10-3

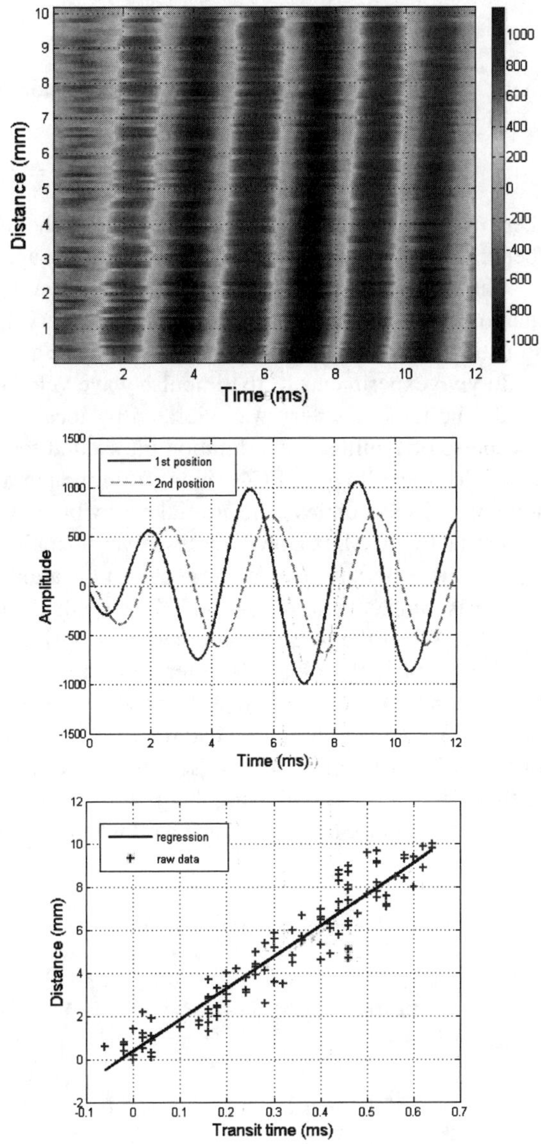

Figure 10-5. (a) Image of a 300-Hz wave propagation in a rubber tube embedded in gelatin; (b) two 300-Hz wave shapes at two positions with 10 mm distance; (c) regression analysis of wave velocity

transit time of the wave velocity. The transit time was measured 0.65 millisecond from the time difference of the first peaks of the two wave shapes. The wave velocity is therefore calculated as 10 mm/0.65 msec = 15.4 m/sec.

A statistical model was developed to measure the wave velocity that uses all the data of the wave image. The regression of the transit time Δt of the wave propagation and the distance Δy the wave takes is analyzed from all the wave curves in the image. The regression line can be obtained by "best fitting" a linear relationship of the transit time Δt and the distance Δy in the least mean squares (LMS) sense

$$\widehat{\Delta y} = a\Delta t + b \qquad (10-8)$$

where $\widehat{\Delta y}$ denotes the value of Δy on the regression for a given value of Δt.

The wave velocity can be estimated by

$$c = \widehat{\Delta y}/\Delta t = a \qquad (10-9)$$

where c is the estimation of wave velocity from the regression analysis that should provide a better result than just using two curves in the image.

The regression analysis for the 300-Hz wave is shown in Fig. 10-5c. The regression equation, with the LMS of a 95% confidence interval, is $\widehat{\Delta y} = 14.5902\Delta t + 0.3918$ and the estimation of the wave velocity is $c = 14.6$ m/sec at 300 Hz.

In vivo experiments of the impulse wave velocity on the femoral artery of pigs were tested. The femoral artery was vibrated by localized ultrasound radiation force using a short pulse of 1-millisecond duration ultrasound tone burst. The excitation was triggered by the electrocardiogram ECG signal from the pig at the peak R-wave of the electrocardiographic QRS complex. The arterial wall vibration velocity resulting from the excitation was sensed by a 6-MHz Doppler transceiver operating in CW mode. The excitation and the detection were both focused at the center of the artery cross section. The vibration velocity signal was sampled at a rate of 100 kHz, amplified and outputted as a voltage, as shown in Fig. 10-6. When the first measurement at a position was done, the force transducer was moved longitudinally along the artery 5 mm away for measuring the resulting vibration at the second position. The time delay was measured 1.1 milliseconds between the first maximum negative peaks of the two impulse curves, and therefore the pulse velocity is calculated as 5 mm/1.1 msec = 4.5 m/sec. The same value of pulse velocity can be measured if the time delay is measured using the first maximum positive peak of the pulse curve. This means that the pulse keeps its shape well when traveling a short distance of 5 mm.

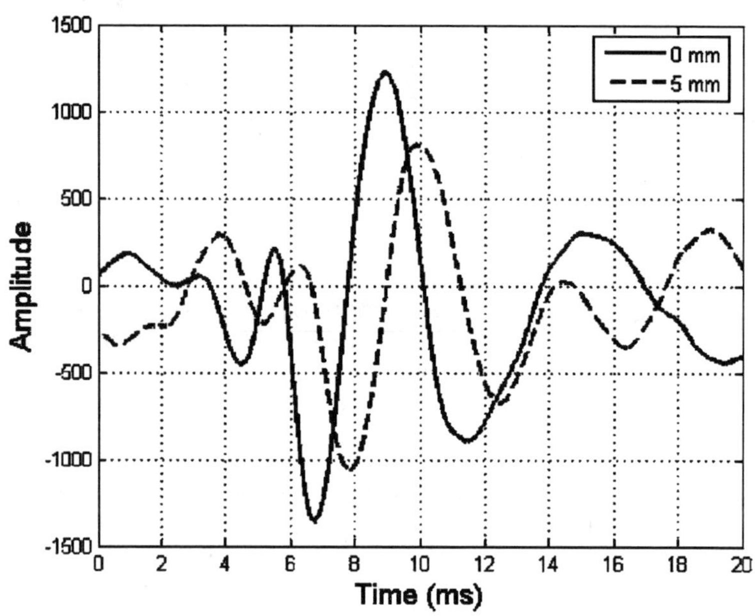

Figure 10-6. The impulse wave velocity is measured 4.5 m/sec from two impulse wave shapes of the in vivo femoral artery of pig at a distance of 5 mm

10.5 CONCLUSIONS

In this paper, an arterial wave is generated non-invasively on the vessel wall by the radiation force of ultrasound. The wave propagation in the artery is measured by either laser or ultrasound method. The wave velocity can be measured with high temporal resolution in a few tens microseconds and fine spatial resolution in a few millimeters. A mathematical model for wave propagation along the artery is developed with which the viscoelasticity of the artery can be evaluated non-invasively and locally. Local arterial material measurement is useful in identifying artery diseases at an early stage, and characterizing the disease and its progression.

Acknowledgment

This study is supported by grant EB 02640 from the National Institutes of Health. The authors thank Mr. R. Kinnick and Dr. C. Pislaru for experimental support.

References

[1] American Heart Association, 2006, *Heart Disease and Stroke Statistics — 2006 Update*, American Heart Association, Dallas.

[2] Hallock, P., 1934, "Arterial Elasticity in Man in Relation to Age as Evaluated by the Pulse Wave Velocity Method," *Archives of Internal Medicine*, **54**, pp. 770–798.

[3] Nichols, W. W., and O'Rourke, M. F., 1990, *McDonald's Blood Flow in Arteries*, 3rd ed., Edward Arnold, London.

[4] Fung, Y. C., 1996, *Biomechanics: Circulation*, 2nd ed., Springer, New York.

[5] Dorbin, P. B., 1978, "Mechanical Properties of Arteries," *Physiological Reviews*, **58**(2), pp. 397–460.

[6] Milnor, W. R., 1989, *Hemodynamics*, 2nd ed., Williams & Wilkins, Baltimore.

[7] O'Rourke, M. F., Staessen, J. A., Vlachopoulos, C., Duprez, D., and Plante, G. E., 2002, "Clinical Application of Arterial Stiffness; Definitions and Reference Values," *American Journal of Hypertension*, **15**, pp. 426–444.

[8] Laurent, S., Katsahian, S., Fassot, C., Tropeano, A., Gautier, I., Laloux, B., and Boutouyrie, P., 2003, "Aortic Stiffness is an Independent Predictor of Fatal Stroke in Essential Hypertension," *Stroke*, **34**(5), pp. 1203–1206.

[9] Brands, P., et al, 1998, "A Noninvasive Method to Estimate Pulse Wave Velocity in Arteries Locally by Means of Ultrasound," *Ultrasound in Medicine & Bio*logy, **24**(9), pp. 1325–1335.

[10] Meinders, J. M., et al, 2001, "Assessment of Local Pulse Wave Velocity in Arteries Using 2D Distension Waveforms," *Ultrasonic Imaging*, **23**, pp. 199–215.

[11] Eriksson, A., et al, 2002, "Arterial pulse wave velocity with tissue Doppler imaging," *Ultrasound in Medicine & Biology*, **28**(5), pp. 571–580.

[12] Steptoe, A., Smulyan, H., and Cribbin, B., 1976, "Pulse Wave Velocity and Blood Pressure Change: Calibration and Applications," *Pyschophysiology*, **13**(5), pp. 488–493.

[13] Pruett, J. D., Bourland, J. D., and Geddes, L. A., 1988, "Measurement of Pulse-Wave Velocity Using a Beat-Sampling Technique," *Annals of Biomechanical Engineering*, **16**, pp. 341–347.

[14] Lehmann, E. D., et al, 1994, "Aortic Compliance Measured by Non-Invasive Doppler Ultrasound: Application of a Personal Computer Based Mk II System and Its Repeatability," *Medical Engineering & Physics*, **16**, pp. 213–221.

[15] Nagai, Y., et al, 1999, "Carotid Arterial Stiffness as a Surrogate for Aortic Stiffness: Relationship Between Carotid Artery Pressure-Strain Elastic Modulus and Aortic Pulse Wave Velocity," *Ultrasound in Medicine & Biology*, **25**(2), pp. 181–188.

[16] Roland, A., et al, 1995, "Assessment of Arterial Distensibility by Automatic Pulse Wave Velocity

Measurement: Validation and Clinic Application Studies," *Hypertension*, **26**(3), pp. 485–490.

[17] Bramwell, J. C., and Hill, A. V., 1922, "The Velocity of the Pulse Wave in Man," *Proceedings of the Royal Society of London, Series B*, **93**(652), pp. 298–306.

[18] Chubachi, N., Kanai, H., Murata, R., and Koiwa, Y., 1994, "Measurement of Local Pulse Wave Velocity in Arteriosclerosis by Ultrasonic Doppler Method," *Ultrasonics Symposium*, pp. 1747–1750.

[19] Zhang, X., Kinnick, R. R., Fatemi, M., and Greenleaf, J. F., 2005, "Noninvasive Method for Estimation of Complex Elastic Modulus of Arterial Vessels," *IEEE Transactions on Ultrasonics, Ferroelectrics, and Frequency Control*, **52**, pp. 642–652.

[20] Zhang, X., and Greenleaf, J. F., 2006, "Measurement of Wave Velocity in Arterial Walls with Ultrasound Transducers," *Ultrasound in Medicine and Biology*, **32**(11), pp. 1655–1660.

[21] Zhang, X., and Greenleaf, J. F., 2006, "Noninvasive Generation and Measurement of Propagating Waves in Arterial Walls," *Journal of the Acoustical Society of America*, **119**(2), pp. 1238–1243.

[22] Zhang, X., and Greenleaf, J. F., 2006, "The Stiffening of Arteries by the Tissue-Mimicking Gelatin," *IEEE Transactions on Ultrasonics, Ferroelectrics, and Frequency Control*, **53**, 1534–1539.

Section 2:

TISSUE CHARACTERIZATION

Chapter 11

TISSUE HARMONIC MOTION ESTIMATION FOR TISSUE CHARACTERIZATION USING ULTRASOUND AND KALMAN FILTER[1]

Yi Zheng[1], Shigao Chen[2], Aiping Yao[1], and James Greenleaf[2]
[1]St. Cloud State University, St. Cloud, MN, USA
[2]Mayo Foundation, Rochester, MN, USA

ABSTRACT

Detection of tissue harmonic motion using ultrasound is introduced for biomedical applications. The harmonic motion induced by the ultrasound radiation force in a tissue region is expressed as a Doppler frequency shift in the fast time and as an oscillatory phase shift in the slow time. A method using pulse echo ultrasound and the Kalman filter is presented for detecting the phase shifts over a short distance in a tissue region. The phase shift over a short distance is used to find the shear wave velocity, which is related to tissue stiffness. The elasticity and viscosity of tissue can be quantitatively calculated with the wave speed dispersion in frequency. Examples and experiments are shown to validate the method.

11.1 INTRODUCTION

Tissue elasticity can be studied by ultrasound. Some of the early studies are reviewed in two comprehensive reports [1, 2]. Those studies include elastography [3], shear wave elasticity imaging [4], sonoelasticity imaging [5], and vibroacoustography [6]. In general, these methods measure elastic deformation of tissue that experiences an external force and are qualitative imaging techniques that give relative mapping of tissue elasticity.

[1]Portions reprinted, with permission, from Zheng Y, Greenleaf JF, Stable and unbiased flow turbulence estimation from pulse-echo ultrasound. *IEEE Transaction on Ultrasonics, Ferroelectrics, and Frequency Control*, Vol. 46, No. 5, pp. 1074 to 1087. Sept. 1999 © 1999 IEEE; and Zheng Y, Chen SC, Tan W, Kinnick R, Greenleaf JF, Detection of tissue harmonic motion induced by ultrasonic radiation force using pulse-echo ultrasound and Kalman filter, *IEEE transaction on Ultrasonics, Ferroelectrics, and Frequency Control*, Vol. 54, No. 2, pp. 1074 to 1087, February 2007 © 2007 IEEE.

One method to quantify both tissue elasticity and viscosity is to evaluate the dispersion of shear wave propagation speed versus its frequency [7]. For a homogeneous medium, the shear wave propagation speed C_s is related to its angular frequency ω_s by [8]

$$c_s(\omega_s) = \sqrt{\frac{2(\mu_1^2 + \omega_s^2\mu_2^2)}{\rho(\mu_1 + \sqrt{\mu_1^2 + \omega_s^2\mu_2^2})}}, \qquad (11-1)$$

where

ρ = density
μ_1 = shear elasticity
μ_2 = shear viscosity of the medium

The speed of shear waves can be estimated from phase shifts of the harmonic motion over the distance propagated [7]:

$$c_s = \omega_s \Delta r / \Delta\phi_s, \qquad (11-2)$$

where $\Delta\phi_s$ = phase change over a propagation distance Δr

Measurements of shear wave speed with Eq. (11-2) at multiple frequencies are then fitted with Eq. (11-1) to solve for the shear elasticity and viscosity of the object. The key component of the method is to use the harmonic radiation force of ultrasound to simulate shear wave within the object. The harmonic radiation force is generated by a confocal transducer having a center disk and an outer ring that introduce two ultrasound beams to the same focal point within the object [6]. These two ultrasound beams have slightly different frequencies and cause the tissue at the focal point to vibrate at the difference frequency. Amplitude modulation waves can also be used to vibrate tissue [7, 9, 10]. The vibration frequency can be easily changed (by altering the ultrasound frequencies) to evaluate the dispersion response of tissue at different frequencies. The typical vibration frequency is between 100 Hz and 500 Hz. A relatively long tone-burst of ultrasound (tens of milliseconds) is needed to generate at least several cycles of low frequency vibrations. Although the results of these studies are promising, observing the shear wave was achieved by an optical probing beam with a laser vibrometry, which limits its medical application because soft tissues are opaque.

A pulse echo ultrasound method is introduced to non-invasively measure tissue harmonic motion for the shear wave resulting from the application of a harmonic radiation force. The tissue harmonic motion is presented as oscillatory phase shifts of ultrasound echoes in slow time. In general, detecting the oscillatory phase shift in slow time is within the frame work of the correlation method [11, 12] using pulse echo ultrasound. The challenge is that shear waves caused by radiation force have low amplitudes that can be overwhelmed by noise and interferences. Because the motion is harmonic and its frequency is known, the motion can be modeled by a second-order differential equation with random constant amplitude and phase, which can be recursively estimated by a Kalman filter. The Kalman filter is an effective tool for estimating a random process in noise with least mean squared (LMS) error [13]. This technique has been validated in various experiments [14]. In this chapter, the keys to the techniques are presented in the following three sections.

11.2 TISSUE MOTION DETECTION USING ULTRASOUND

The correlation method is an effective tool to detect blood flow and tissue motion [11, 12]. The main idea of the correlation method is to measure the phase shift between two consecutive echoes from a moving object. In this section, implementation of the correlation method is presented.

Multiple tone-bursts separated by a period T (pulse repetition period) are transmitted and the mth received echo from the moving tissue can be represented as a band-pass signal [15]:

$$r(t,m) = \Re\{g(t,m)e^{j2\pi f_0 t}\} = |g(t,m)|\cos(2\pi f_0 t + \phi(t,m)), \qquad (11-3)$$

where $g(t,m)$ = complex envelope of $r(t,m)$

$$g(t,m) = |g(t,m)|e^{j\phi(t,m)} = I(t,m) + jQ(t,m) \qquad (11-4)$$

The $|g(t,m)|$, $\phi(t,m)$, $I(t,m)$, and $Q(t,m)$ are magnitude, phase, in-phase, and quadrature components of $g(t,m)$, respectively. f_0 is an effective center frequency.

Applying the quadrature demodulator to two consecutive echoes, two complex envelopes $g(t,m)$ and $g(t,m+1)$ can be obtained as shown in Fig. 11-1. Based on the two complex envelopes, a complex value for tissue motion can be calculated using the complex multiplier, which is

$$\begin{aligned} vv(t,m) &= g(t,m)g^*(t,m+1) = |g(t,m)|e^{j\phi(t,m)}|g(t,m+1)|e^{-j\phi(t,m+1)} \\ &= x(t,m) + jy(t,m) \end{aligned} \qquad (11-5)$$

The real and imaginary components, $x(t,m)$ and $y(t,m)$, are

$$x(t,m) = I(t,m)I(t,m+1) + Q(t,m)Q(t,m+1) \qquad (11-6)$$

and

$$y(t,m) = Q(t,m)I(t,m+1) - I(t,m)Q(t,m+1) \qquad (11-7)$$

So the phase difference of two consecutive echoes can be found by

$$\Delta\phi(t,m) = \tan^{-1}\left(\frac{y(t,m)}{x(t,m)}\right) = \phi(t,m) - \phi(t,m+1) \qquad (11-8)$$

The phase difference of two consecutive echoes can be converted into a time shift that is caused by a moving object between the two echoes with respect to the received instantaneous frequency $f_i(t,m) = f_0 + f_d(t,m)$ at t (with approximation of $f_i(t,m) \approx f_i(t,m+1)$):

$$\Delta T(t,m) = \frac{\Delta\phi(t,m)}{2\pi f_i(t,m)} \qquad (11-9)$$

ΔT is also the round-trip time caused by the displacement of the moving object:

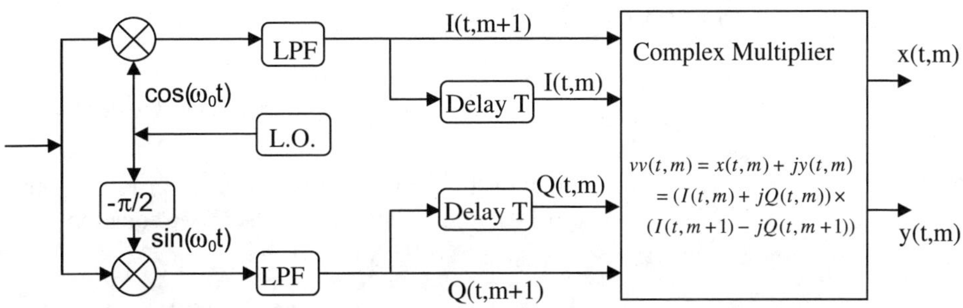

Figure 11-1. Quadrature demodulator and complex multiplier

$$\Delta T(t,m) = \frac{2v(t,m)T\cos\theta}{c} \tag{11 – 10}$$

where

c = propagation speed of ultrasound in tissue

θ = angle between the direction of tissue motion and the axis of the ultrasound beam

Thus, the velocity of the tissue motion can be found by

$$\begin{aligned} v(t,m) &= \frac{\Delta\phi(t,m)c}{4\pi f_i(t,m)T\cos\theta} = \frac{c}{4\pi f_i(t,m)T\cos\theta}\tan^{-1}\left(\frac{y(t,m)}{x(t,m)}\right) \\ &= \frac{c}{4\pi f_i(t,m)T\cos\theta}\tan^{-1}\left(\frac{Q(t,m)I(t,m+1) - I(t,m)Q(t,m+1)}{I(t,m)I(t,m+1) + Q(t,m)Q(t,m+1)}\right) \end{aligned} \tag{11 – 11}$$

If the tissue motion is harmonic, $v(t,m)$ is harmonic in slow time (when m changes). However, it will be very noisy. Therefore, an optimal estimator is needed for the harmonic motion, which will be discussed in the following two sections.

For a constant velocity estimation such as blood flow mapping, the effect of noise on the estimate can be reduced by averaging $x(t,m)$ and $y(t,m)$ over multiple echoes:

$$vv(t) = \sum_{m=1}^{M-1} vv(t,m) = x(t) + jy(t) = \sum_{m=1}^{M-1} x(t,m) + j\sum_{m=1}^{M-1} y(t,m) \tag{11 – 12}$$

Therefore, the mean velocity over a total time interval of $M \times T$ is

$$\begin{aligned} v(t) &= \frac{c}{4\pi f_i(t,m)T\cos\theta}\tan^{-1}\left(\sum_{m=1}^{M-1} y(t,m) \Big/ \sum_{m=1}^{M-1} x(t,m)\right) \\ &= \frac{c}{4\pi f_i(t,m)T\cos\theta}\tan^{-1}\left(\frac{\sum_{m=1}^{M-1}\{Q(t,m)I(t,m+1) - I(t,m)Q(t,m+1)\}}{\sum_{m=1}^{M-1}\{I(t,m)I(t,m+1) + Q(t,m)Q(t,m+1)\}}\right) \end{aligned} \tag{11 – 13}$$

In the above equation, if the mean instantaneous frequency $f_i(t)$ is approximated as the transmitted center frequency ($f_i(t) \approx f_0$ with narrowband assumption), the above equation is identical to Kasai's 1-D autocorrelation result derived from the mean Doppler frequency [11]. However, received echoes from the moving tissue are wideband, and in general, $f_i(t)$ can be significantly different from f_0 even if f_0 is the transmitted center frequency; therefore, the approximation may introduce bias. The instantaneous frequency can be estimated using multiple temporal samples in a sample volume from which a value of effective velocity will be estimated [14]:

$$f_i(k) = f_0 + \frac{1}{2\pi T_s}\tan^{-1}\frac{I(k+1)Q(k) - I(k)Q(k+1)}{I(k+1)I(k) + Q(k)Q(k+1)} \tag{11 – 14}$$

where

f_0 = demodulation frequency and is not necessarily the same as the center frequency of the transmitted wave if the above equation is used

$$f_i(t,m) = f_0 + \frac{1}{2\pi T_s}\tan^{-1}\frac{\sum_{k=1}^{K-1} I(k+1,t)Q(k,t) - I(k,t)Q(k+1,t)}{\sum_{k=1}^{K-1} I(k+1,t)I(k,t) + Q(k,t)Q(k+1,t)} \tag{11 – 15}$$

Example 11-1

The example detailed here verifies the above method by using Field II program [16, 17]. It simulates the ultrasound echoes from a moving scatter. The angle between the direction of the scatter motion and the axis of the ultrasound beam is 45 deg. The velocity of the scatter moving is set as 20 cm/sec. The ultrasound tone-burst is a sinusoid signal with ten cycles and center frequency of 5 MHz. The number of pulses in slow time is 30 ($M = 30$). The sampling frequency is 40 MHz. The pulse repetition frequency is 5 kHz, and the transducer fractional bandwidth is 50% with a Hamming window. The phase shift between two consecutive echoes due to the tissue motion is shown in Fig. 11-2.

Results:

- Estimated velocity using Eq. (11-13) with $f_i(t) = f_0$: $V = 19.78$ cm/sec
- Estimated velocity using Eq. (11-13) with $f_i(t)$ estimated by Eq. (11-15): 19.86 cm/sec.

Comments:

- The correlation method can be used to estimate a constant motion.
- The use of instantaneous frequency reduces the estimation bias.
- The variance σ of the estimated velocity can be derived from the phase variance of the velocity vectors and is called the normalized magnitude difference [15]:

$$\sigma^2(t) = \frac{2}{T^2}\left(1 - \frac{\sqrt{\left(\sum_{m=1}^{M-1} x(t,m)\right)^2 + \left(\sum_{m=1}^{M-1} y(t,m)\right)^2}}{\sum_{m=1}^{M-1}\sqrt{x^2(t,m) + y^2(t,m)}}\right) \qquad (11-16)$$

The estimates of the phase variance can be converted to a variance of velocity:

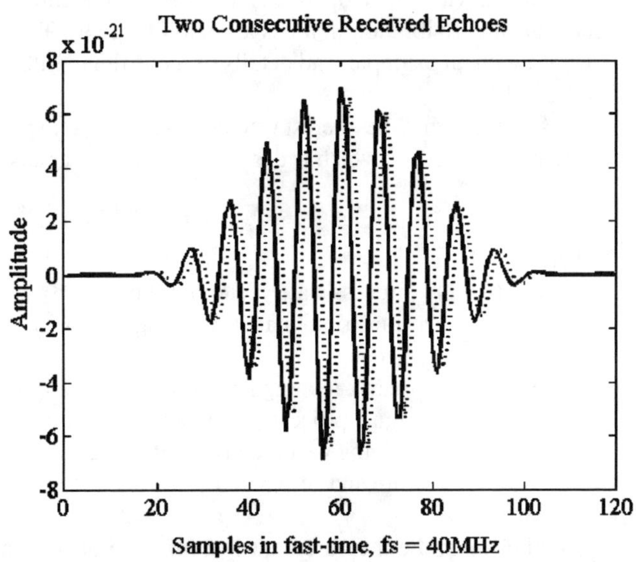

Figure 11-2. Phase shift between two consecutive echoes due to tissue motion

$$\sigma_{\mathrm{v}}^2=\left(\frac{\sqrt{\sigma^2}}{2\pi f_{\mathrm{i}}}\times\frac{c}{2}\right)^2=\sigma^2\left(\frac{c}{4\pi f_{\mathrm{i}}}\right)^2 \tag{11-17}$$

In the above methods, the pulse repetition frequency should be set properly. It cannot be too fast, otherwise, the detection depth is limited. It cannot be too low, which will cause a phase ambiguity problem. The pulse repetition frequency must also satisfy the sampling theorem, which requires the pulse repetition frequency to be at least twice as large as the Doppler frequency caused by tissue motion; otherwise, aliasing will happen. In general, a higher sampling frequency will provide a high signal-to-noise ratio (SNR) in the signal frequency band.

The variance given by Eqs. (11-16) and (11-17) can be used to estimate tissue harmonic motion amplitude in a noisy environment.

11.3 INTRODUCTION TO THE KALMAN FILTER

The Kalman filter is a least squared error filter that recursively estimates a random process from noisy measurement when the spectra of signal and noise are overlapped. It is the optimal estimator among all linear and non-linear estimators when the processes are Gaussian based on the least squared error criteria [13]. It was developed by Rudolph E. Kalman in 1960 to recursively filter discrete data as computers become available for digital filter implementation [18]. The Kalman filter is a set of mathematical equations that efficiently estimate the states of processes with minimized estimated error covariance. The key idea of the Kalman filter is to optimally weight the differences between the predictions using the results of the previous step and the measurement. The prediction uses the state equations that describe the random processes under the consideration. Therefore, a random process can be dealt with using the Kalman filter as long as the process can be described by a difference equation, a differential equation, a power spectral density function, or a correlation function. Since its introduction in 1960, the Kalman filter had been successfully applied in a wide range of studies. Recently, the Kalman filter was applied to estimate the induced small vibration in a tissue region [14, 19]. The Kalman filter is reviewed in this section with an example, and details of the Kalman filter and its applications can be found in Ref. [13].

Assuming that the interest is to estimate a random process that can be represented by a state variable vector, which is described by the linear stochastic difference equation:

$$\mathbf{x}_{k+1}=\boldsymbol{\varphi}_k\mathbf{x}_k+\mathbf{B}\mathbf{u}_{k+1}+\mathbf{w}_k \tag{11-18}$$

where

$\mathbf{x}_k=(n\times 1)$ process state vector at time t_k

$\boldsymbol{\varphi}_k=(n\times n)$ transition matrix relating $\mathbf{x}_k$ to $\mathbf{x}_{k+1}$ in the absence of the driving vector

$\mathbf{B}=(n\times l)$ matrix relating an optional control input u to the state x; in most random signal process estimation applications, $\mathbf{B}$ is zero

$\mathbf{w}_k=(n\times 1)$ driving vector consists of white sequences with known covariance structure and zero mean

In applications, the measurement is a function of the state variables with additive white noise:

$$\mathbf{z}_k=\mathbf{H}_k\mathbf{x}_k+\mathbf{v}_k \tag{11-19}$$

where

$\mathbf{z}_k = (m \times 1)$	measurement vector at time t_k
$\mathbf{H}_k = (m \times n)$	measurement matrix giving the ideal (noiseless) connection between the measurement and the state vector at time t_k
$\mathbf{v}_k = (m \times 1)$	measurement noise vector consisting of white sequences with known covariance structure and zero means. Non-white noise can be represented by state variable

The covariance matrices for the $\mathbf{w}_k$ and $\mathbf{v}_k$ vectors are given by

$$\mathbf{E}[\mathbf{w}_k\mathbf{w}_i^{\mathrm{T}}] = \begin{Bmatrix} \mathbf{Q}_k, & i=k \\ 0 & i \neq k \end{Bmatrix} \qquad (11-20)$$

$$\mathbf{E}[\mathbf{v}_k\mathbf{v}_i^{\mathrm{T}}] = \begin{Bmatrix} \mathbf{R}_k, & i=k \\ 0 & i \neq k \end{Bmatrix} \qquad (11-21)$$

$$\mathbf{E}[\mathbf{w}_k\mathbf{v}_i^{\mathrm{T}}] = 0, \text{ for all } k \text{ and } i \qquad (11-22)$$

The goal of the Kalman filter is to estimate the state variable vector $\mathbf{x}$ from the noisy measurement vector $\mathbf{z}$. It is done by calibrating the prediction based on the results of previous step:

$$\hat{\mathbf{x}}_k = \hat{\mathbf{x}}_k^- + \mathbf{K}_k(\mathbf{z}_k - \mathbf{H}_k\hat{\mathbf{x}}_k^-) \qquad (11-23)$$

where

$\hat{\mathbf{x}}_k$ = estimate of state variable with given measurement $\mathbf{z}_k$ at time step k
$\hat{\mathbf{x}}_k^-$ = prediction for step k with the knowledge from the prior step

$$\hat{\mathbf{x}}_k^- = \varphi_{k-1}\hat{\mathbf{x}}_{k-1} \qquad (11-24)$$

The differences between the measurements and predictions are used to calibrate the estimation. K_k is the filter gain, which is determined by minimizing the mean squared estimation error:

$$\mathbf{P}_k = \mathbf{E}[\mathbf{e}_k\mathbf{e}_k^{\mathrm{T}}] \qquad (11-25)$$

where

$$\mathbf{e}_k = \mathbf{x}_k - \hat{\mathbf{x}}_k$$

The minimization finds,

$$\mathbf{K}_k = \mathbf{P}_k^-\mathbf{H}_k^{\mathrm{T}}(\mathbf{H}_k\mathbf{P}_k^-\mathbf{H}_k^{\mathrm{T}} + \mathbf{R}_k)^{-1}, \qquad (11-26)$$

and the error covariance matrix,

$$\mathbf{P}_k = (\mathbf{I} - \mathbf{K}_k\mathbf{H}_k)\mathbf{P}_k^-, \qquad (11-27)$$

where

$$\mathbf{P}_k^- = \varphi_{k-1}\mathbf{P}_{k-1}\varphi_{k-1}^T + \mathbf{Q}_{k-1} \qquad (11-28)$$

The Kalman filtering process consists of:

- initializing prior estimation of the state vector and its error covariance matrix.
- calculating the filter gain for the current step using Eq. (11-26).
- calculating prior estimation using Eqs. (11-24) and (11-28).
- finding error covariance matrix using Eq. (11-27).
- estimating state variables using the current measurement and Eq. (11-23).
- repeating the process from step 2 to step 5 for the next sample.

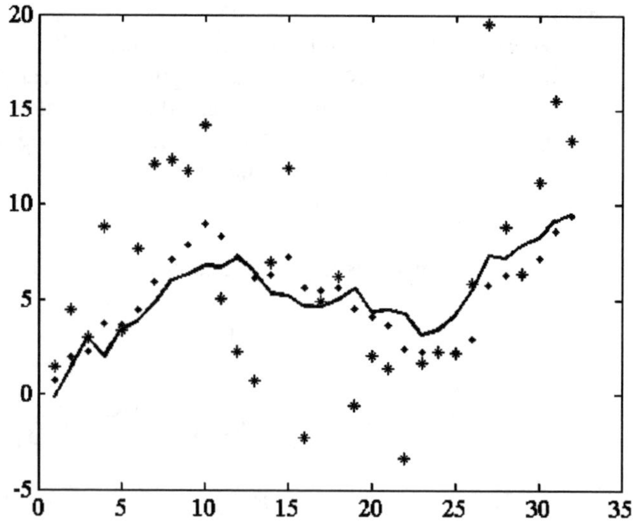

Figure 11-3. Results of estimating a random walking process from noisy measurement. The solid line is a realization of the random walking process. The asterisks are noisy measurements, and the dots are the estimates of the random walk

Example 11-2

Estimate a stochastic process called random walk process (Wiener process) from noisy measurements,

$$z(t) = v(t) + 5n(t) \tag{11-29}$$

where

$n(t)$ = unity Gaussian white noise
$v(t)$ = random walk, which is an integration of a Gaussian white noise $u(t)$:

$$\dot{v}(t) = u(t), \quad v(0) = 0 \tag{11-30}$$

The objective is to estimate $v(t)$ from $z(t)$. Defining as a state variable $x(t) = v(t)$, the discrete-time model of the random walk can be found [13], and then the state model of the system for Eqs. (11-18) and (11-19): $\varphi_k = 1$, $Q_k = 1$, $R_k = 25$, $B = 0$, and $H_k = 1$.

Two white sequences, w_k and v_k are generated first, then Eqs. (11-29) and (11-30) are used to generate x and z recursively. Then the Kalman filter is applied to estimate the random walk $v(t)$ from the noisy measurement $z(t)$. The estimation results are shown in Fig. 11-3. The initial values of $\hat{x}_0^-$ and P_0^- are zeroes. The estimated error standard deviation is 1.27, where the standard deviation of the noise is 5.0. From this example, it can be seen that the Kalman filter gives good estimates if the random process is modeled properly; even if the SNR is very low.

11.4 TISSUE HARMONIC MOTION ESTIMATION

In this section, a method is introduced for detecting the shear wave in a tissue region. The shear wave is induced by ultrasound radiation force, which can be produced from confocal beams [6] or amplitude modulated wave [10]. The frequency of the induced tissue vibration is the same as the frequency difference between the two center frequencies of the confocal beams or the frequency of the modulating signal of the amplitude modulation wave. The shear wave velocity can be utilized to describe tissue stiffness using Eq. (11-1). The shear wave velocity is related to shear wave phase shift over a distance as shown in Eq. (11-2). Therefore, the key is to estimate the shear wave phase shifts at different locations with different vibration frequencies.

11.4.1 Detecting Vibration Information Using Pulse Echo Ultrasound

The tissue vibration displacement caused by the harmonic force can be represented as

$$d(t) = D\sin(\omega_s t + \phi_s) \qquad (11-31)$$

where

ω_s = vibration frequency

The vibration displacement amplitude D and phase ϕ_s are random but constants at a location. The velocity of the motion is

$$v(t) = \frac{dd(t)}{dt} = V\cos(\omega_s t + \phi_s) \qquad (11-32)$$

where

$V = D\omega_s$

When multiple pulses separated by a period T (pulse repetition period) are transmitted to the vibrating tissue through an ultrasound transducer having a center frequency of ω_0, the mth received echo is represented by Eq. (11-3). Because the tissue is moving because of the vibration, there is a travel time shift in the mth echo,

$$\Delta t = 2d(t + mT)\cos(\theta)/c \qquad (11-33)$$

where

c = longitudinal sound propagation speed in the tissue
θ = angle between the ultrasound beam and the tissue vibration direction

This travel time difference can be converted to the phase shift

$$\phi(t,m) = \omega_0 \Delta t = \beta\sin(\omega_s(t + mT) + \phi_s) \qquad (11-34)$$

where

$$\beta = 2D\omega_0 \cos(\theta)/c \qquad (11-35)$$

Therefore, the echo can be represented as,

$$\begin{aligned} r(t,m) &= \Re e\{|g(t,m)|e^{j\beta\sin(\omega_s(t+mT)+\phi_s)}e^{j(\omega_0 t+\phi_0)}\} \\ &= |g(t,m)|\cos(\omega_0 t + \phi_0 + \beta\sin(\omega_s(t + mT) + \phi_s)) \end{aligned} \qquad (11-36)$$

The amplitude and phase of the tissue harmonic motion can be found by estimating β and ϕ_s. But the tissue harmonic motion $\beta\sin(\omega_s(t+mT)+\phi_s$ is a phase term of a cosine function in Eq. (11-36). This is a non-linear estimation problem if β and ϕ_s are directly estimated from Eq. (11-36). To use a linear estimator, the harmonic motion should be extracted from $r(t,m)$. Two methods have been developed to extract the phase term. One of these two methods is simply to take $\tan^{-1}$ of a ratio between the quadrature and in-phase components of demodulation with a procedure to avoid singular division [20]. Another method is to calculate motion variance based on Eq. (11-16).

11.4.2 Extract Vibration Signal from Demodulated Ultrasound Echoes

Applying a quadrature demodulator to $r(t,k)$ in Eq. (11-36), the in-phase and quadrature components are

$$I(t,m) = |g(t,m)|\cos(\beta\sin(\omega_s(t+mT)+\phi_s)+\phi_0 \qquad (11-37)$$

$$Q(t,m) = -|g(t,m)|\sin(\beta\sin(\omega_s(t+mT)+\phi_s)+\phi_0) \qquad (11-38)$$

From Section 11.2, it is known that the phase difference between the two echoes can be found by analyzing a complex velocity quantity,

$$\begin{aligned} v(t,m) &= X(t,m)+jY(t,m) = g(t,m)\,g^*(t,m+1) \\ &= |g(t,m)|\,|g(t,m+1)|\,e^{j(\phi(t,m)-\phi(t,m+1))}, \end{aligned} \qquad (11-39)$$

where the phase difference represents tissue moving velocity and the magnitude represents vibration strength. Because the velocity of the tissue vibration is sinusoidal in slow time, this phase difference is sinusoidal in slow time. Extracting the phase difference, we have

$$\begin{aligned} s(t,m) &= \frac{-1}{2\sin(\omega_s T/2)}\tan^{-1}\left(\frac{Y(t,m)}{X(t,m)}\right) = \frac{-1}{2\sin(\omega_s T/2)}(\phi(t,m)-\phi(t,m+1)) \\ &= \frac{-1}{2\sin(\omega_s T/2)}(\beta\sin(\omega_s(t+mT)+\phi_s)-\beta\sin(\omega_s(t+(m+1)T)+\phi_s)) \\ &= \beta\cos(\omega_s(t+mT)+\phi_s+\omega_s T/2) \end{aligned} \qquad (11-40)$$

This is a direct representation of the vibration signal except that it has an extra phase term, $\omega_S T/2$, which is a constant with fixed values of T and ω_S. This constant phase is removed when we find the phase change ($\Delta\phi_S$) over the propagation distance of the shear wave. The term of $\tan^{-1}$ in Eq. (11-40) is similar to the phase difference for the flow velocity estimation in a medical ultrasound scanner with the capability of color flow mapping (CFM) [11], except the wall filter of CFM must be removed for this application.

This approach uses two complex envelopes to obtain one vibration sample. Thus, it improves performance for noisy data, compared with the method previously used in Refs. [19, 20]. Because the complex velocity quantities in Eq. (11-39) are readily available in a medical ultrasound scanner with CFM capability, this method can be easily implemented in a medical scanner. The difference is that the $s(t,m)$ are averaged for a constant flow velocity estimate in current ultrasound scanners while being used here as inputs to the Kalman filter for estimating the vibration.

11.4.3 Estimate Vibration Displacement and Phase

When vibration is large compared with interference and noise (high SNR) and there are enough samples, the vibration amplitude may be estimated by finding the standard deviation of the $s(t,m)$ in slow time. The phase can be estimated by applying another quadrature demodulation to $s(t,m)$ at the vibration frequency in slow time. A band-pass filter centered at the vibration frequency may be useful if this approach is used. The vibration amplitude can also be estimated by measuring the velocity variance using Eq. (11-16) if the SNR is high [19].

However, when the tissue variation is weak, noisy, and distorted, the direct measurement mentioned above can introduce significant errors with a limited number of samples. When a band-pass filter is applied to noisy data, the amplitude is largely underestimated and the phase may be distorted.

Estimating a harmonic motion with a known frequency and unknown random amplitude and phase is a classical application of the Kalman filter [13]. The Kalman filter recursively extracts the desired information that matches the vibration frequency and provides efficient and real-time solutions. It also allows initial conditions on the state estimation from a Bayesian perspective, which is important for this application. When the state variables are independent and Gaussian distributed, the Kalman filter is optimal among all estimators including linear and non-linear estimators based on the LMS criteria [13]. Therefore, the Kalman filter is selected in this work and has been proven that to be a good selection.

In the Kalman filter modeling, the continue-time model given in Eq. (11-40) in slow time becomes a discrete form including a noise term n_k

$$y_k = \beta_k \cos(\omega_s kT + \phi_{s,k}) + n_k \qquad (11-41)$$

where ϕ_s, $\omega_s t$, and $\omega_s T/2$ are included in $\phi_{s,k}$ for simplicity and $\omega_s t$ and $\omega_s T/2$ will be removed when the change of ϕ_s over a distance is calculated. β_k is a realization of β at kth sample. Because of concern with the estimation in slow time at a particular location in a tissue region, $\omega_s t$ is a constant at the location. With a priori knowledge of the vibration frequency, the Kalman filter estimates the unknown and random phase $\phi_{s,k}$ and amplitude β_k in Eq. (11-41) from noisy measurement y_k based on the LMS criteria. The noise term varies by application. A non-white noise can be modeled as additional state variables in the Kalman filter approach if necessary. In this section, the noise is assumed to mainly come from the electronic system and to be white.

To apply the Kalman filter, the process must be represented in a discrete state model. Equation (11-41) can be rewritten as

$$y_k = [\cos(\omega_s kT), -\sin(\omega_s kT)][\beta_k \cos\phi_{s,k}, \beta_k \sin\phi_{s,k}]^T + n_k \qquad (11-42)$$

We define state variables

$$x_k = \begin{bmatrix} x_k(1) \\ x_k(2) \end{bmatrix} = \begin{bmatrix} \beta_k \cos\phi_{s,k} \\ \beta_k \sin\phi_{s,k} \end{bmatrix} \qquad (11-43)$$

The current state variables are connected to next state variables by a transition matrix $\mathbf{\Phi}$,

$$x_{k+1} = \begin{bmatrix} 1 & 0 \\ 0 & 1 \end{bmatrix} \mathrm{x}_k + w_k = \mathbf{\Phi}\, \mathrm{x}_k + w_k, \qquad (11-44)$$

where amplitude of white sequence w_k represents the variation of the vibration due to other reasons such as background motion. The measurement is related to state variables by a measurement vector $\mathbf{H}$,

$$y_k = \mathbf{H}_k \mathrm{x}_k + n_k \qquad (11-45)$$

where $\mathbf{H}_k = [h_k(1), h_k(2)] = [\cos(\omega_s kT), -\sin(\omega_s kT)]$. Equation (11-42) is obviously equivalent to a state model represented by Eqs. (11-43) to (11-45). The vibration amplitude and phase are related to state variables:

$$\begin{aligned} \beta_k &= \sqrt{\mathrm{x}_k^2(1) + \mathrm{x}_k^2(2)} \\ \phi_{s,k} &= \tan^{-1}(\mathrm{x}_k(2) / \mathrm{x}_k(1)) \end{aligned} \qquad (11-46)$$

If β_k is Rayleigh distributed and $\phi_{s,k}$ is uniformly distributed, $x_k(1)$ and $x_k(2)$ are independent and Gaussian distributed. In this case, the Kalman filter is optimal compared with all other estimators based on the LMS criteria. The filter also provides other important parameters such as estimation error variances. In this case, $\mathbf{P}_k$ is a 2×2 matrix that describes the estimation error variances of the state variables. $\mathbf{P}_k$ can be converted to the error variances of the β and phase estimates [14]:

$$\sigma_\beta^2 = \left(\frac{\partial\beta}{\partial x(1)}\right)^2 P_k(1,1) + \left(\frac{\partial\beta}{\partial x(2)}\right)^2 P_k(2,2) = \frac{x^2(1)P_k(1,1) + x^2(2)P_k(2,2)}{x^2(1) + x^2(2)} \quad (11-47)$$

$$\sigma_{\phi_s}^2 = \left(\frac{\partial\phi}{\partial x(1)}\right)^2 P_k(1,1) + \left(\frac{\partial\phi}{\partial x(2)}\right)^2 P_k(2,2) = \frac{x^2(1)P_k(1,1) + x^2(2)P_k(2,2)}{(x^2(1) + x^2(2))^2} \quad (11-48)$$

Thus, the error variance of displacement amplitude estimates is

$$\sigma_D^2 = \sigma_\beta^2 \left(\frac{c}{2\omega_0 \cos(\theta)}\right)^2 \quad (11-49)$$

Once the state variables are estimated by the Kalman filter outlined in Section 11.2, Eq. (11-46) is used for phase and amplitude estimations. Thus, phase change $\Delta\phi_S$ of the tissue vibration propagating over a known distance Δr is

$$\frac{\Delta\phi_s}{\Delta r} = \frac{\phi_s(r_2) - \phi_s(r_1)}{r_2 - r_1} \quad (11-50)$$

Then, the shear wave speed can be estimated using Eq. (11-2) and the shear wave speed dispersion (c_S changes with vibration frequency) can be used to find the elasticity and viscosity in Eq. (11-1). Note the constant phase $\omega_S t + \omega_S T/2$ is removed from the phase changes by (11-50).

11.4.4 Case Studies

Vibration Estimation. In an in vitro experiment as shown in Fig. 11-4, a freshly excised pig femoral artery (vibrating object in the figure) was embedded in gel containing graphite particles that were used as ultrasonic scatterers to simulate tissue around the artery. Thus, the phantom is opaque. The artery was connected to a water column to provide a fixed static pressure of 70 mmHg. The artery was vibrated at 200 Hz by a 3-MHz transducer using amplitude modulation. The shear wave propagating on the artery wall was detected from 8 mm to 15 mm away from the excitation point.

Pulse echo ultrasound at 5 MHz with 2-kHz pulse repetition frequency was applied to detect the vibration. Applying quadrature demodulation to the RF echoes in fast time and using Eqs. (11-37) to (11-40), vibration signals at locations 8 mm to 15 mm (from the vibration center) were obtained. There were two peaks in RF data in fast time, which indicated the front and posterior walls. The estimations were from locations between the two peaks in the fast time to ensure that they were from the tissue region. The Kalman filter was applied to the extracted vibration in slow time.

The input and output of the Kalman filter are shown in Fig. 11-5. The extract vibration signal represented by Eq. (11-40) is shown by the dotted line in Figure 11-5. It represents the propagated vibration at 15 mm away from the excitation point. The ten cycles of 200-Hz vibration can be clearly seen in the figure. The peak displacement of 600 nm is weak compared with the system noise, as shown by the dotted line. The solid line in Figure 11-5 demonstrates the estimated vibration by the Kalman filter. It is the output of the second pass of the filtering. The first pass establishes initial values for the second pass so that the second pass produces good results from the first few steps. The estimated displacement amplitude is about 600 nm, which represents the vibration strength of the shear wave at 15 mm away from the vibration center.

Tissue Vibration Amplitude Estimation. Experiments were conducted to estimate both amplitude and phase of tissue vibration. The experiment set up is shown in Figure 11-4.

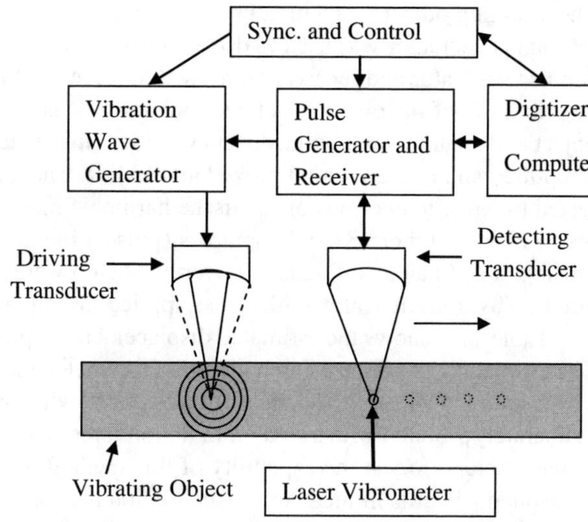

Figure 11-4. Block diagram of experiment setup. A driving transducer induces the radiation force to vibrate the object. The shear wave propagates along the object and is detected by a pulse echo ultrasound system. A laser vibrometer is also used to detect the vibration and validate the pulse echo method. The small dot line circles on the vibration object represent detection locations over a short distance away from the vibration center

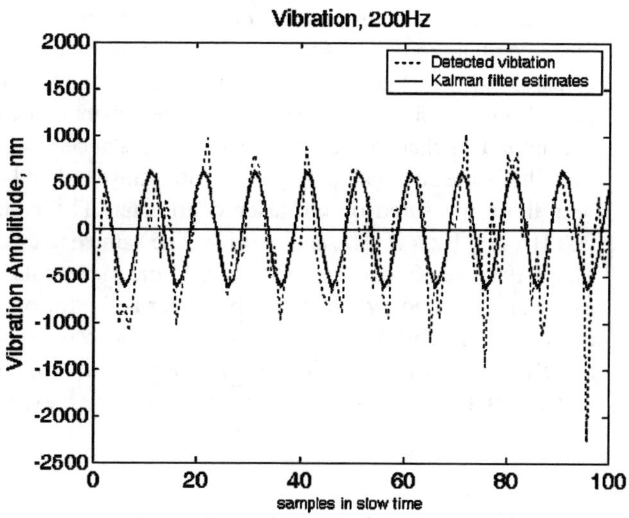

Figure 11-5. Detected vibration of the artery and the estimated vibration by the Kalman filter. The dot line represents the detected vibration and the solid line represents the estimates of the Kalman filter

The vibrating object was a metal beam (1 × 4 × 200 mm) for the amplitude estimation. A mechanical actuator was used to drive the metal beam at a vibration frequency of 500 Hz. A pulse generator and receiver were used to drive a 5-MHz focused transducer and detect the vibration of the object at a fixed location that is away from the vibration center. The object and the ultrasound transducers were within a water tank. The echo was conditioned (amplified and filtered) and digitized at 40 MHz. The number of repetitive pulses was selected to cover ten periods of the tissue harmonic motion. The pulse repetition frequency was 4 kHz, and there were 80 samples (pulses) in slow time. The timing of the driving, pulsing, and Analog-to-Digital Converter (ADC) were synchronized with programmable circuits (Sync). The Kalman filter was applied to estimate the vibration amplitude.

Table 11-1 shows the estimated displacement amplitudes of the vibrating metal beam. The estimation was done using both laser vibrometer and ultrasound with the Kalman filter. The results show that ultrasound estimates match with laser measurements. The object was deliberately vibrated with a smaller radiation force so that the displacement was small (tens of nanometers) to test the capability of this method for future in vivo applications, when ultrasound power is limited and tissue motion is less.

Shear Wave Phase Estimation. The experiment setup was shown in Fig. 11-4, and the distance between the excitation and detection transducers was varied over a short distance to measure the phase and amplitude changes of the shear wave propagating along the arterial wall of an arterial phantom. The object vibrated by the ultrasound radiation force was a rubber tube embedded in a gel phantom. The measured phase changes were used to calculate the phase velocity of the object by Eq. (11-2).

A focused ultrasound transducer introduced oscillatory radiation force confined to a point on the tube using an amplitude modulated wave having a center frequency of 1.3 MHz. Ultrasound pulses were transmitted from the detection transducer (5 MHz center frequency) at a Pulse Repitition Frequency (PRF) of 4 kHz to detect the vibration of the tube. One hundred echoes were digitized at 40 MHz to estimate the vibration phase. The distance between the excitation and detection points was varied from 10 mm to 20 mm away from the vibration center, with an increment of 1 mm, to evaluate shear wave phase and amplitude changes versus propagation distance.

The phases estimated by pulse echo ultrasound and the Kalman filter were compared with those measured by the laser vibrometer. The vibration frequencies were 100 Hz, 200 Hz, 300 Hz, 400 Hz, and 500 Hz. The same excitation energy was used for all vibrations. At a given frequency, there were 11 measurements from 10 mm to 20 mm. The phase estimates provided by the laser vibrometer and ultrasound method with the Kalman filter are shown in Table 11-2. The phase and shear wave speed estimates of the two methods closely match when the vibration frequencies are higher than 200 Hz. The results also match with the earlier study using another model to extract the vibration [20].

TABLE 11-1. Displacement amplitude estimates

Displacements (nm)		Difference
Laser (D_L)	Kalman ($D_k \pm \sigma_D$)	s [(D_k – D_L)/D_L] (%)
28.7	24.5±4.0	−14.6
47.8	43.2±6.0	−9.62
121	122±4.6	0.826
239	242±5.4	1.25
477	471±8.2	−1.26
955	959±18	0.419
1910	1926±30	0.837
3820	3851±60	0.811

TABLE 11-2. Phase change estimates and shear wave speeds

Vibration Frequency (Hz)	Laser Vibrometer		Pulse Echo and Kalman Filter	
	Phase Changes (deg./10 mm)±S.E.	Shear Wave Speed (m/sec)±S.E.	Phase Changes (deg./10 mm)±S.E.	Shear Wave Speed (m/sec) ±S.E.
100	49.9±0.85	7.21±0.12	45.5±2.4	7.91±0.42
200	87.6±2.8	8.22±0.26	79.7±4.5	9.03±0.51
300	107±3.8	10.0±0.36	108±1.5	10.0±0.14
400	127±2.8	11.3±0.25	128±1.8	11.3±0.16
500	144±5.1	12.5±0.44	145±2.2	12.4±0.19

11.5 CONCLUSIONS

The study demonstrates that the Kalman filter applied to pulse echo ultrasound can effectively detect very small harmonic motion induced by ultrasound radiation in a tissue region. The method provides a non-invasive and quantitative technique for tissue characterization.

REFERENCES

[1] Gao, L., Parker, K. J., Lerner, R. M., and Levinson, S. F., 1996, "Imaging of the Elastic Properties of Tissue — A Review," *Ultrasound in Medicine and Biology*, **22**, pp. 959–977.

[2] Ophir, J., Alam, S. K., Garra, B. S., Kallel, F., Konofagou, E. E., Krouskop, T., Merritt, C. R. B., Righetti, R., Souchon, R., Srinivasan, S., and Varghese, T., 2002 (winter), "Elastography: Imaging the Elastic Properties of Soft Tissues With Ultrasound," *Journal Medical Ultrasonics*, **29**, pp. 155–171.

[3] Ophir, J., Cespedes, I., Ponnekanti, H., Yazdi, Y., and Li, X., 1991, "Elastography: A Quantitative Method for Imaging the Elasticity of Biological Tissue," *Ultrasonic Imaging*, **13**, pp. 111–134.

[4] Sarvazyan, A. P., Rudenko, O. V., Swanson, S. D., Fowlkes, J. B., and Emelianov, S. Y., 1998, "Shear Wave Elasticity Imaging: A New Ultrasonic Technology of Medical Diagnostics," *Ultrasound in Medicine and Biology*, **24**, pp. 1419–1435.

[5] Yamakoshi, Y., Sato, J., and Sato, T., 1990, "Ultrasound Imaging of Internal Vibration of Soft Tissue Under Forced Vibration," *IEEE Transactions on Ultrasound, Ferroelectrics, and Frequency Control*, **37**, pp. 45–53.

[6] Fatemi, M., and Greenleaf, J. F., 1998, "Ultrasound-Stimulated Vibro-Acoustic Spectrography," *Science*, **280**, pp. 82–85.

[7] Chen, S., Fatemi, M., and Greenleaf, J. F., 2004, "Quantifying Elasticity and Viscosity From Measurement of Shear Wave Speed Dispersion," *Journal of Acoustical Society of America*, **115**, pp. 2781–2785.

[8] Oestreicher, H. L., 1951, "Field and Impedance of an Oscillating Sphere in a Viscoelastic Medium With an Application to Biophysics," *Journal of Acoustical Society of America*, **23**, pp. 707–714.

[9] Chen, S., 2002, "Shear Property Characterization of Viscoelastic Media Using Vibrations Induced by Ultrasound Radiation Force," Ph.D. thesis, Mayo Graduate School of Medicine, Rochester, MN.

[10] Chen, S., Fatemi, M., and Greenleaf, J. F., 2004, "Comparison of Stress Field Forming Methods for Vibroacoustography," *IEEE Transactions on Ultrasound, Ferroelectrics, and Frequency Control*, **51**, pp. 313–321.

[11] Kasai, C., Namekaw, K., Koyano, A., and Omoto, R., 1985, "Real Time Two-Dimensional Blood Flow Imaging Using an Autocorrelation Technique," *IEEE Transactions on Sonics and Ultrasonics*, **32**, pp. 458–464.

[12] Huang, S., Lerner, R. M., and Parker, J., 1992, "Time Domain Doppler Estimators of the Amplitude and Vibrating Targets," *Journal of Acoustical Society of America*, **91**, pp. 965–974.

[13] Brown, R. G., and Hwang, P. Y. C., 1997, *Introduction to Random Signals and Applied Kalman Filtering*, 3rd ed., John Wiley and Sons, New York.

[14] Zheng, Y., Chen, S., Tan, W., Kinnick, R. R., and Greenleaf, J. F., 2007, "Detection of Tissue Harmonic Motion Induced by Ultrasonic Radiation Force Using Pulse Echo Ultrasound And Kalman Filter," *IEEE Transactions on Ultrasound, Ferroelectrics, and Frequency Control*, **54**, pp. 290–300.

[15] Zheng, Y., and Greenleaf, J. F., 1999, "Stable and Unbiased Flow Turbulence Estimation From Pulse Echo Ultrasound," *IEEE Transactions on Ultrasound, Ferroelectrics, and Frequency Control*, **46**, pp. 1074–1087.

[16] Jensen, J. A., 1996, "Field: A Program for Simulating Ultrasound Systems," *Medical and Biological Engineering and Computing*, **34**, pp. 351–353.

[17] Jensen, J. A., and Svendsen, N. B., 1992, "Calculation of Pressure Fields From Arbitrarily Shaped, Apodized, and Excited Ultrasound Transducers," *IEEE Transactions on Ultrasound, Ferroelectrics, and Frequency Control*, **39**, pp. 262–267.

[18] Kalman, R. E., 1960, "A New Approach to Linear Filtering and Predication Problems," *Transactions of the ASME: Journal of Basic Engineering*, **82**, pp. 35–45.

[19] Zheng, Y., Chen, S., Tan, W., and Greenleaf, J. F., 2003, "Kalman Filter Motion Detection for Vibro-Acoustography Using Pulse Echo Ultrasound," *Proceedings of 2003 IEEE International Ultrasonic Symposium*, pp. 1812–1815.

[20] Zheng, Y., Chen, S., Zhang, X., and Greenleaf, J. F., 2004, "Detection of Shear Wave Propagation in an Artery Using Pulse Echo Ultrasound and Kalman Filtering," *Proceedings of 2004 IEEE International Ultrasonic Symposium*, pp. 1251–1253.

Chapter **12**

CHARACTERIZATION OF TISSUE VISCOELASTICITY FROM SHEAR WAVE SPEED DISPERSION

Shigao Chen and James F. Greenleaf
Department of Physiology and Biomedical Engineering, Mayo Clinic College of Medicine, Rochester, MN, USA

ABSTRACT

Quantitative measurement of tissue elasticity and viscosity has important medical applications because pathologies are often linked to abnormal tissue mechanical properties. Shearwave Dispersion Ultrasound Vibrometry (SDUV) is proposed to quantify tissue viscoelasticity. SDUV uses a focused ultrasound beam *within FDA safety limits* to stimulate formation and propagation of harmonic shear waves in the studied tissue. The propagation speed of induced shear wave is frequency-dependent (dispersive) and relates to the tissue's mechanical properties. Shear wave speeds at a number of frequencies are measured by pulse-echo ultrasound setup and fit with a theoretical dispersion model to inversely solve for tissue elasticity and viscosity. A special pulse sequence is developed to make SDUV compatible with commercial ultrasound scanners. In vitro SDUV measurements in gelatin phantom, liver, and pork and beef muscle show promising results, demonstrating the feasibility of SDUV to quantify tissue elasticity and viscosity.

12.1 INTRODUCTION

Non-invasive measurement of the mechanical properties of soft tissue such as shear elasticity (stiffness) has important medical applications because changes of these properties are often related to pathology. Palpation has long been used by physicians to detect abnormalities by sensing the increased stiffness of the examined tissue. However, palpation is subjective and often fails to find lesions lying deep inside the body. Therefore, elasticity imaging, non-invasive methods that objectively measure tissue mechanical properties, has been a subject of extensive investigation in recent years. The general approach of these elasticity imaging methods is to evaluate tissue response to applied stress. Tissue response can be monitored by magnetic resonance imaging (MRI) [1] or ultrasound. Compared to ultrasound, MRI-based techniques are more expensive and require a longer imaging time. Therefore, the following introduction is focused on ultrasound based elasticity imaging techniques.

Within the ultrasound based techniques, the stress can be applied at the body surface by static compression [2], dynamic vibration [3, 4], or within the studied tissue by ultrasound radiation force [5-8]. Most of these current techniques only provide a *relative* mapping of the tissue elasticity. Although this can be useful for lesion detection, some other medical applications may require quantitative elasticity measurements (i.e., in unit of Pascals). Considerable efforts have been directed to reconstruct true elasticity from tissue displacements [5, 9-12]. However, this estimation is very challenging because the displacement data provided by ultrasound measurement is usually incomplete and noisy. Therefore, these direct inversion approaches usually involve many simplifying assumptions and noise filtering that affect their accuracy [2].

To avoid these limitations, several alternatives for quantitative tissue elasticity measurement, which do not require direct inversion of measured displacement field, have been developed. Girnyk et al [13] showed some promising results of viscoelasticity quantification by studying the displacement magnitude and strain relaxation due to pulsed ultrasound radiation force, although further validations are necessary. Another approach is by measuring the shear wave propagation speed within the studied tissue [7, 12, 14, 15]. However, tissue viscosity is neglected in these shear wave speed methods and this omission may cause significant bias in the estimation of tissue elasticity. In addition, a recent study suggests that viscosity is potentially another useful index of tissue healthiness [16]. Therefore, it is important to investigate techniques that *resolve both tissue elasticity and viscosity quantitatively*.

Toward this aim, a new method called Shearwave Dispersion Ultrasound Vibrometry (SDUV) is proposed to quantify both elasticity and viscosity by characterizing the frequency dispersion of the propagation speed of shear waves within the studied tissue. SDUV uses a focused ultrasound beam within the Food and Drug Administration (FDA) power limits to stimulate (within the studied tissue) cylindrical harmonic shear waves propagating outward from the beam axis. Shear wave propagation is tracked by a separate ultrasound beam in pulse/echo mode. The phase of the shear wave at two locations along its traveling path is used to calculate the shear wave propagation speed. The speed of shear waves at different frequencies is measured and fitted with a theoretical Voigt dispersion model to inversely solve for tissue viscosity and elasticity. In vitro SDUV measurements in gelatin phantom, muscle, and liver tissues will be reported in the following sections. The possibility of implementing SDUV on commercial ultrasound scanners will also be discussed.

12.2 PRINCIPLE OF SDUV

Details about SDUV can be found in the article by Chen et al [17]. A brief description of this method is presented here. For a homogeneous medium, the shear wave propagation speed c_s is related to its angular frequency ω_s by the Voigt model [18]:

$$c_s(\omega_s) = \sqrt{2(\mu_1^2 + \omega_s^2 \mu_2^2) / \rho(\mu_1 + \sqrt{\mu_1^2 + \omega_s^2 \mu_2^2})} \qquad (12-1)$$

where ρ, μ_1, and μ_2 are the density, shear elasticity, and shear viscosity of the medium, respectively. It can be seen that the shear wave propagation speed is frequency-dependent or, in other words, *dispersive*. The density of various soft tissues shows very little variance and can be assumed to be nearly constant (i.e., 1000 kg/m^3). Therefore, if the variation of c_s versus frequency is characterized accurately, it is possible to solve inversely for elasticity and viscosity of the medium by fitting Eq. (12-1) to the dispersion measurements. The

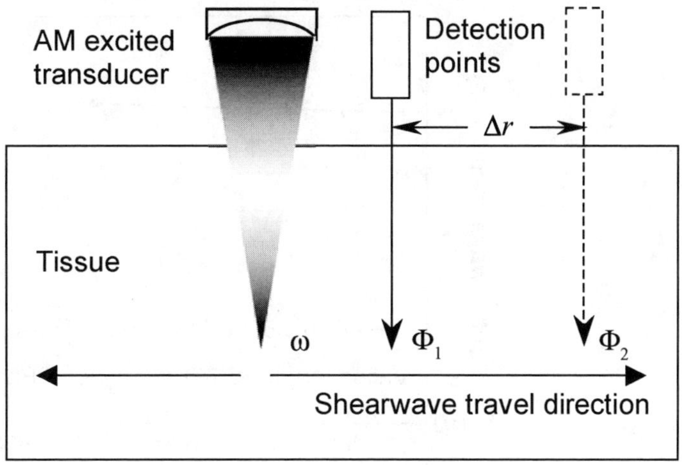

Figure 12-1. Diagram of SDUV. AM ultrasound is used to generate monochromatic shear wave whose propagation speed is estimated by the phase shift of the shear wave over the traveled distance Δr: $c_s(\omega) = \omega\Delta r / (\Phi_1 - \Phi_2)$

implementation of this concept in SDUV is shown in Fig. 12-1. The radiation force of an amplitude-modulated (AM) ultrasound beam from a focused transducer is used to generate a monochromatic shear wave in the studied tissue. The propagation speed of the shear wave for each frequency ω_s can be estimated by the phase difference (Φ_1-Φ_2) of the vibrations at two locations with a known distance Δr:

$$c_s(\omega_s) = \frac{\omega_s \Delta r}{\Phi_1 - \Phi_2} \tag{12 – 2}$$

The frequency of the shear wave (typically between 100 Hz and 1000 Hz) is then varied by changing the modulation frequency of the incident ultrasound to obtain shear speed measurements at multiple frequencies. The dispersion characteristic of the studied tissue is then fitted to Eq. (12-1) to solve for its elasticity and viscosity.

One advantage of using radiation force to produce vibration internally is that the generated shear waves are confined *locally* to a small region because they attenuate quickly and thus are less susceptible to reflections and reverberation interferences. Another more important advantage of this method is that the propagation of the shear wave is solely governed by the local tissue properties once the shear wave is launched. Therefore, measurements based on this method are device independent: measurements do not depend on the beam shape or intensity of the ultrasound. In contrast, tissue response detected at the focus of the ultrasound beam usually depends on these unknown ultrasound beam characteristics and thus can only provide a relative mapping of tissue properties. One should note that SDUV is not a two-dimensional imaging method, but rather provides fast and accurate estimates of elasticity and viscosity at any selected point of measurement within the studied tissue.

Figure 12-2 is the result of an initial validation experiment in a transparent medium where shear wave propagation was monitored by an optical method due to its superior sensitivity. Bars represent ±SD (standard deviation) for six measurements of c_s at each frequency. A least mean square (LMS) fit (the solid line) of Eq. (12-1) to the data gives $\mu_1 = 5.36 \pm 0.13$ kPa and $\mu_2 = 0.38 \pm 0.06$ Pa · sec, which are very close to independent measurement results ($\mu_1 = 5.67$ kPa and $\mu_2 = 0.36$ Pa · sec) [17].

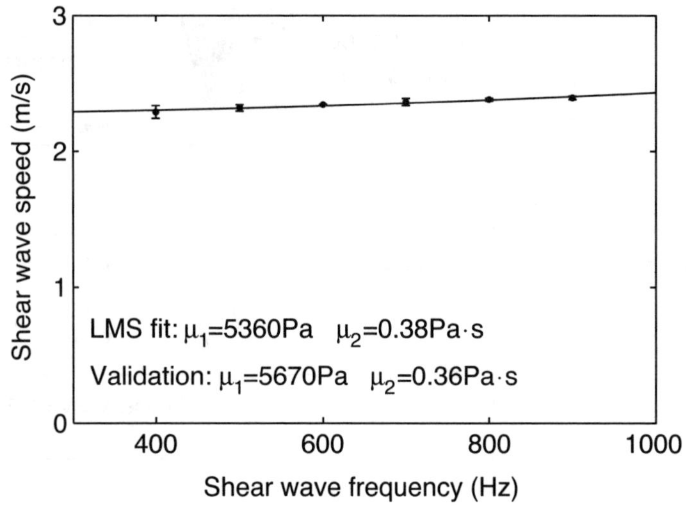

Figure 12-2. Measured shear wave speed dispersion in a gelatin phantom and the inverse fit by Eq. (12-1)

12.3 VIBRATION DETECTION WITH PULSE-ECHO ULTRASOUND

Application of SDUV in tissues requires the use of pulse-echo ultrasound for vibration detection because optical methods are not feasible in such cases. The major challenge is that tissue vibrations resulting from the "pushing" ultrasound beam are usually weak and can be affected by various environmental noises such as body, breathing, and cardiac motions. Therefore, special signal processing is needed to extract these small vibration signals with special emphasis on accurate estimation of vibration phase. The good news is that the vibration is a pure tone and its frequency is known exactly. Thus, strong filtering can be applied at a very specific frequency to extract the useful tissue vibration signal. Kalman filtering has been used to extract the phase of harmonic vibrations detected by pulse-echo ultrasound. In this method, ultrasound pulses are repeatedly transmitted to the same tissue region at a pulse repetition frequency (PRF) of several kilohertz. A fixed time point in the echo corresponding to a fixed tissue region along the ultrasound beam is quadrature demodulated across all echoes to get the tissue motion at that position. Then, a Kalman filter is applied to the tissue motion signal to lock in and extract only the frequency component at that of the oscillatory radiation force. More details can be found in Chapter 11.

Feasibility of using pulse-echo ultrasound for motion detection in SDUV is demonstrated with an in vitro experiment in beef muscle. Instead of the pushing ultrasound beam, a mechanical actuator (Model V203, LDS Test and Measurement Ltd., Herts, England) was used to generate larger vibrations in this experiment. A glass tube with a diameter of 6 mm was glued to a through hole cut at the center of a beef striated muscle (8 cm × 15 cm × 5 cm). The glass tube was driven by the mechanical actuator to introduce monochromatic shear waves ranging from 200 Hz to 500 Hz in the beef. The propagation of the shear wave at different locations separated by 2-mm intervals along and across the fiber of the muscle was tracked by a focused transducer in pulse-echo mode. The echoes received by the transducer were digitized at 100 MHz and processed to estimate the phase of the shear wave. The frequency-dependent propagation speed of the shear wave was estimated by Eq. (12-2) and then fitted to Eq. (12-1) to solve for tissue elasticity and viscosity.

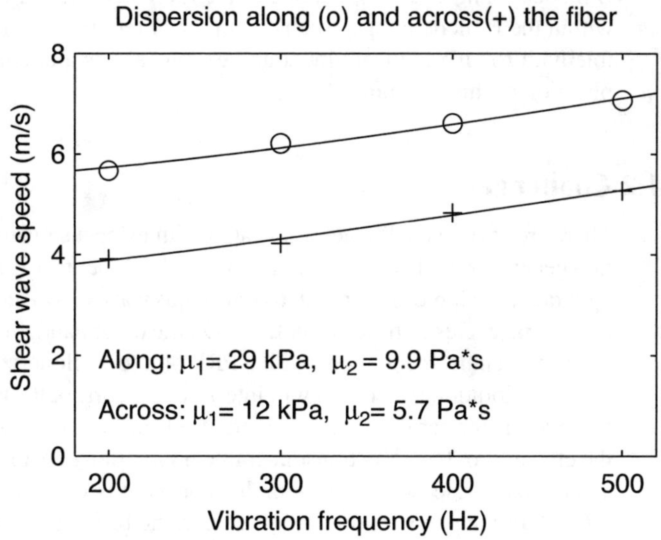

Figure 12-3. Measured dispersion and inversions of viscoelasticity along and across the beef muscle fibers

Figure 12-3 shows the dispersion of shear wave speed measured in the beef muscle. The circles and crosses are frequency-dependent shear wave propagation speeds measured along and across the muscle fibers. The solid lines are the LMS fits from Eq. (12-1) that give a shear elasticity of 29 kPa and a viscosity of 9.9 Pa·sec along the fiber; and a shear elasticity of 12 kPa and a viscosity of 5.7 Pa·sec across the fiber. Independent and accurate measures of elasticity and viscosity in soft tissues are very hard to accomplish and therefore these results can not be validated with a gold standard. However, these measurement show lower elasticity and viscosity for propagation across the fibers than that along the fibers, which is consistent with the trend reported in the literature. The values obtained in this experiment are close to those reported for beef muscle by Catheline et al [19] ($\mu_1 \sim 49$ kPa, $\mu_2 \sim 15$ Pa·sec along the fibers and $\mu_1 \sim 25$ kPa, $\mu_2 \sim 3.3$ Pa·sec across the fibers).

12.4 MOTION GENERATION AND DETECTION WITH A SINGLE-ARRAY TRANSDUCER

12.4.1 Motivation

The SDUV diagram in Fig. 12-1 requires two separate ultrasound transducers: one for tissue vibration and one for pulse-echo motion detection. Although this scheme works well for in vitro experiments, its in vivo application will be challenging because the available acoustic window will become limited. In addition, the alignment of these two transducers would be both time consuming and difficult considering the relatively large physical size of the transducers and the close distance between the excitation location and motion detection locations. Finally, the positioning of SDUV measurement locations requires some kind of imaging guidance, which this two transducer scheme can not provide. To eliminate these limitations, the possibility to implement SDUV is explored in a commercial ultrasound scanner where a single-array transducer will be used for both vibration generation and detection. The same transducer can also provide a guidance image of the studied organ with

B mode scanning. The operator can interactively select the SDUV measurement locations within the B mode image. Such an approach requires no alignment setup, is fully compatible with routine clinical ultrasound exams, and is a crucial step toward the practical application of this technique.

12.4.2 Challenges

There are several challenges associated with using a single-array transducer for both vibration generation and detection. In a typical SDUV setup, the transducer used to vibrate tissue with radiation force is operated in continuous wave (CW) mode for tens of milliseconds (i.e., for several cycles of tissue vibration). On the other hand, the motion detection transducer is operated in pulse-echo mode with a PRF of a few kilohertz. Therefore, the CW ultrasound used for vibration generation will interfere and corrupt the ultrasound echo used for motion detection if the same transducer were used for both purposes. One possibility is to divide the elements of the array transducer into a vibration generation group and a detection group and operate these two groups simultaneously at two separated ultrasound frequencies. But this solution has several limitations. First, the bandwidth of current array transducers may not be wide enough to allow sufficient separation of the vibration ultrasound frequency and the detection ultrasound frequency. Crosstalk between these two groups and tissue echo generated by the vibration ultrasound are likely to be overwhelming and make the detection of small vibrations difficult. Second, the ultrasound intensity required for tissue vibration is quite high. Therefore, the use of only a subset of transducer elements available in the array for vibration generation may not be sufficient. To avoid these limitations, the authors propose to achieve this multiplexing task by operating the array transducer intermittently in sequential vibration and detection modes.

12.4.3 Intermittent Pulse Sequence

As shown in Fig. 12-4, the array transducer focuses ultrasound at one location to vibrate the studied tissue and then steers its focus to another location for vibration detection. This can be achieved with the electronic steering of the ultrasound beam without mechanically moving the array transducer. Phases of the shear wave for different traveled distance r can

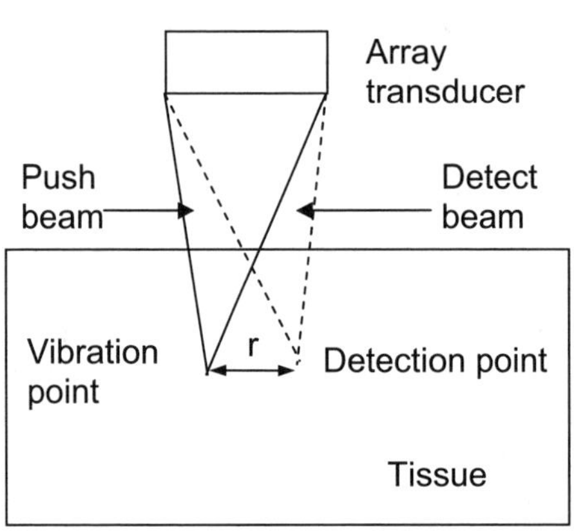

Figure 12-4. Intermittent operation of a single-array transducer for vibration generation and detection

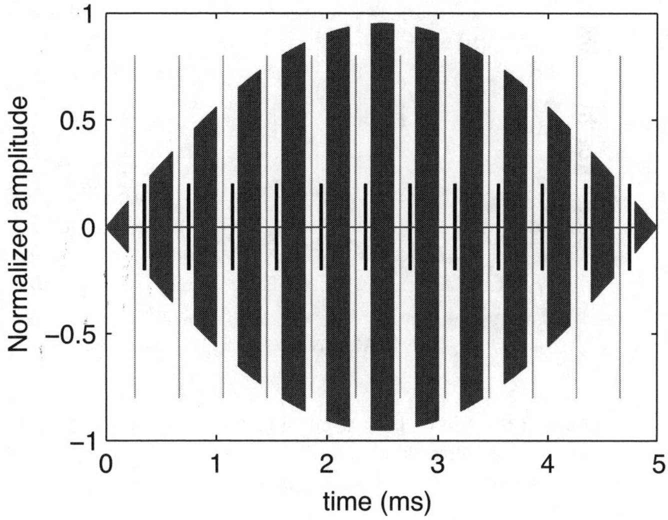

Figure 12-5. Diagram of a pulse sequence that allows the intermittent operation shown in Fig. 12-4. Note that the horizontal time scale is too large for the oscillation of the signal at ultrasound frequency to be seen

be measured with such an array transducer and used in SDUV. Detailed operation of the array transducer is shown in Fig. 12-5. Instead of continuously vibrating the tissue, the AM ultrasound is now gated ON to vibrate the tissue and OFF to provide a time window for pulse-echo detection of tissue motion. Note that the horizontal time scale in this figure is too large for the oscillation of the signal at ultrasound frequency to be seen. When the pushing signal is OFF, a short ultrasound pulse (presented as a high vertical line in Fig. 12-5) is transmitted to the motion detection location and its echo (represented by the low vertical strip) is received and processed to find the tissue vibration. This intermittent pulse sequence allows vibration of a tissue region and at the same time monitoring its response with a single-array transducer. The entire sequence in Fig. 12-5 is repeated several times for a stable vibration state of the tissue to be established and detected.

Figure 12-6 is a simulation of the pushing force resulting from a gated AM ultrasound as shown in Fig. 12-5. The pushing force has a frequency of 200 Hz and is gated ON and OFF every 200 microseconds. Its frequency spectrum has a 200-Hz component and three other components around the gating frequency of 2.5 kHz. To evaluate vibration response to such a gated harmonic force, a laser vibrometer (OFV-303, Polytec, Waldbronn, Germany) was used to study the vibrations in a transparent gelatin phantom induced by a transducer driven at the gated AM mode. The displacement and its frequency spectrum measured with an optical method at the focus of the ultrasound transducer are shown in Fig. 12-7. It can be seen that the primary response of the medium is at 200 Hz, while displacement at the gating frequency is almost invisible (the small jittering at the valleys of the signal corresponds to the gating ON and OFF of the "pushing" ultrasound). This observation is expected if it is assumed (according to Newton's second law) that medium accelerations due to the applied pushing force are comparable at 200 Hz and 2.5 kHz. Then for harmonic motions, displacement of the medium is equal to its acceleration divided by the square of its vibration frequency. Therefore, even though the accelerations are comparable at both frequencies, the displacement at 200 Hz will be at least 100 times higher than that at 2.5 kHz and dominate the displacement spectrum. In addition, higher-frequency shear waves will attenuate more

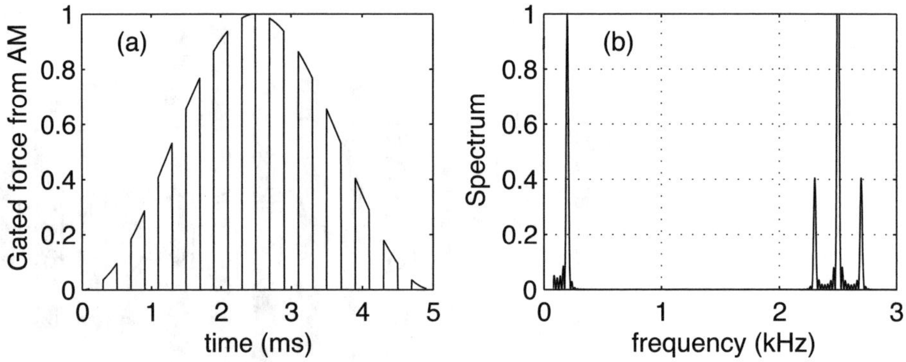

Figure 12-6. Simulation of pushing force (a) and its spectrum (b) resulting from the gated AM ultrasound

quickly when traveling outward from the vibration center. The combined result is that the pushing force from a gated AM ultrasound is still capable of providing pure tone shear waves for SDUV measurements.

The feasibility of the pulse sequence proposed in Fig. 12-5 was tested by a second in vitro experiment in a fresh rabbit liver sample. A single phase array transducer as shown in Fig. 12-4 was not used, because for the time being there is no access to control the operation of an array transducer. Instead, the setup shown in Fig. 12-8 was used. The excitation transducer was operated in gated AM mode. The detection transducer monitored the motion of the studied tissue only when the excitation transducer was OFF. Since the timing of these two transducers followed exactly the sequence shown in Fig. 12-5, results obtained here should be a reliable prediction of the performance of the same intermittent implementation with a single-array transducer in commercial scanners. In Fig. 12-8, the excitation transducer was moved laterally to change the propagation distance of the outgoing shear wave. The excitation transducer and detection transducer could not be positioned on the same side of the tissue because their physical clearance did not allow measurement of tissue motion very close to the excitation center.

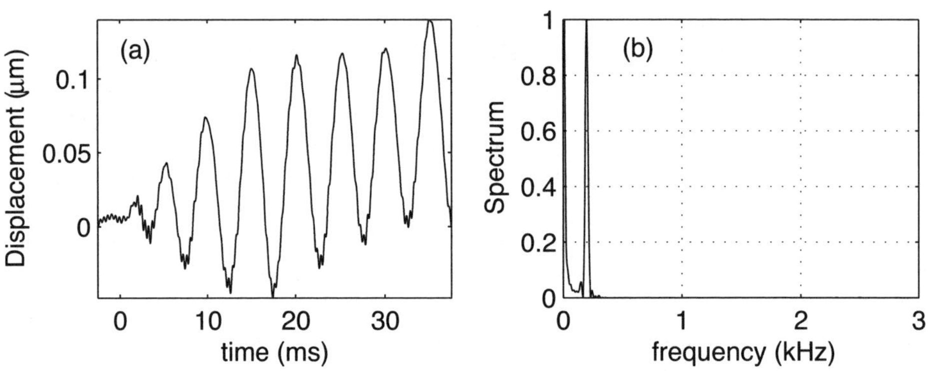

Figure 12-7. Vibration response measured at the focus of a transducer that is driven by the gated AM signal of Fig. 12-5. (a) Displacement measured at the transducer focus showing the primary response of the medium is at 200 Hz. (b) Spectrum of (a)

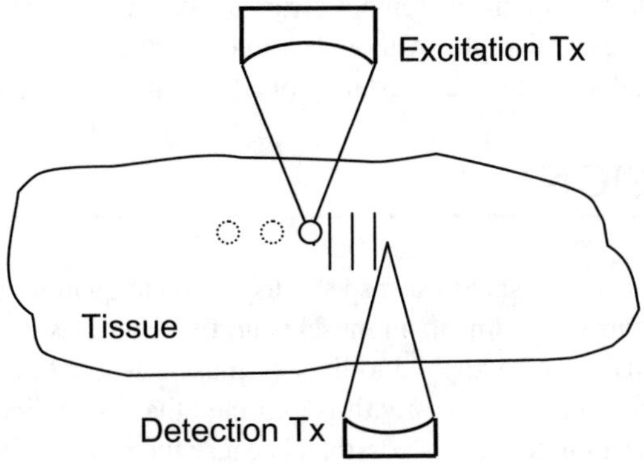

Figure 12-8. Experimental setup used to test feasibility of the intermittent pulse sequence in Fig. 12-5. "Tx" represents transducer

The result of this experiment is shown in Fig. 12-9. The dispersion measured in normal rabbit liver fits well to the theoretical model and gives μ_1 = 1.6 kPa and μ_2= 0.76 Pa · sec. These values are close to results reported for livers in healthy humans (μ_1 = 2.06±0.26 kPa and μ_2= 1.72±0.15 Pa · sec) and normal rats (μ_1 = 1.76±0.37 kPa and μ_2= 0.51±0.04 Pa · sec) [16, 20].

To test method precision, in vitro SDUV measurements were repeated at different locations in a pork tenderloin sample. A two transducer setup as shown in Fig. 12-8 was used, although the pulse sequence is slightly different from Fig. 12-5. Means and standard deviations of SDUV measurements are: μ_1 = 17.5±5.3 kPa, μ_2 = 5.0±1.2 Pa · sec along fibers, and μ_1 = 8.8±1.6 kPa, μ_2 = 1.8±0.6 Pa · sec across fibers. Standard deviations of these

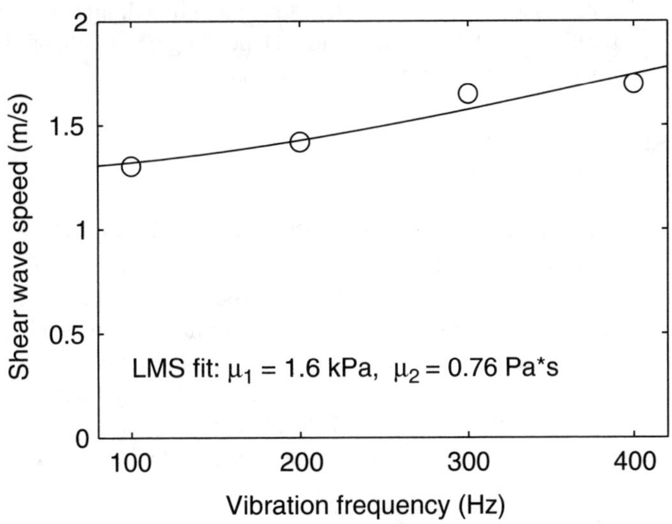

Figure 12-9. Shear wave speed dispersion measured with the intermittent setup of Fig. 12-8 in a normal excised rabbit liver

data contain measurement errors as well as real bio-variations across different locations of the tested sample. Although no gold standard is available in this test, these results give a good idea of the consistency of SDUV measurements.

12.5 DISCUSSION

SDUV uses shear wave speed dispersion to quantify the elasticity and viscosity of a medium. In vitro experiments in muscle and liver samples demonstrate the feasibility of this method. SDUV can be applied to liver or muscle where the requirement that tissue is homogeneous can be met more easily. It is expected that its application in uniform lesions with size larger than 1 cm to 2 cm is possible, because the results of these experiments show that shear wave speed can be reliably estimated within several millimeters from the vibration point. Equation (12-1) used for shear wave speed dispersion fitting is based on the Voigt model. A recent paper by Catheline et al [19] compared the experimental results in agar-gelatin phantom and beef muscles and concluded that the Voigt model is more appropriate than the Maxwell model. Therefore, the use of Eq. (12-1) for dispersion fitting in SDUV is appropriate.

Another concern for SDUV is whether enough tissue displacement can be generated with safe ultrasound power. The ultrasound used to vibrate the tissues in this study had an in situ mechanical index of 1.4, which is below the FDA limit of 1.9. Tissue heating is estimated from the bio-heat transfer equation neglecting heat conduction and perfusion, which will give an overestimation of temperature rise. Under these assumptions, the equation becomes a first-order linear differential equation, whose solution provides a linear relationship between temperature increase ΔT and application time t: $\Delta T = 2\alpha I t / \gamma_V$, where a is absorption coefficient of the tissue, I is temporal average intensity of the acoustic beam in a given location, and γ_v is the volume-specific heat for tissue [21]. Tissue heating was estimated to be about 2°C at the focus of the tissue pushing transducer, which is below the FDA limit of 6°C. Thus, it is concluded that SDUV can be employed at a safe ultrasound power level.

Errors in the shear wave speed measurement using pulse-echo ultrasound will negatively affect the elasticity and viscosity estimation. In vivo applications will also involve gross tissue motion due to heart beat and breathing and thus present additional challenges for ultrasound motion detection. Depending on the site of the application, further optimizations of this method might be necessary.

12.6 CONCLUSIONS

It is demonstrated that both stiffness and viscosity of a viscoelastic medium can be estimated by fitting the measured shear wave speed at multiple frequencies with the Voigt dispersion model. The propagation speed of harmonic shear waves induced by ultrasound radiation force within tissues can be estimated from the phase difference detected by pulse-echo ultrasound over a known propagation distance. A special pulse sequence is developed to allow intermittent operation of a single-array transducer to generate and detect tissue vibrations, which makes this technique compatible with commercial ultrasound scanners. Feasibility of this technique is demonstrated with in vitro tissue experiments.

Acknowledgments

The authors are grateful to Randall Kinnick for his technical assistances on experiments. This work was supported by Grants EB-02640 and EB-02167 from American National Institutes of Health.

References

[1] Muthupillai, R., et al, 1995, "Magnetic Resonance Elastography by Direct Visualization of Propagating Acoustic Strain Waves," *Science*, **269**(5232), pp. 1854–1857.

[2] Ophir, J., Alam, S. K., Garra, B. S., Kallel, F., Konofagou, E. E., Krouskop, T. A., and Varghese, T., 1999, "Elastography: Ultrasonic Estimation and Imaging of the Elastic Properties of Tissues," *Proceedings of the Institution of Mechanical Engineers*, **213**, pp. 203–233.

[3] Sandrin, L., et al, 2002, "Shear Modulus Imaging with 2-D Transient Elastography," *IEEE Transactions on Ultrasonics, Ferroelectrics and Frequency Control*, **49**(4), pp. 426–435.

[4] Parker, K. J., et al, 1998, "Vibration Sonoelastography and the Detectability of Lesions," *Ultrasound in Medicine and Biology*, **24**(9), pp. 1437–1447.

[5] Bercoff, J., Tanter, M., and Fink, M., 2004, "Supersonic Shear Imaging: A New Technique for Soft Tissue Elasticity Mapping," *IEEE Transactions on Ultrasonics, Ferroelectrics and Frequency Control*, **51**(4), pp. 396–409.

[6] Fatemi, M., and Greenleaf, J. F., 1998, "Ultrasound-Stimulated Vibro-Acoustic Spectrography," *Science*, **280**(5360), pp. 82–85.

[7] Sarvazyan, A. P., et al, 1998, "Shear Wave Elasticity Imaging: A New Ultrasonic Technology of Medical Diagnostics," *Ultrasound in Medicine and Biology*, **24**(9), pp. 1419–1435.

[8] Nightingale, K. R., et al, 2001, "On the Feasibility of Remote Palpation Using Acoustic Radiation Force," *Journal of Acoustical Society of America*, **110**(1), pp. 625–634.

[9] Skovoroda, A. R., Emelianov, S. Y., and Odonnell, M., 1995, "Tissue Elasticity Reconstruction Based on Ultrasonic Displacement and Strain Images," *IEEE Transactions on Ultrasonics Ferroelectrics and Frequency Control*, **42**(4), pp. 747–765.

[10] Sumi, C., Suzuki, A., and Nakayama, K., 1995, "Estimation of Shear Modulus Distribution in Soft Tissue from Strain Distribution," *IEEE Transactions on Biomedical Engineering*, **42**(2), pp. 193–202.

[11] Kallel, F., and Bertrand, M., 1996, "Tissue Elasticity Reconstruction Using Linear Perturbation Method," *IEEE Transactions on Medical Imaging*, **15**(3), pp. 299–313.

[12] Nightingale, K., McAleavey, S., and Trahey, G., 2003, "Shear-Wave Generation Using Acoustic Radiation Force: In Vivo and Ex Vivo Results," *Ultrasound in Medicine and Biology*, **29**(12), pp. 1715–1723.

[13] Girnyk, S., et al, 2006, "The Estimation of Elasticity and Viscosity of Soft Tissues in Vitro Using the Data of Remote Acoustic Palpation," *Ultrasound in Medicine and Biology*, **32**(2), pp. 211–219.

[14] Wu, Z., et al, 2004, "Sonoelastographic Imaging of Interference Patterns for Estimation of the Shear Velocity of Homogeneous Biomaterials," *Physics in Medicine and Biology*, **49**(6), pp. 911–922.

[15] Sandrin, L., et al, 2003, "Transient Elastography: A New Noninvasive Method for Assessment of Hepatic Fibrosis," *Ultrasound in Medicine and Biology*, **29**(12), pp. 1705–1713.

[16] Huwart, L., et al, 2006, "Liver Fibrosis: Non-Invasive Assessment with MR Elastography," *NMR in Biomedicine*, **19**(2), pp. 173–179.

[17] Chen, S., Fatemi, M., and Greenleaf, J. F., 2004, "Quantifying Elasticity and Viscosity from Measurement of Shear Wave Speed Dispersion," *Journal of Acoustical Society of America*, **115**(6), pp. 2781–2785.

[18] Yamakoshi, Y., Sato, J., and Sato, T., 1990, "Ultrasonic Imaging of Internal Vibration of Soft Tissue Under Forced Vibration," *IEEE Transactions on Ultrasonics, Ferroelectrics and Frequency Control*, **37,** pp. 45–53.

[19] Catheline, S., et al, 2004, "Measuring of Viscoelastic Properties of Homogeneous Soft Solid Using Transient Elastography: An Inverse Problem Approach," *Journal of Acoustical Society of America*, **116**(6), pp. 3734–3741.

[20] Salameh, N., et al, 2007, "Hepatic Viscoelastic Parameters Measured with MR Elastography: Correlations with Quantitative Analysis of Liver Fibrosis in the Rat," *Journal of Magnetic Resonance Imaging*, **26**(4), pp. 956–962.

[21] Nightingale, K., et al, 2002, "Acoustic Radiation Force Impulse Imaging: In Vivo Demonstration of Clinical Feasibility," *Ultrasound in Medicine and Biology*, **28**(2), pp. 227–235.

Section 3:

BONE CHARACTERIZATION

Chapter 13

APPLICATION OF NON-LINEAR ELASTIC WAVE SPECTROSCOPY (NEWS) TO IN VITRO DAMAGE ASSESSMENT IN CORTICAL BONE

Marie Muller[1], Timothy J. Ulrich[2], Maryline Talmant[1], David Mitton[3], Pascal Laugier[1], and Paul A. Johnson[2]
[1]Université Pierre et Marie Curie-Paris 6, Laboratoire d'Imagerie Paramétrique, UMR CNRS7623, 15 rue de l'Ecole de médecine, F-75006, Paris, France
[2]Geophysics Group, Los Alamos National Laboratory, Los Alamos, NM 87544, USA
[3]Laboratoire de Biomécanique, CNRS, ENSAM, 151 Bd de l'Hôpital, 750013, Paris, France

ABSTRACT

This chapter reviews recent experimental findings applying non-linear dynamical techniques, termed non-linear elastic wave spectroscopy (NEWS), to infer progressive mechanical damage in human bone, in vitro. Standard measures of bone mechanical properties are compared where possible, including the quasi-statically derived stiffness and hysteresis. The goal is to explore whether or not the methods can ultimately be applied in vivo, for applications such as diagnosing and monitoring the progression of osteoporosis, as well as bone healing. It is shown that the NEWS methods are extremely sensitive to progressive damage, far more so than the other methods applied. The results are promising, and current discussion involves how to conduct similar experiments in vivo.

13.1 INTRODUCTION AND BACKGROUND

The non-invasive evaluation of bone biomechanical competence remains a challenge, despite the ever-growing number of physical and imaging tools available. Among existing techniques, vibration and wave propagation analysis have been found to be valuable diagnostic tools to assess bone quality. Assessment of mechanical properties of human bone in vivo by mechanical vibration analysis consists of determining the natural vibration frequencies, mode shapes and/or damping ratio. Two primary vibration techniques have been developed that excite vibrations in the bone to be measured over a frequency range of approximately 50 Hz to 1000 Hz [1]. In the first of these, termed the impulse frequency response (IFR) method, the excitation is provided by a hammer impact. In the second method, termed the bone resonance analysis (BRA) method, the measurement system consists of an electromechanical shaker which applies a sinusoidal driving force to the bone via an impedance head. The resulting vibration is measured with an accelerometer. Using an appropriate model, the vibrational characteristics (e.g., resonance frequencies) can be related to properties of bone that may prove useful as an index of the strength. Jurist [2,3] was among the first investigators to present a theoretical basis and a measurement method for determining the resonance frequency of the ulna in vivo. Bending modes have been identified for long bones (e.g., ulna, tibia, femur) and the resonance frequencies have been related to bending or torsional stiffness using a simple beam model [2-8].

As a quantitative assessment technique, the vibrational response of long bones has been addressed by several researchers for the in vivo determination of bone mechanical characteristics. The potential of vibration analysis applied to bone has been investigated as a non-invasive method to assess the effect that osteoporosis may have on the mechanical properties as well. From in vivo vibration measurements of the tibia in a population of osteoporotic patients and age-matched controls, it was concluded that the bending rigidity (i.e., the product of the Young's modulus and the moment of inertia), calculated from the resonance frequencies in osteoporotic tibiae, was smaller in contrast to the control group [9, 10]. In other studies, the modal damping factor has been found to correlate with fracture energy measured with an impact test [11, 12].

The vibration response has also been used to monitor fracture healing. In particular, the sensitivity of resonance frequency and of different vibration modes to fracture healing has been extensively described [13-17]. Clinical studies reported in, for instance [15, 18], clearly indicate that resonance frequencies can be used clinically to assess the healing state of a fractured long bone. The resonance frequency was found to steadily increase during the process of callus consolidation in both in vitro and in vivo experiments [15, 16]. In 2006, Alizad et al [19] used the acoustic radiation force with an amplitude-modulated signal in order to induce low-frequency vibration in intact, fractured, and bonded excised rat femurs. This study demonstrated that the presence of damage in the samples was responsible for a modification of the vibrational characteristics of the samples, resulting in a decrease of the resonance frequency. A numerical study reported shifting of the vibration spectrum due to fracture, as well as a stronger coupling between lateral and axial vibration spectra [20]. Bending stiffness of long bones was found to be more sensitive to healing than torsional stiffness or resonance frequency [21]. The above studies concur in suggesting that bone macro-damage (fracture) may potentially be assessed by measuring bone vibrational properties.

Another application of bone vibration analysis is the detection of peri-implant conditions. Hip prosthesis loosening can be determined in vivo by the analysis of the vibration of the implant. The excitation vibration is applied to the femoral condyle and the response measured at the great trochanter. The method characterizes changes with implant fixation stability of the complete spectral response of the prosthesis [22, 23]. Rosenstein et al [22] were the first to suggest that loosening may be detectable from the non-linear response of the implant: The vibration of a loose implant exhibits the presence of harmonics of the input frequency when the vibration response of a secure prosthesis matches that of the input vibration. Another resonance technique, which is also used for dental implants, consists in measuring the resonance frequency [24, 25] for the evaluation of the implant stability.

At higher frequencies (approximately 100 kHz to 2 MHz), linear characteristics of ultrasonic wave transmission through bone, such as frequency-dependent attenuation and speed of sound, are widely used to assess skeletal status and predict osteoporotic fracture risk. The use of ultrasonic wave transmission can be classified into two categories according to the experimental setup: namely, transverse transmission with the ultrasonic emitter and receiver placed on opposite sides of a skeletal site to be measured; and axial transmission with the emitter/receiver set placed parallel to the bone longitudinal axis on one side of a long bone [26]. The clinical potential of ultrasound for the investigation of pathological conditions that affect bone strength was recognized as early as in the late 1950s by Siegel et al [27], who reported the first use of axial transmission for monitoring fracture healing at the tibia. Modern bone quantitative ultrasound was described in 1984 by Langton et al [28], who measured the slope of the frequency-dependent attenuation at the calcaneus to discriminate osteoporotic from non-osteoporotic women. Since then, many advances have been achieved and a variety of different sophisticated technologies capable of measuring different skeletal sites such as the heel, fingers, wrist or tibia have been introduced and evaluated [26]. Heel quantitative ultrasound (QUS) parameters are highly correlated with bone mineral density in most cases and predict future fracture risk almost as well as for central dual energy X-ray absorptiometry of the spine or hip [29]. Whereas vibration analysis yields information on whole-bone characteristics (e.g., bending rigidity), ultrasound analysis indicates local changes in bone properties (bone mineral density, microstructure, stiffness) of the segment of bone that is traversed by ultrasound [30-32].

Vibration and ultrasonic analysis of bone have thus been shown to be sensitive to bone conditions that affect bone strength such as osteoporosis or fracture (i.e., macro-damage); however, to the knowledge of the authors, no data is available regarding direct detection of micro-damage using vibrational techniques, likely due to the fact that the elastic parameters assessed in the studies mentioned above fail to be sensitive to discontinuities at the meso or micro scale (e.g., [33]).

Daily physiological loads resulting from walking, lifting, and weight bearing result in bone in local strains in a range of 0.05% to 0.3%, that are responsible for the generation of bone micro-cracking. Following Wolff's law, stating that bone structure is permanently adapting to the applied stresses [34], fatigue damage initiates a bone remodeling process designed to heal micro-damage. In healthy bone, remodeling contributes to bone stiffness and toughness, allowing bone to absorb energy without failing. Fracture risk is increased when an imbalance takes place between micro-damage factors (brittleness, porosity, softness), and bone remodeling [35]. Such imbalances can occur with bone age or with the development of osteoporosis. As a consequence, aged or osteoporotic bone exhibits more micro-damage than young and healthy bone [36]. Micro-crack accumulation has been observed in subjects older than 40 years, and micro-crack density and length show an approximate exponential increase with increasing age [34, 37-39]. Of note is that at the hip, micro-crack density is twice as large for 78-year-old women than for 46-year-old women [40] underscoring the effect of bone age on fragility.

Two hypotheses may explain observed crack accumulation as a function of age. In the first, it may be due to bone age and local hypermineralization of the bone matrix. Although osteoporotic bone is highly porous and globally less dense than normal bone, its mineral content can be locally high, due to bone age. Indeed, in some regions with a low remodeling rate, osteocytes go through their lifetime (around 25 years) without being subject to remodeling [36], and accumulate in the form of necrosed cells in the lacunae, preventing further remodeling. This leads to local hypermineralization and therefore a locally increased brittleness. Thus, aged and osteoporotic bone is locally highly brittle, in addition to being subjected to locally high stresses, because of its increased porosity. Damage can then accumulate in the form of micro-cracks that can lead to the formation of large-scale fractures [38]. Local bone age has been found to be much more important in a population of fractured subjects than in a control (non-fractured) population [36]. In the second hypothesis, damage accumulation may be due to an abnormally high remodeling rate, responsible for a considerably increased porosity, and therefore, for a higher stress concentration [40]. Micro-damage accumulation is thus potentially the result of an ensemble of complex phenomenon.

Bone micro-damage accumulation has important consequences on bone mechanical properties. It has been shown that micro-damage accumulation coincides with a decrease of bone toughness and stiffness [41-43]. Thus, micro-damage accumulation and bone fragility are strongly correlated. The accumulation of bone micro-damage has been proposed as one factor that contributes to increased skeletal fragility with age and that may increase the risk for fracture in older women. Therefore, bone micro-damage characterization in vivo could indeed provide relevant information in terms of bone quality and thus fracture risk and skeletal fragility assessment.

Micro-damage can typically be quantitatively assessed invasively using histomorphometry [44-46]. In addition, newly developed high-resolution imaging techniques such as synchrotron radiation micro-computed tomography (SR-μCT) have been shown to be useful to assess micro-damage [47]. These methods provide very high resolution (up to a few microns using SR-μCT and less than a micron for histomorphometry) and are able to monitor damage accumulation during fatigue, either by placing a fatigue testing device in the synchrotron beam, or by using different fluorochromes for different fatigue states. Clearly, however, these techniques cannot be used in vivo. For instance, the resolution needed for micro-damage imaging requires lethal radiation levels, up to 28 keV in cortical bone. In addition, both techniques provide only local information, not necessarily indicative of the global or volumetric damage state.

Elastic non-linear parameters derived from dynamic wave studies show great sensitivity to damage in a variety of materials [48-52]. Indeed, damaged materials have proved to exhibit characteristic non-linear behaviors that can be used to infer material mechanical integrity. The physical basis of the behavior is related to damage features at scales apparently ranging from 10^{-9} to at least 10^{-1} m (e.g., [53]). Damage, in the form of distributed or localized cracks for instance, results in predictable characteristics of the vibration spectrum. For example, under resonance conditions, the resonance frequency (proportional to the modulus and wave speed) has been shown to shift downward with increasing vibration amplitude, due to the presence of the mechanical damage [48]. The authors note that, in 2002, Lee et al [54] used resonance ultrasound spectroscopy (RUS) to characterize bone linear elastic moduli. However, in this study, only linear parameters were assessed and bone micro-damage was not quantified. Another important manifestation of non-linear dynamic behavior is the fact that in the presence of two pure frequency waves, wave multiplication takes place leading to the presence of sum and difference frequencies (sidebands) as well as harmonics of each wave [49, 50]. The general group of methods, all relying on monitoring spectral components for extracting the elastic non-linear behavior, is termed non-linear elastic wave spectroscopy [49-51].

In the following, feasibility studies conducted by our groups are reviewed, demonstrating the sensitivity of non-linear dynamical parameters to progressively accumulated micro-damage in vitro. In the first part is described the origin of elastic non-linearity followed by a broadly applied theory that describes the elasticity of damaged solids. Following this, experimental protocols and results for the methods presented are shown. The first method, a resonance technique known as non-linear resonant ultrasound spectroscopy (NRUS) [48, 51], is based on the measurement of a resonance frequency shift with increasing vibration amplitude, and is applied to characterizing progressive damage in a population of human femurs. This is followed by progressive damage observations obtained from a modulation method, known as non-linear wave modulation spectroscopy (NWMS) [48, 50], in a single human femur. Following this a discussion and conclusions are presented. The results shown are derived primarily from previous studies by the authors groups [55, 56].

13.2 INTRODUCTION TO ELASTIC NON-LINEARITY

As noted, NEWS methods have their basis in monitoring the material elastic modulus in response to dynamic wave excitation, usually by measurement of wave distortion characteristics and their associated spectral components. In undamaged solids, most materials have small elastic non-linear response due to intrinsic anharmonicity of, for instance, the crystalline lattice-atomic level vibration. This contribution to elastic non-linearity is difficult to measure, and its net effect is to *increase* the material modulus under dynamic oscillatory wave forcing. In a mechanically damaged material there is an additional contribution to the material elastic non-linearity due entirely to the damage itself. It has a different physical origin. The contribution is much larger than that of the anharmonicity, by at least one to two orders of magnitude. Cracks and other soft features in the material cause the material modulus to *decrease* under dynamic forcing: As wave amplitude increases progressively, the material modulus decreases accordingly (i.e., the material softens), depending on the strength or magnitude of the material non-linear parameter(s). The non-linear parameter(s) has/have been empirically related to mechanical damage intensity in a number of studies (e.g., [57-59]). There is no direct analytical relationship between damage intensity and the non-linear parameter(s), however, because the basic physics of dynamic wave/crack interaction is not well understood (e.g., [60]). In fact, there is no direct correlation with mechanical damage in most non-destructive testing methods. Normally, a damage/undamaged result is all that is required, or empirical relations are developed so that correlation between what is measured and damage intensity can be inferred. For the moment, phenomenological models are relied on to describe the elastic behaviors, including the damage characteristics, of these materials. Thus, the non-linear parameters defined in such models are indicative of damage intensity but not a direct measure of them.

13.3 THEORY

Following the phenomenological description of stress-strain developed by Guyer and McCall [61-64], the one-dimensional constitutive relation between the stress σ and the strain ε can be expressed in a first approximation as,

$$\sigma = \int C(\varepsilon, \dot{\varepsilon})\, d\varepsilon \qquad (13-1)$$

where C is the modulus and $\dot{\varepsilon} = d\varepsilon/dt$ is the strain rate. The modulus is strain-amplitude and strain-rate-dependent, and is defined as,

$$C(\varepsilon, \dot{\varepsilon}) = C_0(1 + \beta\varepsilon + \delta\varepsilon^2 - \alpha(\Delta\varepsilon + \varepsilon\,\mathrm{sign}(\dot{\varepsilon})) + \cdots) \qquad (13-2)$$

where C_0 is the linear modulus, ε is the *instantaneous* strain amplitude in an oscillating wave, $\Delta\varepsilon$ is *maximum* (one half-peak) strain amplitude over the previous period ($\Delta\varepsilon = (\varepsilon_{max} - \varepsilon_{min})/2$ for continuous sine wave excitation), and the function $\mathrm{sign}(\dot{\varepsilon}) = 1$ for $\dot{\varepsilon} > 0$ and $\mathrm{sign}(\dot{\varepsilon}) = -1$ for $\dot{\varepsilon} < 0$. The $\mathrm{sign}(\dot{\varepsilon})$ function accounts for hysteresis in the stress-strain response of a propagating or standing wave. The parameter α is a measure of the hysteresis in stress-strain over a half wave cycle. The wave literally follows a small hysteresis loop in stress-strain as a function of time. The parameters β and δ are the classical non-linear coefficients in the Taylor expansion of stress versus instantaneous strain [65]. They indicate that the modulus is strain amplitude dependent (over the wave cycle). In a classical medium like water or many single crystals, only β and δ appear and they are very small in magnitude, meaning there is very little change in the modulus with dynamic strain amplitude.

In a cracked or otherwise damaged material, β and δ are normally at least 1 to 2 orders of magnitude larger than classical media, and are negative in value (in contrast to classical media). This means that the material modulus (and thus wave speed) *decreases* as wave strain amplitude increases, in contrast to classical media, where modulus and wave speed generally increase with strain amplitude. In addition, the presence of hysteresis gives distinctive attributes to the waveform and associated wave spectrum, and to the strain amplitude dependencies of harmonics versus the fundamental drive amplitude, as well as the resonance frequency. Since a wave is normally characterized in terms of its half-peak strain amplitude $\Delta\varepsilon$ (or RMS strain), in a classical non-linear medium, a strain wave of frequency f_1 and amplitude $\Delta\varepsilon$ can transform into a strain wave containing frequency components $2f_1$, $3f_1$, etc., with strain amplitudes proportional to $(\Delta\varepsilon)^2$, $(\Delta\varepsilon)^3$, etc. In the case of a purely hysteretic material, the second harmonic cannot be generated ($\beta = 0$), and the third harmonic is predicted and observed to be quadratic with the strain amplitude of the fundamental, ($\Delta\varepsilon_{3f_1} \propto \alpha\Delta\varepsilon^2_{f_1}$). This implies that hysteresis acts as a second-order non-linearity. In medium where two sine waves $\Delta\varepsilon_1 \cos(2\pi f_1 t)$ and $\Delta\varepsilon_2 \cos(2\pi f_2 t)$ (ignoring the phase $kx=\phi$) are present, the waves multiply leading to new waves that have frequencies $f_1 \pm f_2$ with strain amplitudes proportional to $\beta\Delta\varepsilon_1\Delta\varepsilon_2$. When dynamic hysteresis dominates, due to the presence of mechanical damage in particular, the second-order sidebands ($f_1 \pm 2f_2$) will display strain amplitudes proportional to $\alpha\Delta\varepsilon_1\Delta\varepsilon_2$.

When a pure continuous wave (cw) frequency is applied to a sample simultaneously with an impulsive force that excites a number of the normal modes, the modes multiply with the cw frequency due to the β and α terms in Eqs. (13-1) and (13-2). Many sum and differences are created in this manner, and it tends to be a very robust approach to extracting the non-linear response of a material. The work presented that follows takes advantage of this approach, known as non-linear wave modulation spectroscopy.

In resonance, it can be shown that the resonance frequency f is proportional to the peak strain amplitude $\Delta\varepsilon$ via the non-linear parameter α,

$$\frac{\Delta f}{f_0} = \frac{f - f_0}{f_0} \approx \alpha\Delta\varepsilon \qquad (13-3)$$

where f_0 is the linear (low-amplitude) resonance frequency. This equation is the basis for non-linear resonant ultrasound spectroscopy, where the frequency of the resonance peak is followed with increasing strain amplitude in order to extract α. Note that the fundamental resonance frequency is related to the wave speed and modulus. For instance, in Young's mode resonance, the wave speed c is related to f_0 and sample length L,

$$c = f\lambda = 2f_0 L = \sqrt{\frac{C}{\rho_0}} \qquad (13-4)$$

where ρ_0 is density in the unperturbed state. Thus, as frequency downshifts, the material modulus and wave speed decrease.

13.4 EXPERIMENTS

The overall approach outlined below was to progressively damage bone samples and measure their elastic non-linear response by applying NRUS or using NWMS, as a function of induced damage. In the following the samples and experiment are described.

13.4.1 Samples

The mid-distal part of 18 fresh human femurs obtained from cadavers (5 female and 13 male donors, mean = 81 years, SD = 13 years, range = 47 to 100 years; see Table 13-1) were assessed in the study. Samples were cut to leave 6-cm segments of the diaphysis (see Fig. 13-1 for cross-sectional view). Soft tissue was removed and the specimens were kept frozen at –20°C before measurement sessions. Specimens were then measured at room temperature, and were kept hydrated during the duration of the experiment. The ethical approval for the collection of samples was granted by the Human Ethics Committee of the

TABLE 13-1. Age and gender data for the samples used in the NRUS study. A single sample (70-year-old female) was available for the NWMS measurements

Sample # (*j*)	Age (yr)	Gender (M/F)	Sample # (*j*)	Age (yr)	Gender (M/F)
1	64	M	10	74	M
2	70	M	11	96	F
3	47	M	12	89	F
4	74	M	13	47	M
5	80	M	14	70	M
6	89	F	15	69	M
7	83	M	16	83	M
8	96	F	17	100	F
9	80	M	18	88	M

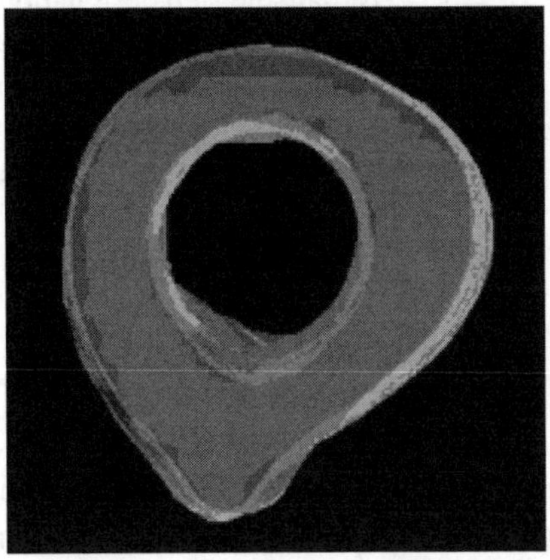

Figure 13-1. Cross section of the sample, a 6-cm tall (out of the plane of the figure) section of cortical bone taken from a human diaphysis (femur). The average outer diameter of the cross-section is approximately 2.5 cm. Similarly shaped and sized samples were used for all NRUS and NWMS experiments presented herein

Institute of Anatomy at the University René Descartes (Paris, France). The tissue donors or their legal guardians provided informed written consent to provide their tissues for investigation, in accord with legal clauses stated in the French Code of Public Health (Code de la Santé Publique Français).

13.4.2. Experimental Protocol

Fatigue damage was progressively induced in the specimens applying compressive mechanical testing (INSTRON 8500, Massachusetts, USA). For the samples subjected to the NRUS experiment, fatigue cycling was performed at 4 Hz (see Table 13-2). The cyclic load was determined for each sample, in order to induce a constant quasi-static strain amplitude: 0.5% for the samples from female donors older than 81 years (this is the median age for hip fracture for women, and thus a lower strain was determined to be appropriate in order to avoid early failure). For the other donors a 0.6% strain was applied. For the progressive damage study using NWMS, the fatigue cycling was done at a strain amplitude of 0.5% at 4 Hz.

TABLE 13.2. Damage step parameters for the progressive damage studies

Study	# Damage Steps (N)	# Cycles per Damage Step	Strain ($\Delta\varepsilon$)	Strain Rate (Hz)
NRUS	10	15,000	0.6%, (0.5%)[1]	4
NWMS	5	30,000	0.5%	4

[1]The quasi-static strain (ε)^s was reduced to 0.5% for female samples age 81 yr and older.

To facilitate the discussion of the details of the experiments and results a notation is defined as follows:

- M = number of samples (M_{NRUS} = 18, M_{NWMS} = 1).
- N = number of damage steps (N_{NRUS} = 10, N_{NWMS} = 5).
- $i = 0,\ldots, N$, damage step designation ($i = 0$ indicates initial undamaged state).
- $j = 1,\ldots, M$, sample designation index.
- $F_{i,j}$ = applied loads during damage step i in sample j.
- $X_{i,j}$ = induced displacements for ith damage step and jth sample.
- $H_{i,j}$ = hysteresis measurements (i.e., area of hysteresis loops).
- $R_{i,j}$ = NRUS measurements.
- $K_{i,j}$ = stiffness measurement (from quasi-static $F_{i,j}$ and $X_{i,j}$ data).
- W_i = NWMS measurement at ith damage step (j is ignored as $j = 1$ for all NWMS experiments).
- $\alpha_{i,j}$, α_i, β_I, and Γ_i are non-linear parameters calculated from the data $R_{i,j}$ or W_i.

The procedure for the NRUS and NWMS studies began with the initial dynamic measurement ($R_{0,j}$ or W_0) for all samples to determine the initial non-linearity of the material. The samples were then taken through a damage step, consisting of cyclic loading as described above (also shown in Tab. 2), during which the quasi-static measurements of $F_{i,j}$ and $X_{i,j}$ were made and the stiffness $K_{i,j}$ determined. The quasi-static hysteresis $H_{i,j}$ was also measured, if observed. After each damage step a non-linear elasticity experiment was conducted ($R_{i,j}$ or W_i).

13.4.3 Non-Linear Resonant Ultrasound Spectroscopy

To perform the resonance measurements a piezo-ceramic source (with an 8-cm-thick steel backload bonded with epoxy) was attached to the samples using phenol salicilate. Frequency sweeps with progressively increasing amplitude were performed using an apparatus from Dynamic Resonance Systems, Inc. (Powell, WY, USA). The resulting displacements at the surface of the sample were measured using a laser interferometer (BMI, SH120, France). The NRUS experimental configuration is shown in Fig. 13-2.

In the NRUS measurements, the $R_{i,j}$ data consists of a suite of resonance curves, e.g., a single resonance peak excited with increasing drive amplitudes. Figure 13-3 displays two suites of resonance curves at two different damage states $i = 0$ and $i = 8$. From the resonance peak data $R_{i,j}$ the non-linear parameter $\alpha_{i,j}$ can be calculated by using Eq. (13-3). Preliminary feasibility studies of the NRUS technique for bone damage assessment were demonstrated in bovine bone in a previous study [66].

13.4.4 Non-Linear Wave Modulation Spectroscopy

This technique has been applied broadly in industrial materials and geomaterials (e.g., [50, 51]). In this feasibility study, only one sample was available for testing. For the NWMS experiments in human cortical bone, the sample was bonded to the piezoelectric source in

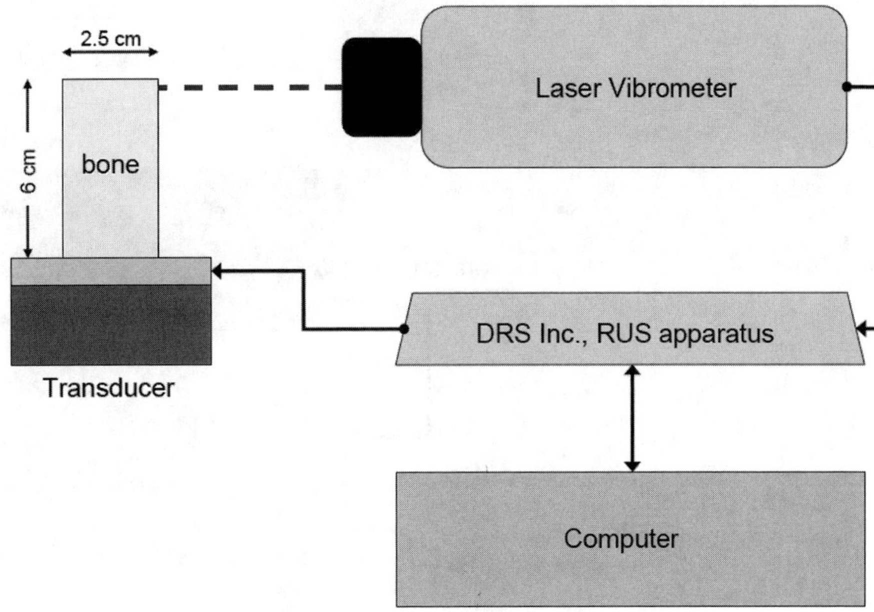

Figure 13-2. Non-linear resonant ultrasound spectroscopy experimental setup

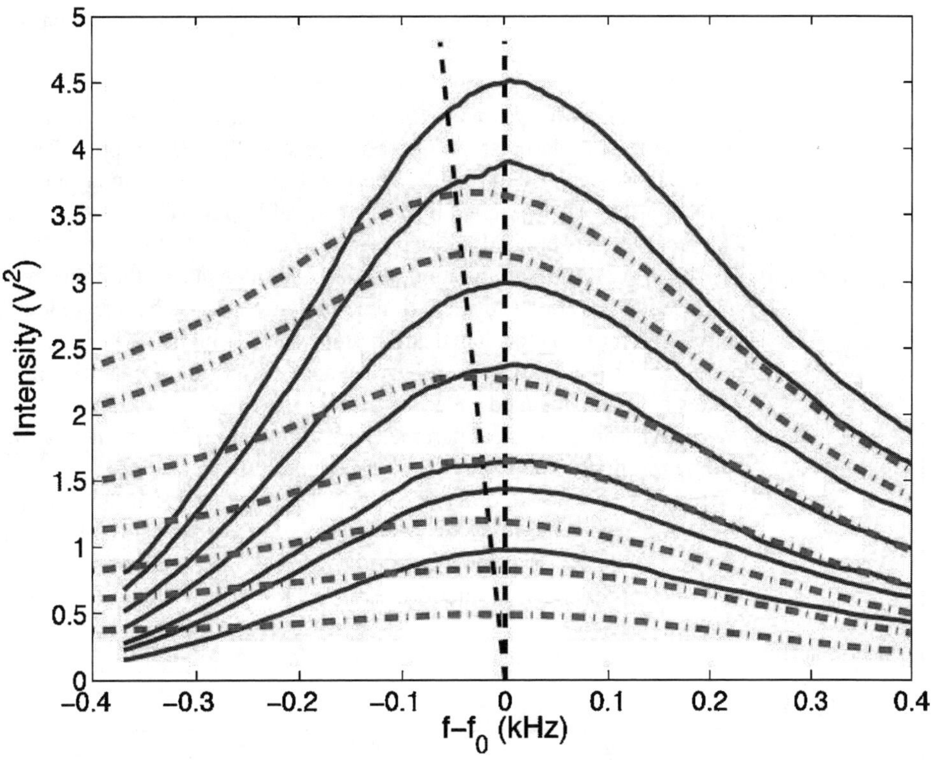

Figure 13-3. Resonance curves for a single sample at damage steps 0 (solid) and 8 (dot-dash). The shift of the resonance frequency is measurably more pronounced in the higher states of damage

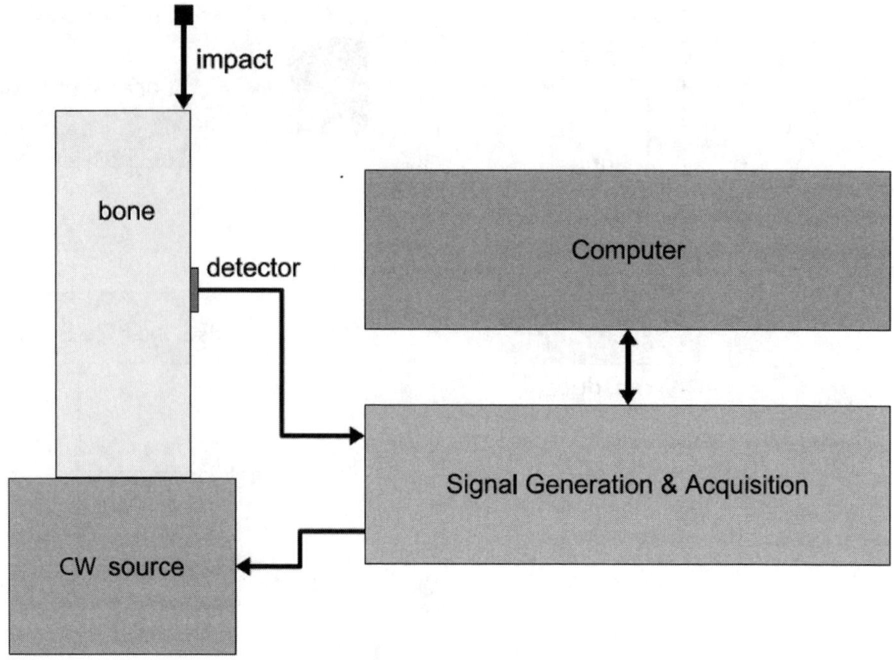

Figure 13-4. Non-linear wave modulation spectroscopy experimental setup. Signal generation and acquisition system consists of a 12-bit D/A converter (max. conversion rate, 150 MS/sec; max. output level, 50 V_{pp}) and a 14-bit A/D acquisition card

the same manner as was done for the NRUS measurements. The piezoelectric source was driven at a continuous amplitude pure tone of 223 kHz to optimize the efficiency of the source and receiver bandwidth. The low-frequency vibrational modes of a bone sample were simultaneously excited by a mechanical impulse (induced by a light tap with a pen). The receiver was bonded midway between the ends of the sample as shown in Fig. 13-4. To obtain various excitation amplitudes, the intensity of the impulse and pure tone were systematically varied (30 excitation combinations at each damage level); however, lack of calibration prevents the actual strain amplitudes from being known. Without calibration, it was necessary to look only at relative changes of the non-linear parameters from their initial values in the undamaged state.

A dynamic non-linear parameter Γ_i, a function of α_i and β_i, was defined and obtained by integrating the frequency spectrum containing the first-order ($f_1 \pm f_2$) and second-order ($f_1 \pm 2f_2$) sidebands in order to include the effects of multiple sidebands simultaneously. The frequency range used for this study was 215 kHz to 231 kHz. In addition, the values of the non-linear parameters β_i and α_i were calculated from the ratio of the first- and second-order sideband amplitudes, respectively, to the product of the source amplitudes (i.e., A_1A_2).

13.5 RESULTS

13.5.1 NRUS Observations

Figure 13-5 shows load versus displacement curves for a test sample taken to failure under larger loading conditions, illustrating the well-known quasi-static non-linear and hysteretic

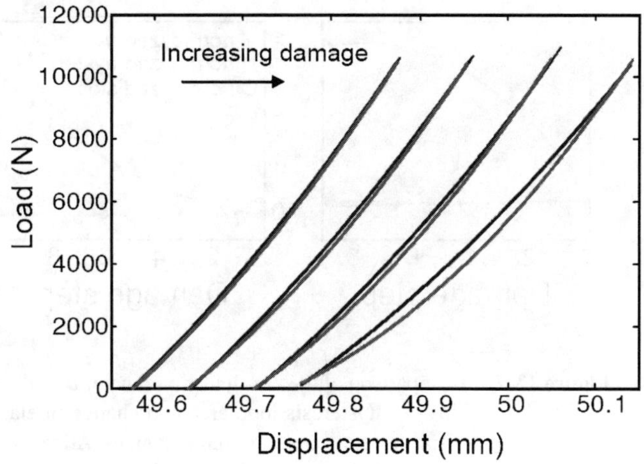

Figure 13-5. Load-displacement curves showing the evolution of the hysteretic elastic behavior with increasing damage

nature of human bone. The stiffness $K_{i,j}$ is extracted from the slope of load-displacement curves. Only very near failure can one easily measure hysteresis.

Figure 13-3 shows sample resonance curves at increasing drive level $R_{i,j}$, at damage step 0 ($R_{0,j}$) before damage cycling commenced, and at damage step, 8 ($R_{8,j}$), well into the damage process. Clearly, Δf is more pronounced at step $i = 8$, meaning the associated $\alpha_{8,j} > \alpha_{0,j}$. This is quantified in Fig. 13-6, which shows the (frequency normalized) dependence of the resonance curve peaks as a function of drive amplitude, from the data shown in Fig. 13-3. What is clear is that at damage step 0, there is little change in the resonance peak frequency with amplitude; however, at step 8, a measurable change can be seen. From such curves, $\alpha_{i,j}$ can be extracted for each damage step (i) and sample (j).

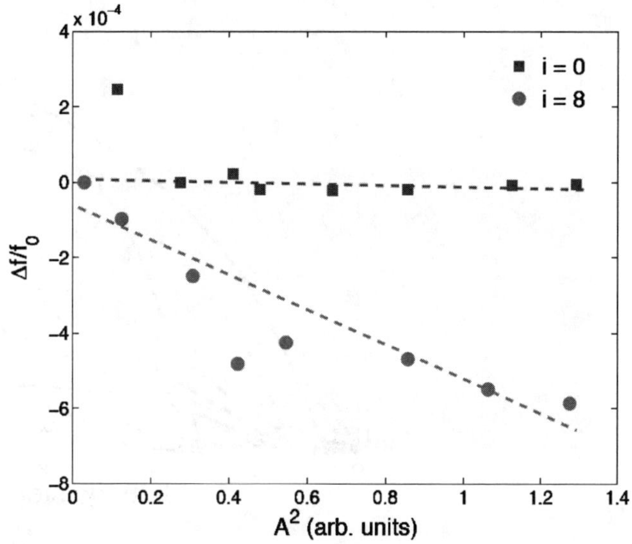

Figure 13-6. Normalized shift of the resonance frequency for the resonance peaks, shown in Fig. 13-3, as a function of the square of the resonance peak amplitude ($A^2 \propto \varepsilon^2$). Dashed lines are shown to more clearly illustrate the trends

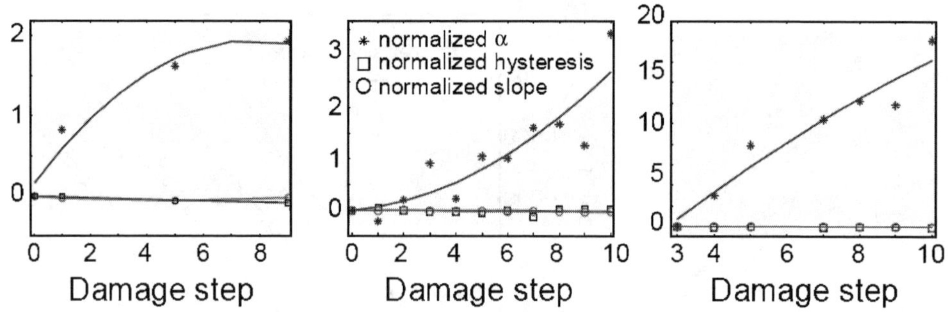

Figure 13-7. Comparison of the non-linear parameter α to the quasi-static load-displacement measures of hysteresis loop area and change in elastic stiffness (i.e., the slope of load-displacement curves) for three samples. All values are normalized to the corresponding value at damage step 0. Note the dramatic increase in the non-linear parameter while the other quantities remain almost unchanged

In Fig. 13-7, normalized α ($\alpha_{i,j}/\alpha_{0,j}$) as a function of damage step i is shown for three samples. The normalized slope of the load-displacement curve, proportional to the stiffness ($K_{i,j}/K_{0,j}$) and the normalized hysteresis ($H_{i,j}/H_{0,j}$) are shown for comparison. The striking result is the fact that there is immeasurable change in the normalized slope and hysteresis, but $\alpha_{i,j}/\alpha_{0,j}$ "sees" the effect of damage cycling very early on, and grows markedly with each damage session. Note that error bars are not shown, but based on repeatability measurements, conducted before the sequence began, and they are on the order of the symbol size. Figure 13-8 shows $\alpha_{i,j}/\alpha_{0,j}$ as a function of damage step for all data collected, in a semi-log plot. Note that there is significant variation in the response, and this is attributed to some or all of the variation to age differences in the sample population (see below). A significant observation is that all samples show $\alpha_{i,j}/\alpha_{0,j}$ increasing as a function

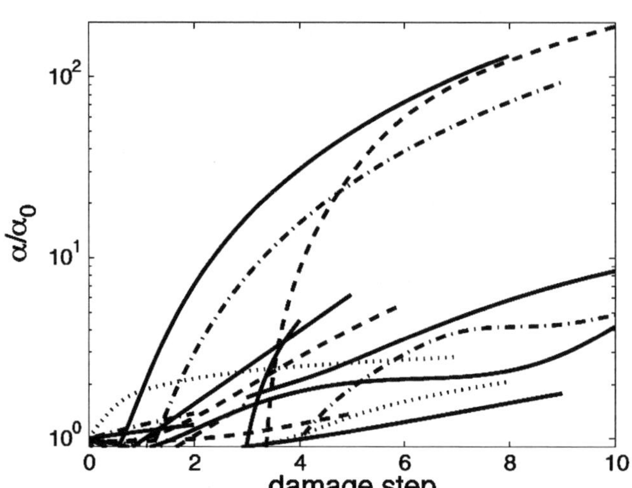

Figure 13-8. Normalized α values for all samples. Trend lines are shown as taken from the original data using smoothing splines. Note the enormity of the variations in the given population

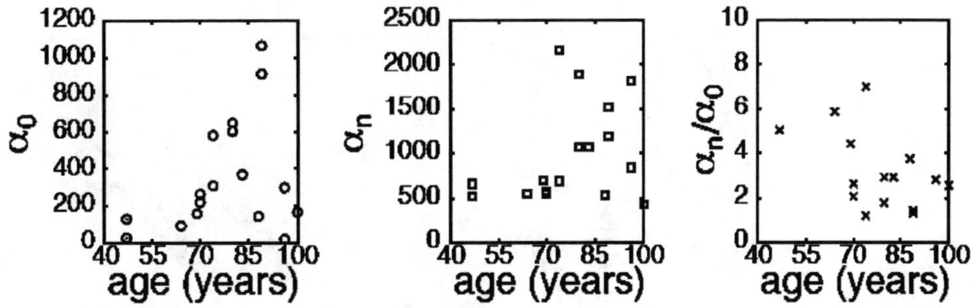

Figure 13-9. Influence of age on the non-linear parameter α (a) before induced fatigue damage, and (b) just before fracture. (c) Ratio of final α_n to initial α_0 values, providing an estimation of the quantity of damage induced before fracture. The few cases where the increase in α was considerably higher (i.e., $\alpha_n/\ \alpha_0 \geq 100$, see Fig. 13-8) are not visible due to the scale

of damage step, and that the stiffness $K_{i,j}/K_{0,j}$ and hysteresis $H_{i,j}/H_{0,j}$ show very little, if any, change with increased damage.

Figure 13-9 shows absolute values of $\alpha_{0,j}$ and $\alpha_{n,j}$ as a function of the sample age, here $n = 7,\ldots, 10$ depending on the sample j and indicates the final damage step completed before failure of the sample occurred. In the event that the sample did not reach catastrophic failure, the value of n was taken as the final damage state, i.e., $n = 10$. The spread in the data is large, and the authors infer that this reflects initial damage state in the population. The spread apparently increases with age; however, since the number of young donors is small, this observation should be treated with caution. Also shown in this figure are the ratios $\alpha_{n,j}/\alpha_{0,j}$ as a function of donor age. This plot shows how much α changed over the duration of the experiment. On average, $\alpha_{n,j}/\ \alpha_{0,j}$ increased by about 2.5 times, but in some cases, far more (see caption).

13.5.2 NWMS Observations

In Fig. 13-10, the observations of loading for all damage sessions in the sample used for the NWMS study are shown [56]. It can be observed that an evolution of the sample length is present (i.e., a permanent compressional deformation); however, due to imperfections of the sample reinsertion into the load frame, as well as the possible presence of the bonding material (a residual from bonding to the transducer) remaining on the samples, this change in sample length was not used in any way for analysis. The load-displacement data was used solely for the extraction of linear fits to extract the stiffness K_i (in an identical fashion as was done for the NRUS study) after each damage session from which normalized parameters are calculated and displayed. The imprecision of the quasi-static displacement measurements X_i prevented the observation of any quasi-static hysteresis.

In Fig. 13-11, an expanded view of the frequency spectrum about the continuous wave frequency is shown. Spectra from two steps are shown, $i = 0$ and 5. The notable difference between the two spectra is the emergence of sideband energy (solid line) around the continuous wave frequency. In contrast, at $i = 0$ there is no measurable sideband energy. This observation is indicative of the enhanced non-linear response with damage cycling. The non-linear parameter Γ_i is obtained from the integral in the frequency range shown in Fig. 13-11, excluding the continuous wave frequency peak (~221 kHz to 225 kHz).

Figure 13-12 shows comparison of the normalized Γ (Γ_i/Γ_0) with normalized K (K_i/K_0) obtained from all damage sessions. Clearly, the non-linear response is far more sensitive

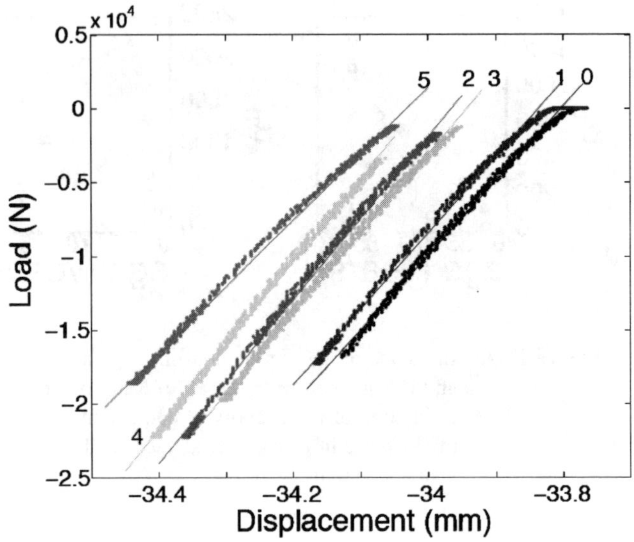

Figure 13-10. Load vs. displacement for each damage step. A linear fit has been applied to extract relative changes in the stiffness K (proportional to Young's modulus) as the fatigue damage progressed

to mechanical property change than the stiffness measure, as was also seen in the NRUS experiments (see Fig. 13-7). The respective changes are about 700% and 10%, respectively. Error bars are not noted, but are much smaller than the overall changes. The normalized α and β parameters follow a similar trend to Γ (see Fig. 13-13), as would be expected since Γ

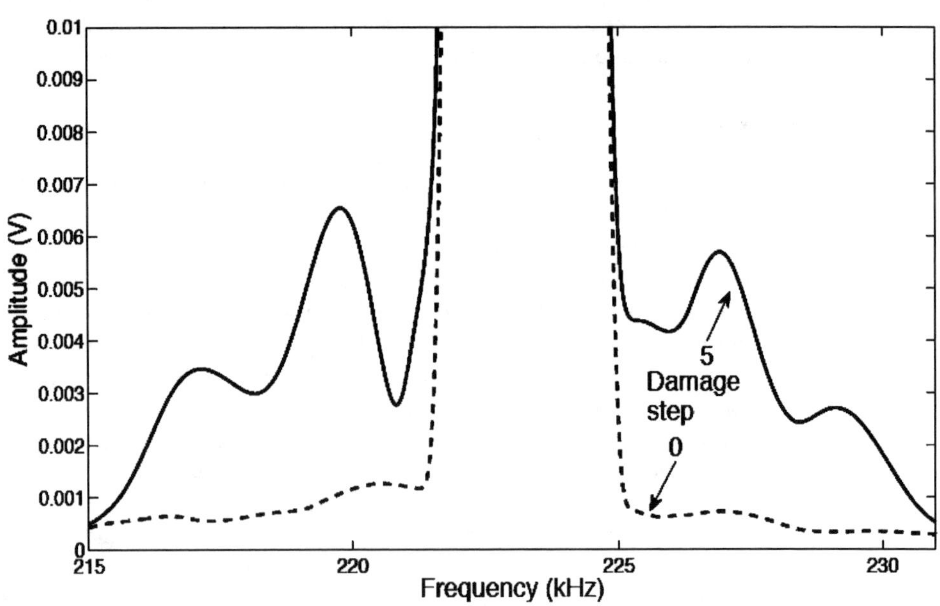

Figure 13-11. Power spectra of sideband frequency range for two damage steps (0 and 5). The increase in sideband energy after fatigue damage has been introduced is clearly visible. The calculation of Γ is done in the frequency range shown (i.e., 215–231 kHz). For the calculation, the central peak at 223 kHz (i.e., ~221–225 kHz) is removed

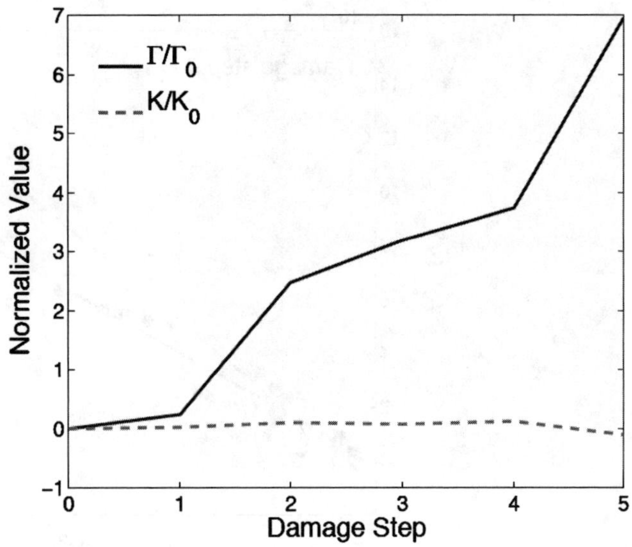

Figure 13-12. Comparison of the relative changes of the non-linear parameter Γ with the relative change in modulus (extracted from linear elastic measurements). Note the almost 700% change in the non-linear parameter Γ, compared to the 10% change in *K*

must be a function of both α and β. Alone, α and β are more sensitive to damage by roughly an order of magnitude; however, they are more difficult to measure precisely and therefore the errors can be much larger.

Figure 13-14 illustrates how the values in Fig. 13-13 are obtained. Equations (13-1) and (13-2) state that the first- and second-order sidebands, associated with β_i and α_i, respectively, are linearly proportional to the product of the fundamental frequency amplitudes, e.g., $A_{f1 \pm f2} \propto \beta A_{f1} A_{f2}$ and $A_{f1 \pm 2f2} \propto \alpha A_{f1} A_{f2}$. Here, A is used instead of absolute ε, as absolute

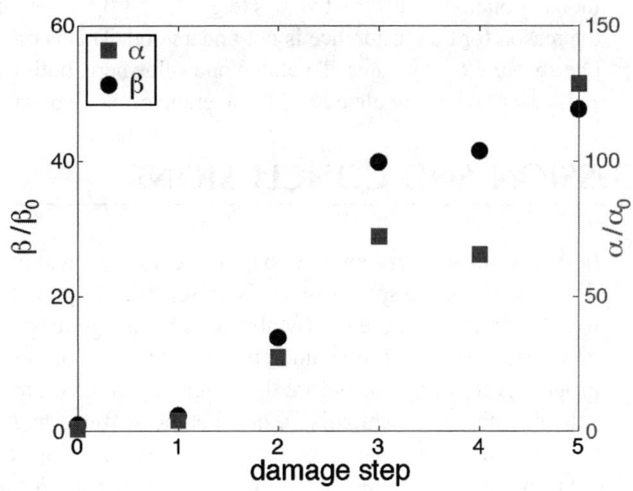

Figure 13-13. Relative changes in the non-linear parameters α (squares) and β (circles). Both non-linear parameters are more sensitive than the parameter Γ introduced above; however, the ease of extracting either α or β from experimental data is for more difficult to do with adequate precision and certainty compared to the calculation of Γ

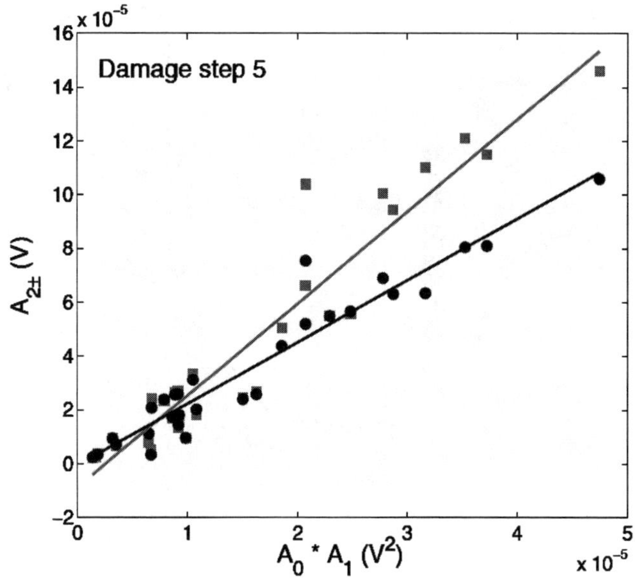

Figure 13-14. Example of second-order sideband amplitudes (e.g., f_{--} = 217 and f_{++} = 230 kHz in Fig. 13-11) as a function of the product of the source amplitudes after the final damage step. The non-linear parameters β and α are extracted from linear fits of such data sets of the first-order and second-order sidebands, respectively

values were not measured. Using all 30 excitation combinations available in each W_i, it is possible to extract the non-linear parameters from a linear fit of the appropriate sideband amplitude versus the product of the excitation amplitudes. As there is no mathematical distinction between amplitudes of sum and difference frequencies (e.g., $f_+ = f_1 + f_2$ and $f_- = f_1 - f_2$, respectively), it should be possible to extract two identical values for α and β from, i.e., one from the sum frequency f_+ and one from the difference frequency f_-. In reality it is not uncommon to obtain two different values (e.g., β_+ and β_-), depending on the use of f_+ or f_-, although the reason for this difference is not understood. This is one of the difficulties in determining one absolute α or β value. To obtain one value here, both values (e.g., α_+ and α_-, as shown in Fig. 13-14) were calculated and the average of the two was taken as the "single" value.

13.6 DISCUSSION AND CONCLUSIONS

In this work, an overview of two non-linear dynamical methods that are derived from non-linear elastic wave spectroscopy is presented, as applied to human bone samples in vitro, in order to demonstrate their viability as a damage diagnostic. The first method, known as non-linear resonant ultrasound spectroscopy, was applied to 18 donor samples that were progressively fatigued. Before the fatigue cycling commenced the non-linear parameter α was obtained. For comparison, quasi-static stiffness and, where possible, hysteresis were measured from the load-displacement curves, quantities that are commonly applied for the measurement of the mechanical state of the material. These measurements provided a baseline in order to characterize the initial state of the samples. The samples were then subject to ten damage sessions in a mechanical load frame, in order to progressively induce mechanical damage. Between each damage session, the same measurements of α, quasi-static stiffness and hysteresis were made. A number of samples failed before the ten sessions were applied. However, despite this, the data trends are clear.

In Figs. 13-7 and 13-8, it has been shown that the non-linear parameter α is extremely sensitive to progressive damage, in contrast to quasi-static stiffness and hysteresis. Figure 13-7 shows that α responds dramatically to the increased damage state within the first two cycling sessions, and by the last session, may increase many hundreds of times. In contrast, the quasi-static stiffness and hysteresis remain essentially unchanged, insensitive to the increasing damage intensity. Figure 13-8 shows results of α from the entire suite of samples. From this plot, it is clear that, (a) α increases as a function of induced damage intensity for all samples, and (b) that there exists a large range in individual responses. The results suggest that the non-linear response *must* be related to the damage intensity. In Fig. 13-9, α has been shown as a function of donor age. The authors infer that these data illustrate that the population distribution of damage intensity before damage cycling commenced, suggesting that there is a large variation in the "damage state" of human bone that may depend on age and other factors. The observations of α as function of age and as a function of progressively induced damage underscore the differences in bone integrity in the population. Confidence in relating non-linear response to progressive damage, and in fact, correlating non-linear response with damage in general, is based on numerous studies relating these quantities in non-biologic materials. Studies that show the same trends as those measured here, include work by Morris et al [58], where progressive damage was induced by bending in high-strength aluminum; Nagy [57], who observed progressive damage from bending in samples of aluminum alloy, polymers, titanium matrix composite, and adhesive layers; Van Den Abeele and De Visscher [59], who observed progressive damage induced by three-point bending in concrete; as well as work that the authors have conducted in, for instance, progressive damage in tension-cycled samples of steel (unpublished). These and many other studies support this interpretation.

This chapter has also described a second NEWS method, known as non-linear wave modulation spectroscopy. This method relies on the non-linear-induced wave multiplication of two or more elastic waves at different frequencies. When the vibration spectrum, induced by an impulsive forcing, is mixed with a pure frequency, continuous wave, the method is perhaps the most robust NEWS method in existence. A hybrid of this method has been shown, where the sidebands in the frequency domain are integrated in order to obtain Γ a function of α and β from Eqs. (13-1) and (13-2). This result is shown in Fig. 13-12, where the result is compared with the stiffness obtained quasi-statically. Like the NRUS results comparing α to stiffness for three samples shown in Fig. 13-7, the non-linear response shows the sensitivity to damage early on (after the first damage session), changing by nearly 700% in contrast to the 10% change in the stiffness.

The observations suggest that if the method is to be successfully applied in vivo, statistics relating Γ, α, and/or β to bone damage will be very important. In future studies, it is clear that empirical relations between measured crack characteristics (e.g., crack density, length, orientation) and non-linear response will have to be developed in order to quantitatively assess damage for both in vitro and in vivo studies. The results also suggest that the method may be of value in following one individual as a function of time if in vivo application is realized.

The samples used in the work presented here had fluids removed from the samples before measuring. It is known from previous work that fluids have a large influence on elastic non-linear response (e.g., [64]), so different magnitudes of non-linear parameters than those shown here may be expected in fluid saturated material. Furthermore, inducing sufficient strain amplitudes in order to induce non-linear response in bone in vivo will be challenging. Separating bone non-linearity from tissue non-linearity should not be a serious issue because tissue non-linearity is significantly smaller than that observed in bone, and exhibits no hysteretic term (α). That is, unless the bone non-linear response is much diminished in vivo. Moreover, measuring the signals in vivo presents its own difficulties due to wave dissipation through tissue. The authors are currently are designing methods to overcome these problems, and intend to begin in vivo studies in the near future.

References

[1] Cornelissen, P., Cornelissen, M., Van der Perre, G., Christensen, A. B., Ammitzboll, F., and Dyrbye, C., 1986, "Assessment of Tibial Stiffness by Vibration Testing in situ — II. Influence of Soft Tissues, Joints and Fibula," *Journal of Biomechanics*, **19**, pp. 551–561.

[2] Jurist, J. M., 1970, "In Vivo Determination of the Elastic Response of Bone. I. Method of Ulnar Resonance Frequency Determination," *Physics in Medicine and Biology*, **15**, pp. 417–426.

[3] Jurist, J. M., 1970, "In Vivo Determination of the Elastic Response of Bone. II. Ulnar Resonant Frequency in Osteoporotic, Diabetic and Normal Subjects," *Physics in Medicine and Biology,* **15**, pp. 427–434.

[4] Van der Perre, G., Van Audekercke, R., Martens, M., and Mulier, J. C., 1983, "Identification of in Vivo Vibration Modes of Human Tibiae by Modal Analysis," *Journal of Biomechanical Engineering*, **105**, pp. 244–248.

[5] Cornelissen, P., Cornelissen, M., Van der Perre, G., Christensen, A. B., Ammitzboll, F., and Dyrbye, C., 1987, "Assessment of Tibial Stiffness by Vibration Testing in Situ — III. Sensitivity of Different Modes and Interpretation of Vibration Measurements," *Journal of Biomechanics*, **20**, pp. 333–342.

[6] Thomas, A. M., Luo, D. Z., and Dunn, J. W., 1991, "Response of Human Femur to Mechanical Vibration," *Journal of Biomedical Engineering*, **13**, pp. 58–60.

[7] Lowet, G., Van Audekercke, R., Van der Perre, G., Geusens, P., Dequeker, J., and Lammens, J., 1993, "The Relation Between Resonant Frequencies and Torsional Stiffness of Long Bones in Vitro. Validation of a Simple Beam Model," *Journal of Biomechanics*, **26**, pp. 689–696.

[8] Couteau, B., Hobatho, M. C., Darmana, R., Brignola, J. C., and Arlaud, J. Y., 1998, "Finite Element Modelling of the Vibrational Behaviour of the Human Femur Using CT-Based Individualized Geometrical and Material Properties," *Journal of Biomechanics*, **31**, pp. 383–386.

[9] Van der Perre, G., and Lowet, G., 1994, "Vibration, Sonic and Ultrasonic Wave Propagation Analysis for the Detection of Osteoporosis," *Clinical Rheumatology*, **13** (Suppl. 1), pp. 45–53.

[10] Van Der Perre, G., and Lowet, G., 1996, "In Vivo Assessment of Bone Mechanical Properties by Vibration and Ultrasonic Wave Propagation Analysis," *Bone*, **18** (Suppl. 1), pp. 29S–35S.

[11] Panteliou, S. D., Abbasi-Jahromi, H., Dimarogonas, A. D., Kohrt, W., and Civitelli, R., 1999, "Low-Frequency Acoustic Sweep Monitoring of Bone Integrity and Osteoporosis," *Journal of Biomechanical Engineering*, **121**, pp. 423–431.

[12] Christopoulou, G. E., Stavropoulou, A., Anastassopoulos, G., Panteliou, S. D., Papadaki, E., Karamanos, N. K., and Panagiotopoulos, E., 2006, "Evaluation of Modal Damping Factor as a Diagnostic Tool for Osteoporosis and Its Relation with Serum Osteocalcin and Collagen I N-Telopeptide for Monitoring the Efficacy of Alendronate in Ovariectomized Rats," *Journal of Pharmaceutical and Biomedical Analysis*, **41**, pp. 891–897.

[13] Singh, V. R., Yadav, S., and Adya, V. P., 1989, "Role of Natural Frequency of Bone as a Guide for Detection of Bone Fracture Healing," *Journal of Biomedical Engineering,* **11**, pp. 457–461.

[14] Tower, S. S., Beals, R. K., and Duwelius, P. J., 1993, "Resonant Frequency Analysis of the Tibia as a Measure of Fracture Healing," *Journal of Orthopaedic Trauma*, **7**, pp. 552–557.

[15] Benirschke, S. K., Mirels, H., Jones, D., and Tencer, A. F., 1993, "The Use of Resonant Frequency Measurements for the Noninvasive Assessment of Mechanical Stiffness of the Healing Tibia," *Journal of Orthopaedic Trauma*, **7**, pp. 64–71.

[16] Nakatsuchi, Y., Tsuchikane, A., and Nomura, A., 1996, "The Vibrational Mode of the Tibia and Assessment of Bone Union in Experimental Fracture Healing Using the Impulse Response Method," *Medical Engineering & Physics*, **18**, pp. 575–583.

[17] Lowet, G., Dayuan, X., and Van der Perre, G., 1996, "Study of the Vibrational Behaviour of a Healing Tibia Using Finite Element Modelling," *Journal of Biomechanics*, **29**, pp. 1003–1010.

[18] Christensen, A. B., Ammitzboll, F., Dyrbye, C., Cornelissen, M., Cornelissen, P., and Van der Perre, G., 1986, "Assessment of Tibial Stiffness by Vibration Testing in Situ — I. Identification of Mode Shapes in Different Supporting Conditions," *Journal of Biomechanics*, **19**, pp. 53–60.

[19] Alizad, A., Walch, M., Greenleaf, J. F., and Fatemi, M., 2006, "Vibrational Characteristics of Bone Fracture and Fracture Repair: Application to Excised Rat Femur," *Journal of Biomechanical Engineering*, **128**, pp. 300–308.

[20] Nikiforidis, G., Bezerianos, A., Dimarogonas, A., and Sutherland, C., 1990, "Monitoring of Fracture Healing by Lateral and Axial Vibration Analysis," *Journal of Biomechanics*, **23**, pp. 323–330.

[21] Roberts, S. G., and Steele, C. R., 2000, "Efficacy of Monitoring Long-Bone Fracture Healing by Measurement of Either Bone Stiffness or Resonant Frequency: Numerical Simulation," *Journal of Orthopaedic Research*, **18**, pp. 691–697.

[22] Rosenstein, A. D., McCoy, G. F., Bulstrode, C. J., McLardy-Smith, P. D., Cunningham, J. L., and Turner-Smith, A. R., 1989, "The Differentiation of Loose and Secure Femoral Implants in Total Hip Replacement Using a Vibrational Technique: An Anatomical and Pilot Clinical Study," *Proceedings of the Institution of Mechacnical Engineers [H]*, **203** (2), pp. 77–81.

[23] Rowlands, A., Duck, F. A., and Cunningham, J. L., 2007 (Jun 21), "Bone Vibration Measurement Using Ultrasound: Application to Detection of Hip Prosthesis Loosening," *Medical Engineering & Physics*, [Epub ahead of print].

[24] Denayer, I., Jaecques, S. V., Burny, F., Puers, R., Borgwardt, A., and Van der Perre, G., 2000, "Measurement of Implant Stability and Osseointegration: Non-Linearities and Resonance Frequencies," *12th Conference European Society of Biomechanics*, Dublin, p. 151.

[25] Cawley, P., Pavlakovic, B., Alleyne, D. N., George, R., Back, T., and Meredith, N., 1998, "The Design of a Vibration Transducer to Monitor the Integrity of Dental Implants," *Proceedings of the Institution of Mechacnical Engineers [H]*, **212**, pp. 265–272.

[26] Njeh, C. F., Hans, D., Fuerst, T., Glüer, C. C., and Genant, H. K. 1999, *Quantitative Ultrasound: Assessment of Osteoporosis and Bone Status*, Martin Dunitz, London, United Kingdom, 1999.

[27] Siegel, I. M., Anast, G. T., and Fields, T., 1958, "The Determination of Fracture Healing by Measurement of Sound Velocity Across the Fracture Site," *Surgery, Gynecology & Obstetrics*, **107**, pp. 327–332.

[28] Langton, C. M., Palmer, S. B., and Porter, R. W., 1984, "The Measurement of Broadband Ultrasonic Attenuation in Cancellous Bone," *Engineering in Medicine*, **13**, pp. 89–91.

[29] Marin, F., Gonzalez-Macias, J., Diez-Perez, A., Palma, S., and Delgado-Rodriguez, M., 2006, "Relationship Between Bone Quantitative Ultrasound and Fractures: A Meta-Analysis," *Journal of Bone and Mineral Research*, **21**, pp. 1126–1135.

[30] Laugier, P., Droin, P., Laval-Jeantet, A. M., and Berger, G., 1997, "In Vitro Assessment of the Relationship Between Acoustic Properties and Bone Mass Density of the Calcaneus by Comparison of Ultrasound Parametric Imaging and Quantitative Computed Tomography," *Bone*, **20**, pp. 157–165.

[31] Nicholson, P. H., Muller, R., Cheng, X. G., Ruegsegger, P., Van Der Perre, G., Dequeker, J., and Boonen, S., 2001, "Quantitative Ultrasound and Trabecular Architecture in the Human Calcaneus," *Journal of Bone and Mineral Research*, **16**, pp. 1886–1892.

[32] Raum, K., Leguerney, I., Chandelier, F., Bossy, E., Talmant, M., Saied, A., Peyrin, F., and Laugier, P., 2005, "Bone Microstructure and Elastic Tissue Properties Are Reflected in QUS Axial Transmission Measurements," *Ultrasound in Medicine and Biology*, **31**, pp. 1225–1235.

[33] Nicholson, P. H., Bouxsein, M. L., 2000, "Quantitative Ultrasound Does Not Reflect Mechanically Induced Damage in Human Cancellous Bone," *Journal of Bone and Mineral Research*, **15**, pp. 2467–2472.

[34] Martin, B., 1993, "Aging and Strength of Bone as a Structural Material," *Calcified Tissue International*, **56** (Suppl 1) S34–S39; discussion S39–S40.

[35] Turner, C. H., 2002, "Biomechanics of Bone: Determinants of Skeletal Fragility and bone quality," *Osteoporosis International*, **13**, pp. 97–104.

[36] Parfitt, A. M., 1993, "Bone Age, Mineral Density, and Fatigue Damage," *Calcified Tissue International*, **53** (Suppl. 1) S82–S85; discussion S85–S86.

[37] Norman, A. L., and Wang, Z., 1997, "Microdamage of Human Cortical Bone: Incidence and morphology in Long Bones," *Bone*, **20**, pp. 375–379.

[38] Schaffler, M. B., Chio, K., and Milgrom, C., 1995, "Aging and Matrix Microdamage Accumulation in Human Compact Bone," *Bone*, **17**, pp. 521–525.

[39] Martin, R., 2003, "Fatigue Microdamage as an Essential Element of Bone Mechanics and Biology," *Calcified Tissue International*, **73**, pp. 101–107.

[40] Burr, D. B., Forwood, M. R., Fyhrie, D. P., Martin, R. B., Schaffler, M. B., and Turner, H. C., 1997, "Bone Microdamage and Skeletal Fragility in Osteoporotic and Stress Fractures," *Journal of Bone and Mineral Research*, **12** (1), pp. 9–15.

[41] Zioupos, P., 2001, "Accumulation of *in Vivo* Fatigue Microdamage and Its Relation to Biomechanical Properties in Aged Human Cortical Bone," *Journal of Microscopy*, **201** (Pt2), pp. 270–278.

[42] Schaffler, M. B., Radin, E. L., and Burr, D. B., 1989, "Mechanical and Morphological Effects of Strain Rate on Fatigue of Compact Bone," *Bone*, **10**, pp. 207–214.

[43] Fazzalarin, N. L., Forwood, M. R., Smith, K., Manthey, B. A., and Herreen, P., 1998, "Assessment of Cancellous Bone Quality in Severe Osteoarthrosis: Bone Mineral Density, Mechanics, and Microdamage," *Bone*, **22**, pp. 381–388.

[44] Burr, D. B., and Stafford, T., 1990, "Validity of the Bulk Staining Technique to Separate Artifactual from *in Vivo* Bone Microdamage," *Clinical Orthopaedics*, **260**, pp. 305–308.

[45] Lee, T. C., Myers, E. R., and Hayes, W. C., 1998, "Fluorescence Aided Detection of Microdamage in Compact Bone," *Journal of Anatomy*, **193**(Pt 2), pp. 179–184.

[46] Lee, T. C., Arthur, T. L., Gibson, L. J., and Hayes, W. C., 2000, "Sequential Labeling of Microdamage in Bone Using Chelating Agents," *Journal of Orthopaedic Research*, **18**, pp. 322–325.

[47] Thurner, P. J., Wyss, P., Voide, R., Stauber, M., Stampanoni, M., Sennhauser, U., and Muller, R., 2006, "Time-Lapsed Investigation of Three-Dimensional Failure and Damage Accumulation in Trabecular Bone Using Synchrotron Light," *Bone,* **39**, pp. 288–299.

[48] Johnson, P. A., Van Den Abeele, K., TenCate, J., and Guyer, R., Resonant Nonlinear Ultrasound Spectroscopy, US Patent Number 6,330,827, Dec 18, 2001.

[49] Johnson, P. A., "The New Wave in Acoustic Testing," 1999, *Materials World, the J. Inst. Materials*, **7**, pp. 544–546.

[50] Van Den Abeele, K. E.- A., Johnson, P. A., and Sutin, A., 2000, "Nonlinear Elastic Wave Spectroscopy (NEWS) Techniques to Discern Material Damage. Part I: Nonlinear Wave Modulation Spectroscopy (NWMS)," *Research on Nondestructive Evaluation*, **12**, pp. 17–30.

[51] Van den Abeele, K. E.- A., Carmeliet, J., Ten Cate, J. A., and Johnson, P. A., 2000, "Nonlinear Elastic Wave Spectroscopy (NEWS) Techniques to Discern Material Damage, Part II: Single-Mode Nonlinear Resonance Acoustic Spectroscopy," *Research in Nondestructive Evaluation*, **12**, pp. 31–42.

[52] Ostrovsky, L., and Johnson, P. A., 2001, "Dynamic Nonlinear Elasticity in Geomaterials," *Rivista del Nuovo Cimento*, **24**, pp. 1–46.

[53] Johnson, P. A., and Sutin, A., 2005, "Slow Dynamics and Anomalous Nonlinear Fast Dynamics in Diverse Solids," *Journal of the Acoustical Society of America*, 117, pp. 124–130.

[54] Lee, T., Lakes, R. S., and Lal, A., 2002, "Investigation of Bovine Bone by Resonant Ultrasound Spectroscopy and Transmission Ultrasound," *Biomechacnics and Modelling in Mechanobiology*, **1**, pp. 165–175.

[55] Muller, M., Mitton, D., Talmant, M., Laugier, P., and Johnson, P. A., 2007, "Nonlinear Ultrasound Can Detect Accumulated Damage in Human Bone," *J. Biomechanics*, in review.

[56] Ulrich, T. J., Johnson, P., Muller, M., Mitton, D., Talmant, M., and Laugier, P., 2007, "Application of Nonlinear Dynamics to Progressive Fatigue Damage in Human Cortical Bone," *Applied Physics Letters*, in press.

[57] Nagy, P., 1998, "Fatigue Damage Assessment by Nonlinear Ultrasonic Materials Characterization," *Ultrasonics*, **36**, pp. 375–381.

[58] Morris, W. L., Buck, O., and Inman, R. V., 1979, "Acoustic Harmonic Generation Due to Fatigue Damage in High Strength Aluminum," *Journal of Applied Physics*, **50**, pp. 6737–6741.

[59] Van Den Abeele, K., and De Visscher, J., 2000, "Damage Assessment in Reinforced Concrete Using Spectral and Temporal Nonlinear Vibration Techniques," *Cement and Concrete Research*, **30**, pp.1453–1464.

[60] Guyer, R. A., and Johnson, P. A., 1999, "Nonlinear Mesoscopic Elasticity: Evidence for a New Class of Materials," *Physics Today*, **52**, pp. 30–35.

[61] McCall, K. R., and Guyer, R. A., 1994, "Equation of State and Wave Propagation in Hysteretic Nonlinear Elastic Materials," *Journal of Geophysical Research,* **99**, pp. 23887–23897.

[62] Guyer, R. A., McCall, K. R., and Boitnott, G. N., 1995, "Hysteresis, Discrete Memory and Nonlinear Wave Propagation in Rock: A New Paradigm," *Physical Review Letters,* **74**, pp. 3491–3494.

[63] McCall, K. R., and Guyer, R. A., 1996, "Hysteresis, Discrete Memory and Nonlinear Elastic Wave Propagation in Rock: A New Theoretical Paradigm," *Nonlinear Processes in Geophysics*, **3**, pp. 89–101.

[64] Van Den Abeele, K. E.- A., Carmeliet, J., Johnson, P. A., and Zinzsner, B., 2002, "The Influence of Water Saturation on the Nonlinear Mesoscopic Response of Earth Materials, and the Implications to the Mechanism of Nonlinearity," *Journal of Geophysical Research*, **107**, pp. 10,1029–10,1040.

[65] Landau, L. D., and Lifshitz, E. M., 1980, *Theory of Elasticity*, Pergamon, Oxford, United Kingdom, 1980.

[66] Muller, M., Sutin, A., Guyer, R., Talmant, M., Laugier, P., and Johnson, P. A., 2005, "Nonlinear Resonant Ultrasound Spectroscopy (NRUS) Applied to Damage Assessment in Bone," *Journal of the Acoustical Society of America*, **118**, pp. 3946–3952.

Appendix

INTERNATIONAL STANDARDS AND GUIDELINES FOR THE SAFE USE OF DIAGNOSTIC ULTRASOUND IN MEDICINE

Kwan-Hoong Ng and Muhammad Shahrun Nizam
Department of Biomedical Imaging, University of Malaya, Kuala Lumpur, Malaysia

A.1 ULTRASOUND BIOPHYSICS AND BIOEFFECTS

Diagnostic ultrasound has been experiencing a widespread trend toward increased output with the introduction of color flow imaging, pulse Doppler, and greater usage of real-time imaging. In response to this trend, recommendations for the safe use of ultrasound have been issued by several international organizations (see Table A-1). In addition, recent regulations have changed the emphasis of responsibility so that more onus is now placed on the operators to ensure that ultrasound is used safely on patients. This chapter summarizes the biophysics and bioeffects and the standards issued, and outlines recommendations for safe use in diagnostic ultrasound.

There has been very extensive research aimed at understanding the basic biophysics and evaluation of potential bioeffects. When ultrasound propagates through human tissue, there are potential biological effects. Many studies are dose-effect studies and virtually all ultrasound-induced adverse bioeffects have occurred at higher intensities than diagnostic ultrasound.

The mechanisms by which ultrasound interacts with tissue are generally classified as:

- Thermal effects — heating of tissue as ultrasound is absorbed by tissue. Heat is also produced at the transducer surface.
- Cavitation — the formation of gas bubbles at high negative pressure.
- Other mechanical effects — radiation forces leading to streaming in fluids and stress at tissue interfaces.

There are excellent review articles, reports [1-18], and books [19-27] on the biophysics, bioeffects, and safety of diagnostic ultrasound.

TABLE A-1. Published recommendations by major ultrasound organizations on the safe use of diagnostic ultrasound (modified from [6]).

	WFUMB	AIUM	EFSUMB	ODS/FDA
B-mode	Not contraindicated on thermal grounds when no gas present			
Doppler	Use lowest available power consistent with obtaining good diagnostic information Minimize time beam passes through one point		Prudent use involves minimum output levels and exposure times	
Exposure levels		Refer to ODS		FDA upper intensity limit Ophthalmic exposures: <50 mW/cm^2 Others: <720 mW/cm^2
Thermal effects	Temperature rises ≤1.5°C (38°C) can be used without reservation. Obstetric exposures resulting in a temperature increase of 4°C for 5 min are potentially hazardous.	At the FDA regulatory limit ($I_{SPTA.3}$ = 720 mW/cm^2) max. temperature increase in the conceptus can exceed 2°C.		FDA does not regulate TI. For general use, TI should be <6.
Non-thermal effects	When gas (include contrast agents) is present exposure levels and duration should be reduced to the min. to obtain required information.	The threshold value of MI for extravasation of blood cells in mouse lung is approximately 0.3.		For general use, MI should be <1.9.
First trimester exposure			Absence of knowledge about first trimester US exposures means that care is required in the application of transvaginal US in early pregnancy.	
Epidemiology		Insufficient evidence of a causal relationship between diagnostic US and adverse effects.		

A.1.1 Thermal Effects

As the ultrasound waves are absorbed in tissue, their energy is converted into heat. Ultrasound-induced temperature rise varies with tissue properties such as absorption coefficient, thermal conductivity, density, and blood perfusion [18]. The ultrasound field parameters to consider include wave form (pulse or continuous), average, and spatial intensity (W/cm^2), exposure time, duty cycle, frequency, pulse duration, pulse repetition frequency (PRF), and beam/scanning configuration. Generally, pulsed waves produce less heat than continuous waves, while intensity parameters determine the total energy transferred to the tissue. This induced temperature rise is focal, which may not trigger systemic heat dissipating thermoregulatory mechanisms.

Temperature rises are steeper in tissues that are poorly vascularized (tendon, fat) and in tissues that conduct heat (bone). Tissues adjacent to bone are particularly susceptible to heat increase via conduction. Biological tissue absorption is directly related to the protein content (collagen has very high absorption). Generally, absorption is greatest in mineralized bone followed by skin/tendon, spinal cord, and finally, brain/liver/kidney.

A.1.2 Cavitation

The main non-thermal (mechanical) interactions deal with the generation, growth, vibration, and possible collapse of microbubbles within the tissue. This phenomenon is known as cavitation. Cavitation essentially describes the biophysical interaction of gaseous inclusions (bubbles) within tissues when exposed to an incident waveform (e.g., ultrasound). Typically, bubbles will expand and contract as the peak positive and negative pressures propagate through the tissue and it is this oscillatory behavior of the bubble that induces local biophysical effects.

There are two types of cavitation: stable (non-inertial) cavitation refers to the creation of bubbles that oscillate with the sound beam; transient (inertial) cavitation refers to the process in which the oscillations grow so strong that the bubbles collapse violently, producing very intense, localized effects.

Stable cavitation has beneficial influence(s) on local soft tissues. It is described as the phasic oscillation of gaseous inclusions, resulting in a mechanical vibration within the tissue in the absence of deleterious free radical formation. This vibration creates eddy current flow around the oscillating bubble. Unstable (inertial) cavitation is described as a rapid collapse of the bubble which produces very high local temperatures and/or pressures resulting in the production of toxic "free radicals" and eventual tissue damage.

Cavitation is rare in vivo owing to the lack of naturally occurring gas inclusions in living biological tissues [2]. The use of gas-based contrast agents for imaging studies increases the potential for cavitation [10].

A.1.3 Other Mechanical Effects

When ultrasound waves pass through tissue, they cause a low-level radiation force on the tissue. This force produces a pressure in the direction of the beam and away from the transducer and should not be confused with the oscillatory pressure of the ultrasound itself. This results in shearing forces. The pressure that results and the pressure gradient across the beam are very low, even for intensities at the higher end of the diagnostic range [23].

For a ultrasound wave, the radiation pressure varies with the local intensity across the wavefront. When the propagation medium is an absorbing fluid, the radiation force exerted when the ultrasound energy is absorbed causes the fluid to move in the direction of wave propagation. This effect is known as streaming. Even though water has a small coefficient of absorption, it can still be caused to stream at velocities of several cm per second by a pulsed ultrasound beam [19].

A.2 IS THERE A RISK?

The question of safety has received a lot of international attention because technically sophisticated equipment with higher outputs are being used clinically. Just because bioeffects on human at diagnostic levels have not been observed does not mean that they do not exist. It is known that the potential for risk exists. It is important for practitioners to be aware of biophysics and bioeffects so that they can make informed decisions to safeguard patients.

There are three sources of information on ultrasound bioeffects to help determine the risk factors: epidemiology, *in vitro* cell studies, and animal studies.

- Epidemiology — no adverse effects, including no evidence of low birth weights from diagnostic ultrasound have been demonstrated. The American Institute of Ultrasound in Medicine (AIUM) [17] evaluated epidemiological studies and concluded that: "Widespread clinical use over 25 years has not established any adverse effect arising from exposure to diagnostic ultrasound." Randomized clinical studies are the most rigorous method for assessing potential adverse effects of diagnostic ultrasound. Studies using this methodology show no evidence of an effect on birth weight in humans. Other epidemiology studies have shown no causal association of diagnostic ultrasound with any of the adverse fetal outcomes studies.
- *In vitro* cell studies — *in vitro* studies generally expose macromolecules, membrane transport systems, cells, or clumps of cells suspended in liquid. AIUM [17] stated that: "Although the exposure conditions and mechanisms are different from *in vivo* situations, an *in vitro* effect must be regarded as a real effect of ultrasound". *In vitro* studies also provide the capability to control experimental variables. This enables researchers to explore and evaluate mechanisms for damage. One should be cautious with reports of *in vitro* studies that claim direct clinical significance.
- Animal studies — most of the studies were done at spatial peak-time average intensities and exposure times that exceed diagnostic values. At high time average intensity levels, fetal weight reductions in rats, death of rat fetuses, and altered mitotic rates were observed. For these effects to be produced, animals are exposed to some minimal time average intensity for a given period. If the intensity is reduced, the exposure time had to be increased to compensate for the reduced acoustic energy.

Figure A-1 shows the intensity versus time curve relevant to the production of bioeffects in mammalian tissue [17]. The line (below 100 mW/cm^2) divides the data into a region of positive biological effects and a no-effect observed region. The line follows a curve for intensity greater than 100 mW/cm^2 in which $I \cdot T = 50$ J/cm^2. AIUM has claimed that no bioeffects were observed for intensities as high as 1 W/cm^2 for highly focused beams (dotted line).

A.3 DEVELOPMENT OF SAFETY GUIDELINES AND REGULATION

Unlike ionizing radiation, there has been little international safety standard on the clinical use of ultrasound or standard for the calibration of output from diagnostic equipment. The FDA introduced application-specific limits on acoustic output for the United States only. The permissible limit was lowest for ophthalmic (17 mW/cm^2 spatial peak temporal

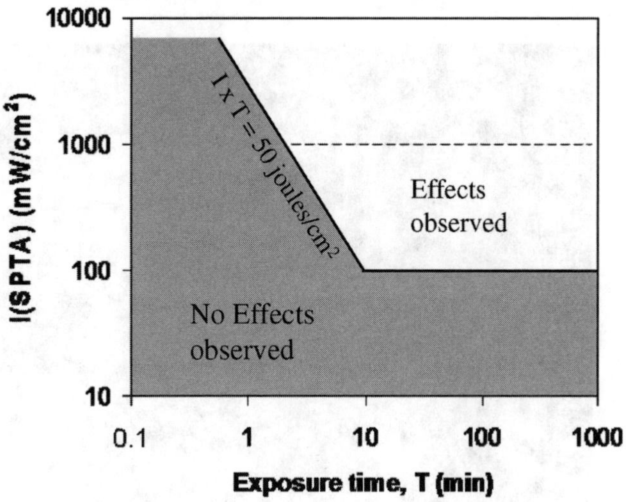

Figure A-1. Intensity vs. time curve relevant to the production of bioeffects in mammalian tissue. (Data derived from [17])

average intensity [I_{SPTA}]) and fetal (94 mW/cm^2 [I_{SPTA}] exposures, where the tissues are particularly sensitive to damage [28].

There is a large body of scientific literature on bioeffects but it is difficult to interpret much of the early work in the context of the safety of diagnostic ultrasound as the exposure conditions used were not reported accurately nor was it clinically relevant. It was often difficult to find biological endpoints that were sufficiently sensitive to respond since modest acoustic outputs were used. The Bioeffects Committee of AIUM has stated that there is no evidence of independently confirmed adverse significant biological effects in mammalian tissue exposed *in vivo* to intensities (I_{SPTA}) below 100 mW/cm^2 [17]. The situation became more complicated when it was first reported that bleeding could be induced in lung capillaries in animal following exposure to diagnostic levels of ultrasound at average intensities far below 100 mW/cm^2. This finding has subsequently been verified in other animal species [10].

A significant development in safety is the introduction of the output display standard (ODS) providing real-time acoustical output labeling developed in 1992 [28, 29]. Two quantities related to acoustical output levels are defined: Thermal Index (TI) and Mechanical Index (MI). The TI estimates the potential for producing thermally-induced bioeffects in soft tissue and bones and the MI estimates the potential for producing non-thermal/mechanical bioeffects in tissue. An example of the ODS being implemented in clinical ultrasound scanner is shown in Fig. A-2.

TI is the ratio of acoustical power produced by the transducer to the power required to raise the temperature in tissue by 1°C. A TI value of 1 means that under tissue conditions assumed in the algorithm, a 1°C elevation of temperature is possible. A thermal index of 2 would be twice that power but does not necessarily indicate a peak temperature rise of 2°C.

Because temperature rise is dependent on tissue type and is particularly dependent on the presence of bone, the TI is subdivided into three indices:

(1) TIS: thermal index for soft tissue
(2) TIB: thermal index with bone at/near the focus
(3) TIC: thermal index with bone at the surface (e.g., cranial examination)

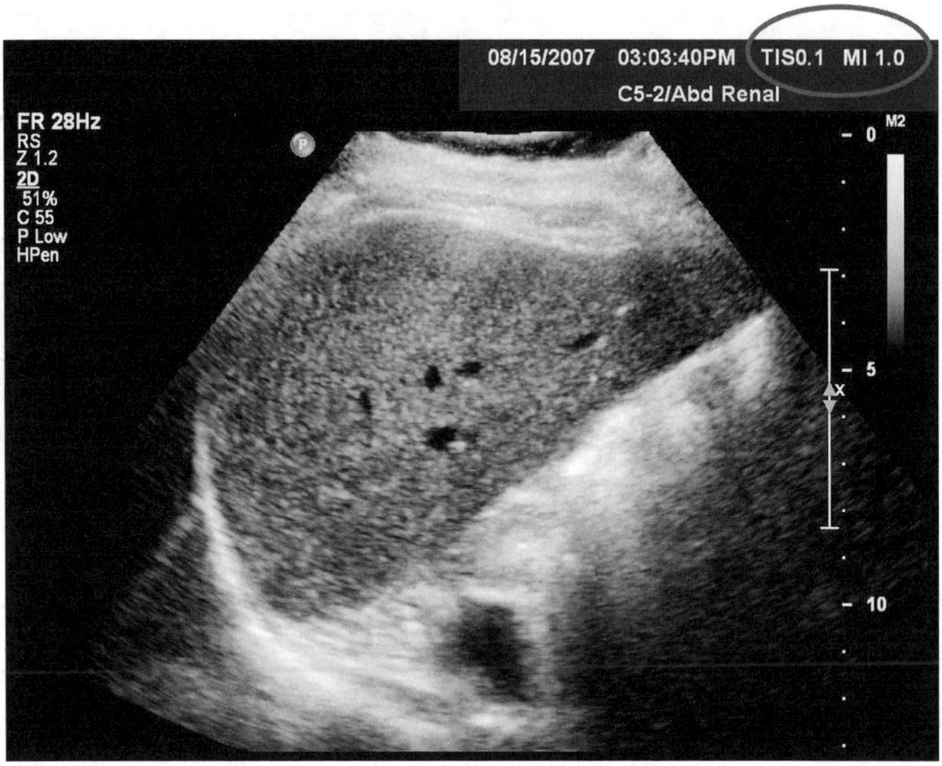

Figure A-2. An abdomen ultrasound scan. Note the TIS and MI values being displayed at the top right-hand corner

The MI value is computed from the peak rarefactional pressure and the frequency, and is intended to estimate the potential for mechanical bioeffects. The higher the index value, the higher the probability for bioeffects occurring; values less than "1" are generally considered "safe." Due to the difficulties of estimating tissue conditions, it is necessary to bear in mind that these indices provide indicators of risks rather than quantifiable values. They do not take into account factors such as dwell time, examination time, patient *in vivo* temperature, or the presence of contrast agents [30].

An IEC standard (Draft IEC 61681) has been drawn up to establish a safety classification for ultrasound equipment based on its ability to produce cavitation or a temperature rise. The standard proposes two classifications of equipment: class A, which has a lower output and for which no output display is required, and class B, which has a higher output and for which an output display is required. The draft is currently undergoing review [31].

A.4 BENEFITS AND RISKS — THE ALARA

The extensive clinical application of ultrasound over many decades has proved conclusively its diagnostic benefit. However, the potential risks must not be overlooked. But there is another risk that must be considered: the risk of not doing the ultrasound examination or choosing a less desirable or invasive way. However, attention must now be focused on evaluating the cost of morbidity resulting from sub-optimal equipment that give poor diagnostic studies.

How can the benefits and risks be balanced? Prudent use can be achieved by applying the concept of ALARA (As Low As Reasonably Achievable). Following the ALARA principle means that total ultrasound exposure is kept as low as reasonably achievable, while optimizing diagnostic information.

There are operator controls for improving image quality and minimizing output intensity. First are those controls that directly affect output intensity. Then are those controls that change the characteristics of the transmitted ultrasound field and indirectly affect intensity: system mode, pulse repetition frequency, focusing depth, pulse length, and transducer choice. The "receiver controls" that help to improve image quality are: receiver gain, time gain compensation (TGC), video dynamic range, and post-processing. There is also the philosophical aspect of ALARA. This includes minimizing scan time, justifying the clinical examination, and not compromising image quality by rushing through an examination.

Because the threshold of bioeffect intensity is not known, it is the responsibility of the operator/s to use his/her/their judgment and insight to adjust the intensity output of the equipment so as to get the most information at the lowest output power.

The AIUM's complete statement on clinical safety is quoted here [16]:

"Diagnostic ultrasound has been in use since the late 1950s. Given its known benefits and recognized efficacy for medical diagnosis, including use during human pregnancy, the American Institute of Ultrasound in Medicine herein addresses the clinical safety of such use:

No confirmed biological effects on patients or instrument operators caused by exposure at intensities typical of present diagnostic ultrasound instruments have ever been reported. Although the possibility exists that such biological effects may be identified in the future, current data indicate that the benefits to patients of the prudent use of diagnostic ultrasound outweigh the risks, if any, that may be present."

A.5 INTERNATIONAL SAFETY STANDARDS AND GUIDELINES

Ultrasound organizations have been issuing statements on the safe use of ultrasound. These are not for regulatory purpose but are intended to educate and advise. WFUMB guidelines have been issued in two special issues of *Ultrasound in Medicine and Biology*. Statements and recommendations are given on B-mode scanning, Doppler imaging, transducer heating, thermal effects. The AIUM have produced statements on the safety of ultrasound. The European Committee for Ultrasound Radiation Safety has published statements on the use of pulsed Doppler measurement in fetuses, stating that its use in routine examinations during the period of organogenesis is considered inadvisable at present.

Selected safety statements from international organizations are listed in the appendices at the end of this chapter.

Acknowledgment

The authors wish to thank Ms. Chai-Hong Yeong for helping with the literature review and research.

REFERENCES

[1] O'Brien, W. D. 2007, "Ultrasound-Biophysics Mechanisms," *Progress in Biophysics and Molecular Biology*, **93**(1–3), pp. 212–255.

[2] Dalecki, D., 2004, "Mechanical Bioeffects of Ultrasound," *Annual Review of Biomedical Engineering*, **6**: pp 229–248.

[3] Ng, K. H., 2002, "International Guidelines and Regulations for the Safe Use of Diagnostic Ultrasound in Medicine," *Journal of Medical Ultrasound*, **10**(1), pp. 5–9.

[4] Barnett, S. B., and Maulik, D., 2001, "Guidelines and Recommendations for Safe Use of Doppler Ultrasound in Perinatal Applications," *Journal of Maternal-Fetal Medicine*, **10**(2), pp. 75–84.
[5] Nyborg, W. L., 2001, "Biological Effects of Ultrasound: Development of Safety Guidelines. Part II: General Review," *Ultrasound in Medicine and Biology*, **27**(3), pp. 301–333.
[6] Barnett, S. B., 2000, "Biophysical Aspects of Diagnostic Ultrasound," *Ultrasound in Medicine and Biology*, **26**(Supplement 1), pp. 68–70.
[7] Barnett, S. B., Ter Haar, G. R., Ziskin, M.C., Rott, H.D., Duck, F.A., and Maeda, K., 2000, "International Recommendations and Guidelines for the Safe Use of Diagnostic Ultrasound in Medicine," *Ultrasound in Medicine and Biology*, **26**(3), pp. 355–366.
[8] Fowlkes, J.B., and Holland, C.K., 2000, "Mechanical Bioeffects from Diagnostic Ultrasound: AIUM Consensus Statements," *Journal of Ultrasound in Medicine*, **19**(2), pp. 69–72.
[9] Nyborg, W.L., 2000, "Biological Effects of Ultrasound: Development of Safety Guidelines. Part I: Personal Histories," *Ultrasound in Medicine and Biology*, **26**(6), pp. 911–964.
[10] Barnett, S.B. (ed), 1998, "World Federation for Ultrasound in Medicine and Biology Symposium on Safety of Ultrasound in Medicine; Conclusions and Recommendations on Thermal and Non-Thermal Mechanisms for Biological Effects of Ultrasound," *Ultrasound in Medicine and Biology*, 24(Supplement 1).
[11] Barnett, S. B., Rott, H. D., ter Haar, G. R., Ziskin, M. C., Maeda, K., 1997, "The Sensitivity of Biological Tissue to Ultrasound," *Ultrasound in Medicine and Biology*, **23**(6), pp. 805–812.
[12] European Federation of Societies for Ultrasound in Medicine and Biology, 1996, "Clinical Safety Statement for Diagnostic Ultrasound; Report from the European Committee for Ultrasound Radiation Safety," *European Journal of Ultrasound*, **3**, pp. 283.
[13] Miller, M. W., 1996, "A Review of In Vitro Bioeffects of Inertial Ultrasonic Cavitation From a Mechanistic Perspective," *Ultrasound in Medicine and Biology*, **22**(9), pp. 1131–1154.
[14] European Federation of Societies for Ultrasound in Medicine and Biology, 1995, "Guidelines for the Safe Use of Doppler Ultrasound for Clinical Applications; Report from the European Committee for Ultrasound Radiation Safety," *European Journal of Ultrasound*, **2**, pp. 167–168.
[15] Barnett, S. B., ter Haar, G. R., Ziskin, M. C., Nyborg, W. L., Maeda, K., and Bang, J., 1994, "Current Status of Research on Biophysical Effects of Ultrasound," *Ultrasound in Medicine and Biology*, **20**(3), pp. 205–218.
[16] American Institute of Ultrasound in Medicine, 1993, "Bioeffects Committee: Bioeffects and Safety of Diagnostic Ultrasound," Laurel MD, AIUM, 1993 (Re-approved 2003).
[17] American Institute of Ultrasound in Medicine, 1992, "Bioeffects and Safety of Diagnostic Ultrasound," Laurel, MD, AIUM, 1992.
[18] Barnett, S. B., and Kossoff, G. (eds), 1992, "Issues and Recommendations Regarding Thermal Mechanisms for Biological Effects of Ultrasound" in "WFUMB Symposium on Safety and Standardization in Medical Ultrasound," *Ultrasound in Medicine and Biology*, Vol. 18, Special Issue.
[19] Pye, S. D., and Zeqiri, B., 2002, "Ultrasound," *Practical Radiation Protection in Healthcare*, C. J. Martin, D. G. Sutton, eds., Oxford University Press, Oxford, UK, pp. 393–407.
[20] National Council on Radiation Protection and Measurements (NCRP), 2002, *Exposure Criteria for Medical Diagnostic Ultrasound: II. Criteria Based On Mechanical Mechanisms*, NCRP, Bethesda, MD.
[21] Ter Haar, G., and Duck, F. A., 2000, *The Safe Use of Ultrasound in Medical Diagnosis*, The British Institute of Radiology, London, UK.
[22] Wells, P. N. T., 1999, "Safety Issues in Diagnostic Ultrasound in Baxter," *Clinical Diagnostic Ultrasound*, 2nd edn, G. M. Baxter, P. Allan, and P. Morley, eds, Blackwell Science, Oxford, UK, pp. 73–81.
[23] Duck, F. A., 1998, "Acoustic Streaming and Radiation Pressure in Diagnostic Applications: What are the Implications?" *Safety of Diagnostic Ultrasound*, S. B. Barnett, G. Kossoff, eds., Parthenon Publishing, Carnforth, UK, pp. 87–98.
[24] Barnett, S. B., and Kossoff, G. (eds), 1998, *Safety of Diagnostic Ultrasound, Progress in Obstetric and Gynecological Sonography Series*, Parthenon, New York.

[25] Zagzebski, J. A., 1996, *Essential of Ultrasound Physics*, Mosby, St. Louis, 1996.

[26] Nyborg, W. L.. 1985, "Mechanisms for Biological Effects of Ultrasound," *Clinics in Diagnostic Ultrasound Vol. 16: Biological Effects of Ultrasound*, W. L. Nyborg and M. Ziskin, eds., Churchill Livingston, New York.

[27] Ziskin, M., 1985, "Epidemiology and Human Exposure," *Clinics in Diagnostic Ultrasound Vol. 16: Biological effects of Ultrasound*, W. L. Nyborg and M. Ziskin, Churchill Livingston, New York.

[28] American Institute of Ultrasound in Medicine, 1992, Standard for Real-time Display of Thermal and Mechanical Acoustic Indices on Diagnostic Ultrasound Equipment, AIUM, Laurel, MD, 1992.

[29] Abbott, J. G., 1999, "Rationale and Derivation of MI and TI — A Review," *Ultrasound in Medicine and Biology*, **25**, pp. 431–442.

[30] Jago, J. R., Henderson, J., Whittingham, T.A., and Willson, K., 1995, "How Reliable Are Manufacturer's Reported Acoustic Output Data?," *Ultrasound in Medicine and Biology*, **12**, pp. 135–136.

[31] FDA, 1993, Revised 510(k) Diagnostic Ultrasound Guidance for 1993, Food and Drug Administration, Center for Devices and Radiological Health, Rockville, MD.

SELECTED SAFETY STATEMENTS FROM INTERNATIONAL ORGANIZATIONS

AA.1 WFUMB STATEMENTS ON THE SAFETY OF DIAGNOSTIC ULTRASOUND (1998)

Ultrasound in Medicine & Biology, Vol. 24, Supplement 1, 1998 [10]
http://www.wfumb.org/default.asp?ID=139&pID=99

AA.1.1 B-Mode Imaging (1992)

Known diagnostic ultrasound equipment as used today for simple B-mode imaging operates at acoustic outputs that are not capable of producing harmful temperature rises. Its use in medicine is therefore not contraindicated on thermal grounds. This includes endoscopic, transvaginal, and transcutaneous applications.

AA.1.2 Doppler (1992)

It has been demonstrated in experiments with unperfused tissue that some Doppler diagnostic equipment has the potential to produce biologically significant temperature rises, specifically at bone/soft tissue interfaces. The effects of elevated temperatures may be minimized by keeping the time during which the beam passes through any one point in tissue as short as possible. Where output power can be controlled, the lowest available power level consistent with obtaining the desired diagnostic information should be used. Although the data on humans are sparse, it is clear from animal studies that exposures resulting in temperatures less than 38.5°C can be used without reservation on thermal grounds. This includes obstetric applications.

AA.1.3 Transducer Heating (1992)

A substantial source of heating may be the transducer itself. Tissue heating from this source is localized to the volume in contact with the transducer.

AA.1.4 Recommendations on Thermal Effects (1997)

A diagnostic exposure that produces a maximum temperature rise of no more than 1.5°C above normal physiological levels (37°C) may be used without reservation on thermal grounds.

A diagnostic exposure that elevates embryonic and fetal *in situ* temperature to 4°C (4°C above normal temperature) for 5 minutes or more should be considered potentially hazardous.

AA.1.5 Non-Thermal Effects

Thresholds for non-thermal biological effects are lowest in:

- tissues that naturally contain gas bodies, e.g., post-natal lung and intestines;
- all tissue with the presence of introduced gas bodies, e.g., ultrasonic contrast agents.

AA.1.6 Contrast Agents

Gas bodies introduced by a contrast agent increase the probability of cavitation. A physician should take this into account when considering the benefit/risk ratio of an examination.

AA.2 SELECTED AIUM STATEMENTS ON THE SAFETY OF DIAGNOSTIC ULTRASOUND (1992)

AA.2.1 Prudent Use and Clinical Safety *(Approved March 19, 2007)*

http://www.aium.org/publications/statements/_statementSelected.asp?statement=34

Diagnostic ultrasound has been in use since the late 1950s. Given its known benefits and recognized efficacy for medical diagnosis, including use during human pregnancy, the American Institute of Ultrasound in Medicine herein addresses the clinical safety of such use:

No independently confirmed adverse effects caused by exposure from present diagnostic ultrasound instruments have been reported in human patients in the absence of contrast agents. Biological effects (such as localized pulmonary bleeding) have been reported in mammalian systems at diagnostically relevant exposures but the clinical significance of such effects is not yet known. Ultrasound should be used by qualified health professionals to provide medical benefit to the patient.

AA.2.2 Mammalian in Vivo Ultrasonic Biological Effects *(Approved October 20, 1992)*

http://www.aium.org/publications/statements/_statementSelected.asp?statement=9

Information from experiments utilizing laboratory mammals has contributed significantly to our understanding of ultrasonically induced biological effects and the mechanisms that are most likely responsible. The following statement summarizes observations relative to specific ultrasound parameters and indices. The history and rationale for this statement are provided in *Bioeffects and Safety of Diagnostic Ultrasound* (AIUM, 1993).

In the low megahertz frequency range, there have been no independently confirmed adverse biological effects in mammalian tissues exposed in vivo under experimental ultrasound conditions, as follows.

(a) When a thermal mechanism is involved, these conditions are unfocused-beam[1] intensities below 100 mW/cm^2, focused-beam[2] intensities below 1 W/cm^2, or thermal index values less than 2. Furthermore, such effects have not been reported for higher values of thermal index when it is less than 6 (log 10*t*/0.6), where *t* is exposure time ranging from 1 minute to 250 minutes, including off-time for pulsed exposure.

(b) When a non-thermal mechanism is involved[3], in tissues that contain well-defined gas bodies, these conditions are in situ peak rarefactional pressures below approximately 0.3 MPa or mechanical index values less than approximately 0.3. Furthermore, for other tissues no such effects have been reported.

AA.3 SELECTED EFSUMB STATEMENTS ON THE SAFETY OF DIAGNOSTIC ULTRASOUND (1995-1996)

AA.3.1 Guidelines for the safe use of Doppler Ultrasound for clinical applications[4]

European Federation of Societies for Ultrasound in Medicine and Biology. *Clinical safety statement for diagnostic ultrasound.* Report from the European Committee for Ultrasound Radiation Safety. *Eur J Ultrasound* 1996;3:283. [12]

http://www.efsumb.org/efsumb/Guidelines/safe%20use%20doppler%20clinical.pdf

Quantitative pulsed Doppler measurements involve stationary beams of high output levels (SPTA intensities may be 100 times those used in B-mode imaging). Color flow imaging employs swept beams and lower acoustic outputs, which are intermediate between those of pulsed Doppler and B-mode imaging. Continuous wave Doppler devices use intensities that are sufficiently low to give minimum concern about potential heating effects.

The differences between the various Doppler applications arise because high intensities are needed to obtain strong echoes from poorly reflecting blood cells for flow studies. Strongly reflecting moving structures, such as those found in the heart, return strong echoes from low intensity pulses. These guidelines have been drawn up on the basis that there is the need to minimize any potential adverse thermal effects from Doppler applications. Although other effects (such as cavitation) may occur, it is considered that these do not give serious cause for concern in diagnostic equipment used clinically at present. The absorption of ultrasound by different tissues varies greatly, with fully calcified bone absorbing most. Thus, the temperatures that result from exposure to a given ultrasound beam will vary according to the tissue type exposed. It should be borne in mind that tissues with low perfusion rates will experience higher temperatures than their well-perfused counterparts. When considering the thermal implications of a given exposure, it is necessary not only to consider temperature increases within tissue structures lying in the path of the ultrasonic beam, but also their sensitivity to thermal damage. On this basis, safe clinical practice, with regard to thermal effects, should be conducted according to the following guidelines:

[1] Free-field spatial peak, temporal average (SPTA) for continuous wave and pulsed exposures.

[2] Quarter-power (–6 dB) beam width smaller than four wavelength or 4mm, whichever is less at the exposure frequency.

[3] For diagnostically relevant ultrasound exposures.

[4] Drawn up by the European Committee for Ultrasound Radiation Safety.

AA.3.1.1 Output conditions and exposure time

i. *New Patient Investigation.* For each new patient investigation, set the machine output conditions to a minimum, and receiver settings to a maximum. This should ideally be achieved by a default (start-up) setting on the machine each time a new patient is examined. Where this is not available, the manufacturer should be consulted.

ii. *Doppler Exposure Times and Levels.* Keep Doppler exposure times and output levels to the minimum consistent with good clinical practice. Choose the region for Doppler interrogation using B-mode imaging. Where possible, set the Doppler gate to the required position under B-mode guidance before turning on the Doppler beam.

iii. *Continued Acoustic Output.* The user must know whether there is continued acoustic output when the image is frozen. If the acoustic output is not interrupted, then the probe should be removed from the patient during freeze frame conditions. Before it is replaced, its front surface temperature should be checked, since some transducers may self heat to unacceptable levels when operating in air. A hot transducer will cool when the pulsed Doppler is switched off.

iv. *Bone.* Special care should be taken when bone falls within the pulsed Doppler field as rapid heating may result in biologically hazardous temperatures. Patients should be encouraged to report bone pain.

v. *Pediatrics.* In pediatric applications, where tissues that may be especially sensitive to temperature elevations lie very close to, or within, bone, extra care must be taken to minimize pulsed Doppler exposures, The tissue of most concern here is the neonatal brain lying adjacent to the skull. Care should also be exercised when pulsed Doppler exposures are undertaken in the vicinity of the epiphyses of growing bone.

vi. *Obstetrics.* In view of the possibility of significant temperature elevation in tissues in the path of the pulsed Doppler beam, routine examination with Doppler ultrasound of the first trimester embryo is considered inadvisable at present. Investigations for which the embryo does not lie in the path of the pulsed Doppler beam, such as studies of the vessels of the uterine wall, may be undertaken.

vii. Pulsed Doppler exposures become increasingly likely to produce biologically significant local heating as the pregnancy proceeds through the second and third trimesters because the acoustic absorption of fetal bone increases with its progressive mineralization. This is of particular importance when brain or growing epiphyses lie in the path of the pulsed Doppler beam.

viii. It should be remembered that sensitive structures lying in front of, or behind, the "measurement target" will also be in the pulsed Doppler beam and thus will be exposed to ultrasound. This is of particular importance in obstetric applications.

ABOUT THE EDITORS

Ahmed Al-Jumaily holds a PhD and MSc from the Ohio State University, and BSc from the University of Baghdad, Iraq (all in mechanical engineering). He has over 30 years of teaching and research experience in two American universities, four Middle Eastern universities and the Auckland University of Technology. Currently he is a Professor of Mechanical Engineering and the Director of the Institute of Biomedical Technologies at the Auckland University of Technology, Auckland, New Zealand. He is a member of the American Society of Mechanical Engineering, Acoustical Society of America, American Engineering Education, International Society of Optical Engineering, International Institute of Acoustics and Vibration and Institute of Professional Engineers New Zealand. He is also a registered professional engineer. He has supervised more than 80 graduate students in vibrations, acoustics, system dynamics and control, biomechanics and biomedical devices. He has published more than 170 papers in international journals and conference proceedings and acted as a referee for several international journals. Professor Al-Jumaily current research interest focuses on the application of vibration and acoustics for diagnostics and therapy of the respiratory and cardiovascular systems.

Mostafa Fatemi received his PhD degree in Electrical Engineering from Purdue University, West Lafayette, IN. Dr. Fatemi joined Mayo Clinic College of Medicine in 1994. His current research interests include ultrasound imaging and evaluating the elastic properties of biological tissues by means of ultrasound radiation force. Dr. Fatemi's pioneering work include a novel imaging method called vibro-acoustography, which was presented in two papers in Science (1998) and Proceedings of the National Academy of Sciences (1999). He has been awarded six patents in various aspects of vibro-acoustography and fetal stimulation by ultrasound. Dr. Fatemi has served at the editorial boards of a number of journals in his field, including IEEE Transactions on Medial Imaging. Dr. Fatemi is a fellow of the American Institute for Medical and Biological Engineering and a member of the American Institute of ultrasound in Medicine, Institute of Electrical and Electronics Engineers, and the Acoustical Society of America. Currently, Dr. Fatemi is a Professor of Bioengineering at Mayo Clinic College of Medicine, where he teaches at the Mayo Graduate School and leads research programs in the areas of breast and prostate vibro-acoustography.

INDEXING WORDS